Springer Proceedings in Complexity

Springer Complexity

Springer Complexity is an interdisciplinary program publishing the best research and academic-level teaching on both fundamental and applied aspects of complex systems-cutting across all traditional disciplines of the natural and life sciences, engineering, economics, medicine, neuroscience, social, and computer science.

Complex Systems are systems that comprise many interacting parts with the ability to generate a new quality of macroscopic collective behavior the manifestations of which are the spontaneous formation of distinctive temporal, spatial, or functional structures. Models of such systems can be successfully mapped onto quite diverse "real-life" situations like the climate, the coherent emission of light from lasers, chemical reaction-diffusion systems, biological cellular networks, the dynamics of stock markets and of the Internet, earthquake statistics and prediction, freeway traffic, the human brain, or the formation of opinions in social systems, to name just some of the popular applications.

Although their scope and methodologies overlap somewhat, one can distinguish the following main concepts and tools: self-organization, nonlinear dynamics, synergetics, turbulence, dynamical systems, catastrophes, instabilities, stochastic processes, chaos, graphs and networks, cellular automata, adaptive systems, genetic algorithms, and computational intelligence.

The three major book publication platforms of the Springer Complexity program are the monograph series "Understanding Complex Systems" focusing on the various applications of complexity, the "Springer Series in Synergetics", which is devoted to the quantitative theoretical and methodological foundations, and the "SpringerBriefs in Complexity" which are concise and topical working reports, case-studies, surveys, essays, and lecture notes of relevance to the field. In addition to the books in these two core series, the program also incorporates individual titles ranging from textbooks to major reference works.

More information about this series at
http://www.springer.com/series/11637

Hideki Takayasu • Nobuyasu Ito • Itsuki Noda
Misako Takayasu
Editors

Proceedings of the International Conference on Social Modeling and Simulation, plus Econophysics Colloquium 2014

Springer Open

Editors
Hideki Takayasu
Sony Computer Science Laboratories, Inc.
Shinagawa
Tokyo, Japan

Nobuyasu Ito
Department of Applied Physics
The University of Tokyo
Bunkyo
Tokyo, Japan

Itsuki Noda
Center for Service Research
National Institute of Advanced Industrial Science and Technology
Tsukuba
Ibaraki, Japan

Misako Takayasu
Department of Computational Intelligence and Systems Science
Tokyo Institute of Technology
Yokohama
Kanagawa, Japan

ISSN 2213-8684 ISSN 2213-8692 (electronic)
Springer Proceedings in Complexity
ISBN 978-3-319-20590-8 ISBN 978-3-319-20591-5 (eBook)
DOI 10.1007/978-3-319-20591-5

Library of Congress Control Number: 2015947289

Springer Cham Heidelberg New York Dordrecht London

Printed on acid-free paper

Springer International Publishing AG Switzerland is part of Springer Science+Business Media (www.springer.com)

Preface

Big data analyses have uncovered many empirical laws hidden in our society and economy. Mathematical models have been introduced successfully explaining those empirical laws as typically seen in the new field of econophysics. One of the goals of this trend of research may be modeling and simulations of the whole society, which can directly contribute to the industry as well as help in decision making.

This book is the proceedings of the international conference, SMSEC2014, which was held on 4-6 November 2014 in Kobe, Japan, as a joint conference of the first "Social Modeling and Simulations" and the 10th "Econophysics Colloquium" (http://aph.t.u-tokyo.ac.jp/smsec2014/). It consisted of 21 invited talks, 77 oral and 53 poster presentations, with 174 participants. A variety of problems in wide fields, such as financial markets, traffic systems, epidemic contagion, and social media, were the subjects of intensive discussion. Data analysis, agent-based modeling, complex networks, and supercomputers were the examples of methods.

The conference was supported by many organizations: Tateishi Science and Technology Foundation, Kobe Convention and Visitors Association "MEET IN KOBE21", the Japanese Society for Artificial Intelligence, Society for Serviceology, the Japanese Association of Financial Econometrics and Engineering JAFEE, Center for Cooperative Work on Computational Science in University of Hyogo, the Physical Society of Japan, and RIKEN Advanced Institute for Computational Science. On behalf of all the participants, we would like to thank those supporters, as well as the following companies, without whose financial support the workshop would not have been possible: Hottolink, Sony CSL, and EBS.

As organizers, we are grateful for the cooperation of the steering committee: Kiyoshi Izumi (Univ. Tokyo), Yukie Sano (Univ. Tsukuba), Takahiro Sasaki (Sony CSL), Takashi Shimada (Univ. Tokyo), Kenta Yamada (Univ. Tokyo), and Naoki Yoshioka (Univ. Tokyo). Finally, we would like to thank all the authors for their contributions to this volume.

Shinagawa, Tokyo, Japan
Bunkyo, Tokyo, Japan
Tsukuba, Ibaraki, Japan
Yokohama, Kanagawa, Japan
April 2015

Hideki Takayasu
Nobuyasu Ito
Itsuki Noda
Misako Takayasu

Contents

Part I
Financial Market

Chapter 1
Influence Networks in the Foreign Exchange Market

Arthur M.Y.R. Sousa, Hideki Takayasu, and Misako Takayasu

Abstract The Foreign Exchange Market is a market for the trade of currencies and it defines their relative values. The study of the interdependence and correlation between price fluctuations of currencies is important to understand this market. For this purpose, in this work we search for the dependence between the time series of prices for pairs of currencies using a mutual information approach. By applying time shifts we are able to detect time delay in the dependence, what enable us to construct a directed network showing the influence structure of the market. Finally, we obtain a dynamic description of this structure by analyzing the time evolution of the network. Since the period of analysis includes the great earthquake in Japan in 2011, we can observe how such big events affect the network.

1.1 Introduction

The Foreign Exchange Market is a market in which currencies are traded; it is continuously open during the weekdays and it has the largest transaction volume among the financial markets (average of \$5.3 trillion/day in April 2013 [1]). The importance of this market is that it defines the relative values of currencies and affects other markets, such as the stock markets [2].

In this market, traders can make orders for buying and selling which are organized in the order book according to their corresponding prices. The highest price of the buy orders in a given time is called best bid and the lowest price of the sell orders, best ask, and their average defines the mid-quote; a deal occurs when the best bid meets the best ask.

A.M.Y.R. Sousa (✉) • M. Takayasu
Department of Computational Intelligence and Systems Science, Interdisciplinary Graduate School of Science and Engineering, Tokyo Institute of Technology, G3-52 4259 Nagatuta-cho, Yokohama 226-8502, Japan
e-mail: yamashita.a.ai@m.titech.ac.jp

H. Takayasu
Sony Computer Science Laboratories Inc., 3-14-13 Higashigotanda, Shinagawa-ku, Tokyo 141-0022, Japan

H. Takayasu et al. (eds.), *Proceedings of the International Conference on Social Modeling and Simulation, plus Econophysics Colloquium 2014*, Springer Proceedings in Complexity, DOI 10.1007/978-3-319-20591-5_1

Information about dependence between price fluctuations of currencies is important to understand the foreign exchange market. Several studies try to model this market and access those dependences [3–5]. However there are no studies on the influence structure in this market and the time evolution of the dependences. To contribute to fill this gap, we analyse the dependences in foreign exchange data during a period of 3 weeks using the mutual information, a non-linear dependence measure from the information theory [6, 7]. By doing a time shift analysis we can infer temporal dependence between markets making possible the construction of directed networks that show the influence structure of the foreign exchange market.

1.2 Data and Method

We analyze the foreign exchange data of the Electronic Broking Services (EBS) by ICAP. This data contains the orders for pairs of currencies in a resolution of 0.1 s. Here we use the 6 currencies with the largest transaction volume: USD (United States dollar), EUR (Euro), JPY (Japanese yen), GBP (Pound sterling), AUD (Australian dollar) and CHF (Swiss franc) in the period between 2011, March, 07th and 2011, March, 25th, each day from 22:00:00 to 21:59:59 GMT. The chosen period is a special one because it includes the great earthquake in Japan on 2011, March, 11th and the announcement of the intervention in the foreign exchange market as a response to the effects of the earthquake on 2011, March, 17th [8]. For this data we define the price $P(t)$ as the last mid-quote, where t is the real time in intervals of 0.1 s. As an example of the data, Fig. 1.1 shows the price $P(t)$ for the market USD/JPY on 2011, March, 09th, before the great earthquake in Japan.

We work with the sign of the difference of price $P(t)$ [9]:

$$S(t) = sign[P(t) - P(t-1)], \tag{1.1}$$

so that we obtain a time series for each pair of currencies with the symbols + (price increasing), − (price decreasing) and 0 (price unchanged). By comparing two of these time series, we can identify 4 states not containing 0: (+, +), (+, −), (−, +) and (−, −). The removal of the states with 0, e.g. (+, 0), is an important step because then we compare the series only when there is activity in both of them, avoiding issues regarding the volume difference and the time zone difference. Table 1.1 illustrates the number of occurrence of each state when comparing the EUR/USD with other markets on 2011, March, 07th (time series of each market with 863,999 points).

Studies in financial markets commonly use the Pearson correlation coefficient as a measure to infer dependence [5, 10]. But the correlation coefficient detects only linear correlation between two variables, not having information about the dependence. The mutual information on the other hand deals direct with the probability distributions being a measure not only for linear and non-linear correlations, but also for dependence. The mutual information is zero if and only if the random variables

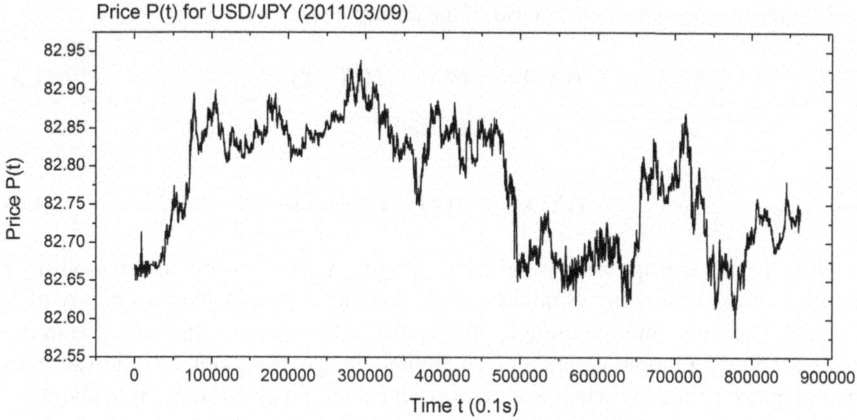

Fig. 1.1 Price P(t) for the market USD/JPY on 2011, March, 09th. Here we work with the sign of the difference of the price P(t)

Table 1.1 Number of states for EUR/USD and other markets on 2011, March, 07th (no time shift)

Market	(+, +)	(+, −)	(−, +)	(−, −)	0[a]
AUD/JPY	3256	2904	2941	3303	851,595
AUD/USD	2425	1707	1591	2332	855,944
CHF/JPY	125	129	184	184	863,377
EUR/AUD	55	59	66	48	863,771
EUR/CHF	3817	3061	3160	3895	850,066
EUR/GBP	3956	3305	3272	4086	849,380
EUR/JPY	5351	3918	3956	5202	845,572
GBP/AUD	53	47	45	53	863,801
GBP/CHF	43	47	56	52	863,801
GBP/JPY	4791	4431	4238	4807	845,732
GBP/USD	3088	2359	2533	3134	852,885
USD/CHF	2874	3656	3689	3032	850,748
USD/JPY	5822	7131	7081	5743	838,222

[a](+, 0), (−, 0), (0, 0), (0, −), (0, +)

are independent. There are evidences that mutual information can reveal aspects ignored by the correlation coefficient and studies comparing both measures [11–13]. Another reason for using mutual information in this work is that we are dealing with symbolic series, in which the numerical values that are taken in account for the correlation coefficient have no meaning.

The mutual information $I(X;Y)$ between two random variables X and Y:

$$I(X;Y) = \sum_{x}\sum_{y} p(x,y) log \frac{p(x,y)}{p(x)p(y)}, \tag{1.2}$$

which can also be expressed in term of the entropies H:

$$I(X;Y) = H(X) - H(X \mid Y) \tag{1.3}$$

or

$$I(X;Y) = H(Y) - H(Y \mid X). \tag{1.4}$$

$H(X)$ is the entropy of the random variable X and can be understood as a measure of its uncertainty. Similarly, $H(X \mid Y)$ can be seen as the uncertainty of X given Y. Thus, one interpretation for the mutual information is the reduction in the uncertainty of a random variable given the knowledge of the other. If the variables are independent, the knowledge of one variable does not give information about the other and then the mutual information is zero.

The final dependence measure we use is the global coefficient:

$$\lambda(X;Y) = \sqrt{1 - e^{-2I(X;Y)}}, \tag{1.5}$$

This quantity has desired characteristics for a dependence measure, as taking value zero for independent variables and being in the range [0;1] [14], and has been used in financial data [12].

In order to compute the global coefficient of the financial series, we estimate the probability of each state using the relative frequency in a time window of 1 day. We also determine a significance level to decide if the computed coefficient is significantly different from the one of a random series; we randomize the analysed series and calculate the global coefficient until it reaches a stationary value which corresponds to the coefficient for the corresponding random series and we take this value as the significance level.

1.3 Results and Discussion

For each two pairs of currencies we compute the global coefficient for their sign time series as function of the time shift between them. For this data, we find four general types of structures according to the presence of peaks that represent dependence between the markets, as illustrated in Fig. 1.2.

- No peak: no dependence between markets.
- Peak at time shift zero: both markets are synchronized. External influences (e.g. economic news) make the markets to have similar behaviour, the change in the price occurs simultaneously in both markets.

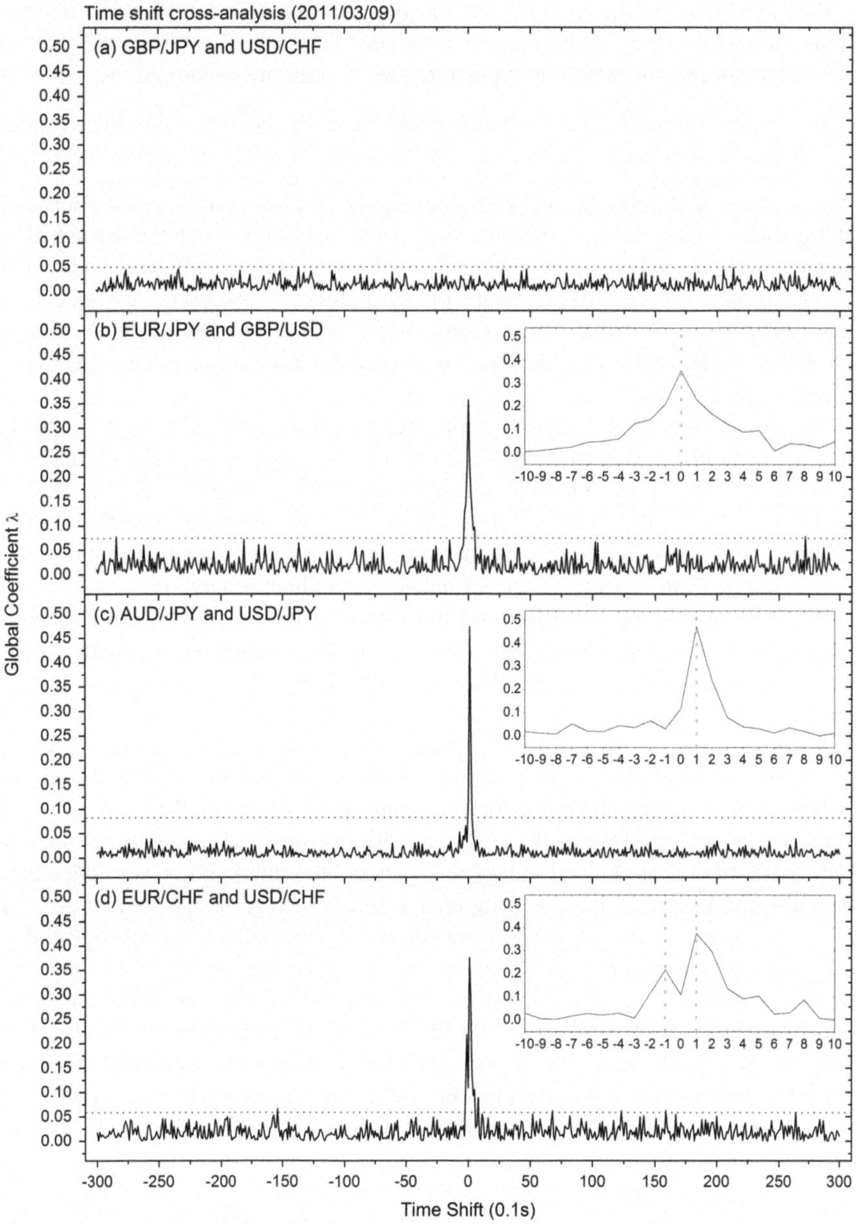

Fig. 1.2 Examples of results for the time shift cross-analysis. (**a**) GBP/JPY and USD/CHF on 2011, March, 09th: no dependence between the markets, same result for random time series. (**b**) EUR/JPY and GBP/USD on 2011, March, 09th: dependence at time shift 0. (**c**) AUD/JPY and USD/JPY on 2011, March, 09th: dependence when the USD/JPY series is shifted 0.1 s forward in relation to the AUD/JPY series. (**d**) EUR/CHF and USD/CHF on 2011, March, 09th: dependence at time shift 0.1 s in both directions. *Dotted lines* indicate the significance level

- Peak at a time shift different of zero: one market influences the other, i.e., there is an internal influence. This means that the past of one market affects the present of the other market, which could be interpreted as an information flow.
- Two peaks at time shifts in both directions: there are also internal influences, but in this case both markets affect each other during the analysed period.

We can build an influence network defining the pairs of currencies as nodes and adding the links according to the time shift cross-analysis between the markets that correspond to the nodes: (a) no peak: no link; (b) peak at time shift zero: undirected link; (c) peak at a time shift different from zero: directed link from the market that influences the other one, i.e., the market that goes ahead, whose past values affects the present values of the other market; (d) two peaks at time shifts in both directions: extraverted link.

We proceed with this analysis for all weekdays from 2011, March, 07 to 2011, March, 25. In this period two important events took place: the great earthquake in Japan on March, 11 and the announcement intervention in the foreign exchange market on March, 17. Figures 1.3, 1.4 and 1.5 show the time evolution of the influence network with day resolution during those 3 weeks. Figure 1.6 shows the time evolution of the different types of links in the influence network.

We observe that the structure does not present major changes within the first week from March, 07th to March, 10th, before the earthquake in Japan. Some characteristic features are: (a) EUR/USD and USD/JPY are the nodes with higher out-degree, meaning those are the markets that always go ahead being followed by the others, and (b) almost no extraverted links (with exception of link between USD/CHF and EUR/CHF, which is always present), i.e., information flows only in one direction, creating a hierarchy of importance between the markets.

From March, 11th (first week) to March, 17th (second week), which corresponds to the period between the earthquake in Japan and the intervention, we notice that the influence network changes compared to the structure in the first week. An important change is the increase in the number of directed and extraverted links, suggesting the interdependence between markets becomes stronger (not only due external influences, but internal ones). The new extraverted links that appeared involve the nodes EUR/USD and USD/JPY, that continue being the most important nodes (highest out degree), but now they are also influenced by other markets. One possible interpretation is that the players of these important markets are now being more careful, waiting for the information of other markets to decide to change the price.

After the announcement of the intervention on March, 17th, we observe another change in the structure, specially the disappearance of the extraverted link between EUR/USD and USD/JPY. Gradually the influence network returns to a structure similar to the one of the first week (before the earthquake).

Those results suggest that the event of the earthquake affected the dependence between markets and the event of the announcement of the intervention contributed for the return of the market to a state previous the earthquake, i.e., it was efficient in the sense of reversing the changes caused by the earthquake in the foreign exchange

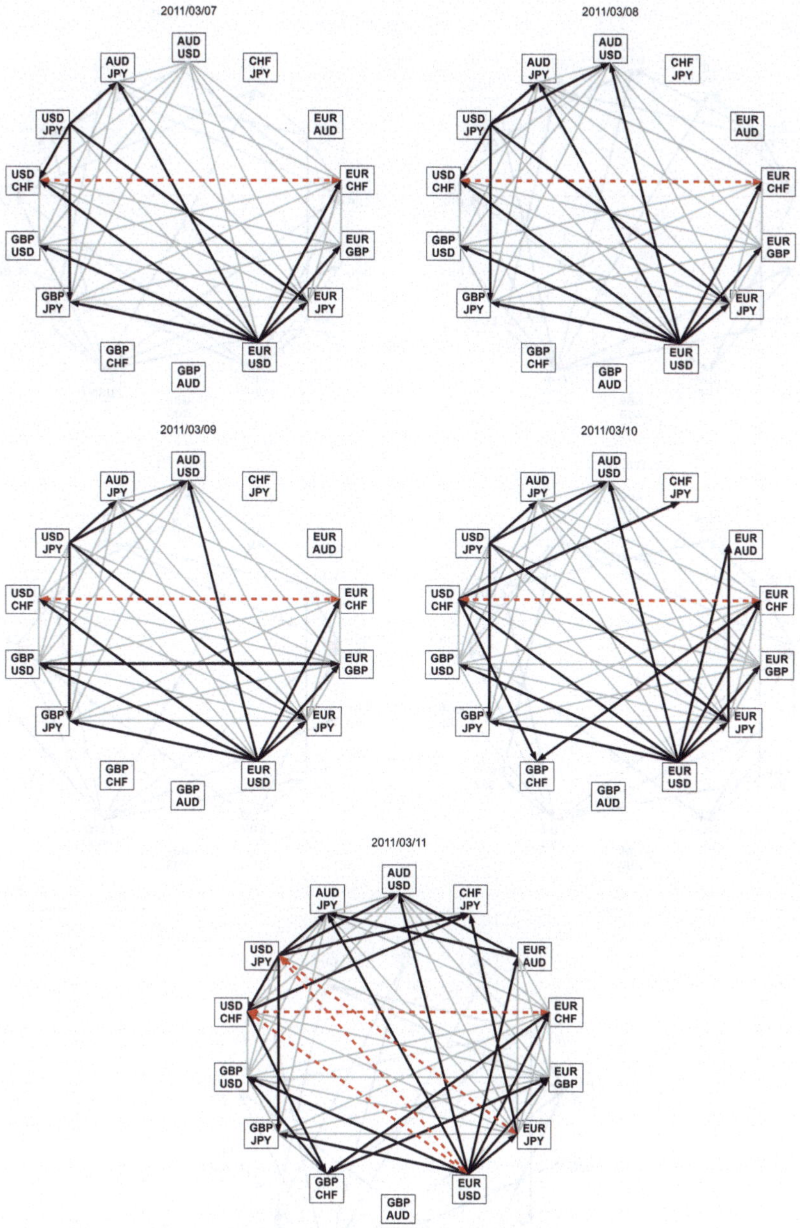

Fig. 1.3 Influence Networks of the Foreign Exchange Market for the currencies USD, EUR, JPY, GBP, AUD and CHF from 2011, March, 07th to 2011, March, 11th. The Great Earthquake in Japan took place on 2011, March, 11th. In this network nodes represent the pairs of currencies and there are three types of links according to the time shift cross-analysis: (i) undirected link (*gray*) corresponding to peak at time shift zero; (ii) directed link (*black*), peak at a time shift different from zero, in this case 0.1 s, from the market that influences the other one; (iii) extraverted link (*red*), two peaks at time shifts, also 0.1 s, in both directions

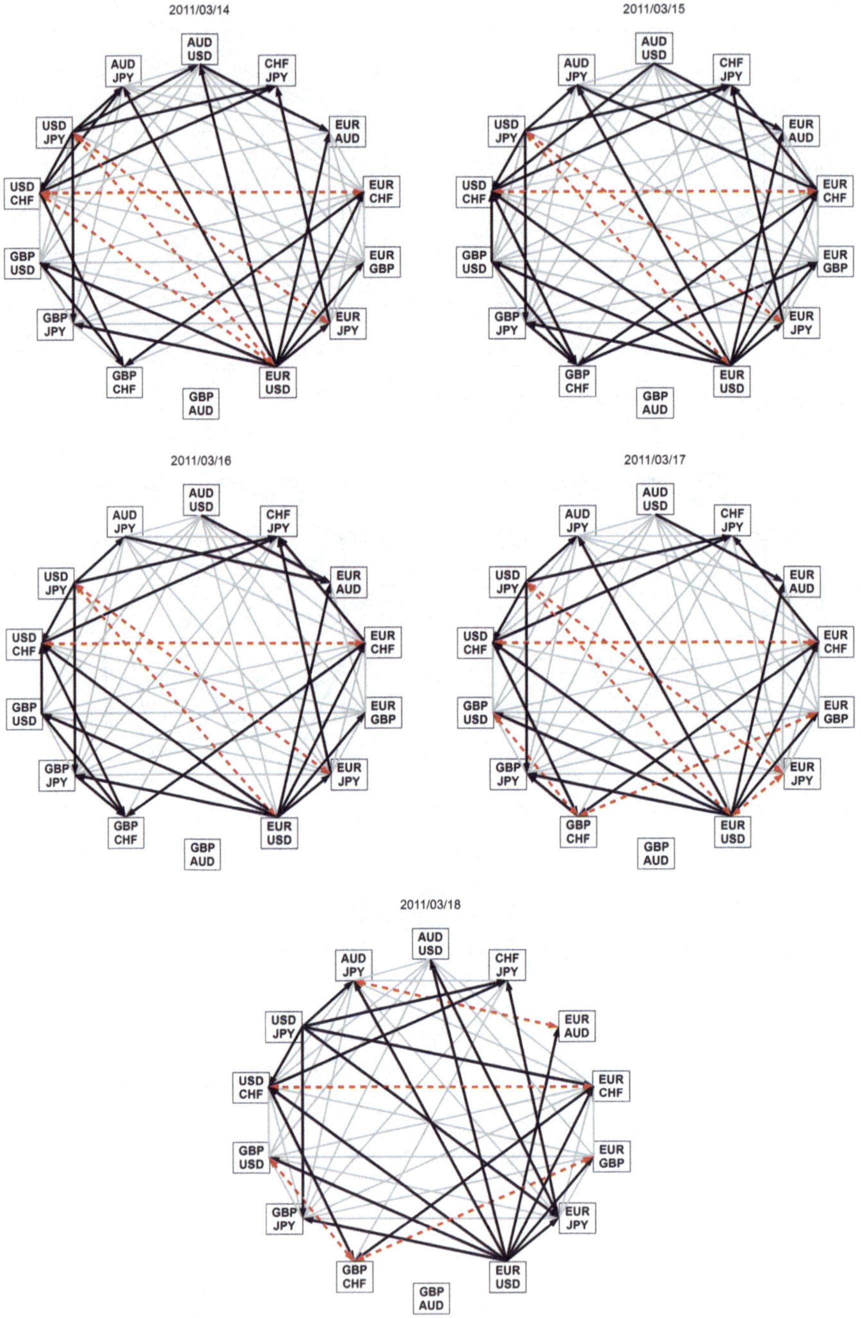

Fig. 1.4 Influence Networks of the Foreign Exchange Market for the currencies USD, EUR, JPY, GBP, AUD and CHF from 2011, March, 14th to 2011, March, 18th. The Intervation in the Foreign Exchange Market was announced in the end of 2011, March, 17th

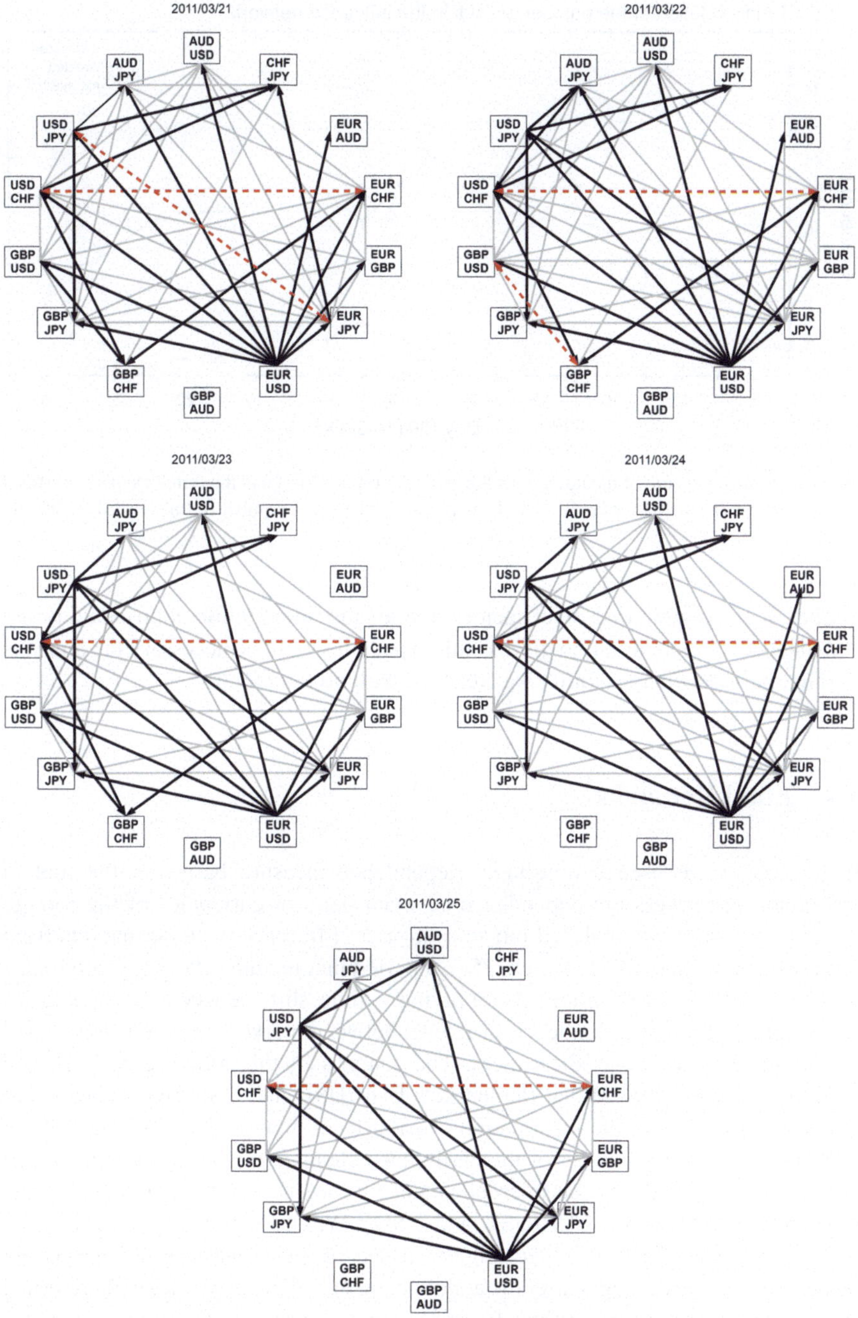

Fig. 1.5 Influence Networks of the Foreign Exchange Market for the currencies USD, EUR, JPY, GBP, AUD and CHF from 2011, March, 21st to 2011, March, 25th

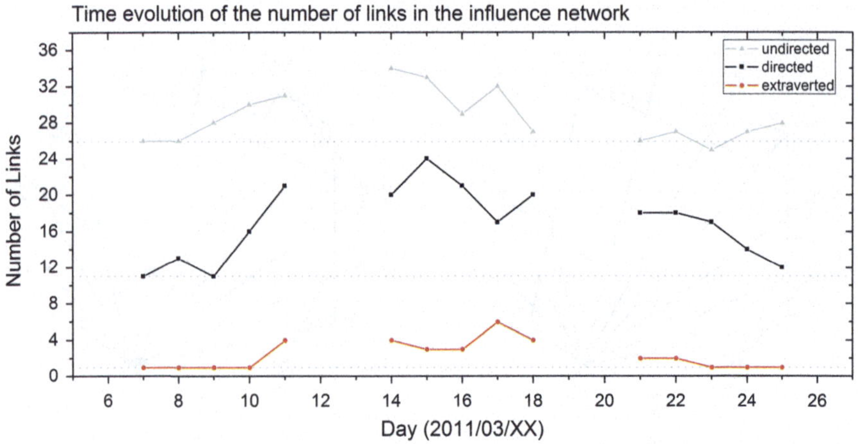

Fig. 1.6 Time evolution of the number of the different types of links in the influence network from 2011, March, 07th to 2011, March, 25th. *Dotted lines* indicate the number of links on 2011, March, 07th

market. It is possible that other factors besides the intervention contributed to the stabilization of the market; to discuss this aspect, it would be necessary the analysis of other periods where stability was reached with no intervention.

1.4 Final Remarks

In this paper we used a non-linear dependence measure based on the mutual information to access the dependence between pairs of currencies of the foreign exchange market. We analysed the sign of price difference of these markets from 2011, March, 07th to 2011, March, 25th, a period that includes the great earthquake in Japan and the intervention. By applying a time shift between the sign series we obtained different dependence structures between markets and then constructed an influence network based on them. The analysis of the influence network and its time evolution showed that the markets EUR/USD and USD/JPY are the most important nodes, with the information flowing from them to the other markets. It also suggested that the event of the earthquake changed the influence structure of the network, intensifying the interdependence between markets and changing the dynamics of the markets EUR/USD and USD/JPY; and the announcement of the intervention was effective in reverting the effects of the earthquake: changes could be observed in the day right after the announcement and the network totally returned to the state previous the earthquake in less than 1 week. The results represent a contribution to understand how the foreign exchange market reacts to big events and thus what can be done in periods of crisis. The analysis can also be useful to predict the behavior of one market based on the past behavior of another, if there is an influence relationship between them.

One important observation is that in the time shift cross-analysis the typical time shift is 0.1 s, i.e., when we have a market influencing another the time delay is 0.1 s. This fact is possibly related to the resolution of the data, also 0.1 s. We analysed the same data but with resolution 1s and could not detect time delay between markets as we found for resolution 0.1 s. We still need to study if we can detect the directionality between markets in other time resolution data or if the resolution 0.1 s is essential to detect such feature. Further researches also should include other currencies, a larger period of analysis and the possibility of time windows smaller than 1 day.

References

1. Bank for International Settlements (BIS) (2013) Triennial Central Bank Survey of foreign exchange turnover in April 2013—preliminary results released by the BIS. http://www.bis.org/press/p130905.htm.Accessed21Dec2014
2. Levinson M (2006) Guide to financial markets, 4th edn. Profile Books, London
3. Yura Y, Takayasu H, Sornette D, Takayasu M (2014) Financial brownian particle in the layered order-book fluid and fluctuation-dissipation relations. Phys Rev Lett 112(9):098703.doi:10.1103/PhysRevLett.112.098703
4. Mizuno T, Kurihara S, Takayasu M, Takayasu H (2003) Analysis of high-resolution foreign exchange data of USD-JPY for 13 years. Physica A 324(1):296–302.doi:10.1016/S0378-4371(02)01881-2
5. Mizuno T, Takayasu H, Takayasu M (2006) Correlation networks among currencies. Physica A 364:336–342.doi:10.1016/j.physa.2005.08.079
6. Shannon CE (1948) Bell Syst Tech J 27:379–423
7. Cover TM, Thomas JA (2006) Elements of information theory, 2nd edn. Wiley, Hoboken
8. Neely CJ (2011) Economic Synopses, 23
9. Ohira T, Sazuka N, Marumo K, Shimizu T, Takayasu M, Takayasu H (2002) Predictability of currency market exchange. Physica A 308(1):368–374.doi:10.1016/S0378-4371(02)00561-7
10. Kullmann L, Kertész J, Kaski K (2002) Time-dependent cross-correlations between different stock returns: a directed network of influence. Phys Rev E 66(2):026125.doi:10.1103/PhysRevE.66.026125
11. Fiedor P (2014) Networks in financial markets based on the mutual information rate. Phys Rev E 89(5):052801.doi:10.1103/PhysRevE.89.052801
12. Dionisio A, Menezes R, Mendes DA (2004) Mutual information: a measure of dependency for nonlinear time series. Physica A 344(1):326–329.doi:10.1016/j.physa.2004.06.144
13. Li W (1990) J Stat Phys 60:823–837
14. Granger C, Lin JL (1994) J Time Ser Anal 15:371–384

Chapter 2
Entropy and Transfer Entropy: The Dow Jones and the Build Up to the 1997 Asian Crisis

Michael Harré

Abstract Entropy measures in their various incarnations play an important role in the study of stochastic time series providing important insights into both the *correlative* and the *causative* structure of the stochastic relationships between the individual components of a system. Recent applications of entropic techniques and their linear progenitors such as Pearson correlations and Granger causality have included both *normal* as well as *critical* periods in a system's dynamical evolution. Here I measure the entropy, Pearson correlation and transfer entropy of the intra-day price changes of the Dow Jones Industrial Average (DJIA) in the period immediately leading up to and including the Asian financial crisis and subsequent mini-crash of the DJIA on the 27th October 1997. I use a novel variation of transfer entropy that dynamically adjusts to the arrival rate of individual prices and does not require the binning of data to show that quite different relationships emerge from those given by the conventional Pearson correlations between equities. These preliminary results illustrate how this modified form of the TE compares to results using Pearson correlation.

2.1 Introduction

One of the most pressing needs in modern financial theory is for more accurate information on the structure and drivers of market dynamics. Previous work on correlations [1] has lead to a better understanding of the topological structure of market correlations and mutual information [2] has been used to extend an earlier notion [3, 4] of a market crash as analogous to the phase-transitions studied in physics. These studies are restricted to static market properties in so far as there is no attempt to consider any form of causation. However, one of the goals of econophysics is to gain a better understanding of market dynamics and the drivers of these dynamics need to be extended to trying to measure causation. This is

M. Harré (✉)
Faculty of Engineering and IT, Centre for Research in Complex Systems, Sydney University, Sydney, NSW, Australia
e-mail: michael.harre@sydney.edu.au

H. Takayasu et al. (eds.), *Proceedings of the International Conference on Social Modeling and Simulation, plus Econophysics Colloquium 2014*, Springer Proceedings in Complexity, DOI 10.1007/978-3-319-20591-5_2

extremely difficult, strongly non-linear systems such as financial markets have feedback loops where the most recent change in price of equity a influences the price of b which in turn influences the price of a. This can make extracting causation relationships exceptionally difficult: the empirical distributions need to accurately reflect the temporal order in which price changes in the equities occur, and the time between these changes is itself a stochastic process. The goal of this paper is to introduce a (non-rigourous) heuristic that addresses these concerns using a modification to the conventional definition of the Transfer Entropy (TE) applied to the intraday tick data of the equities that make up the Dow Jones Industrial Average (DJIA) in the tumultuous build up of the Asian Financial Crisis (AFC) that culminated in the crash of the DJIA on the 27th October 1997. This article is arranged in the following way: Sect. 2.2 introduces the linear Pearson correlations I use as a comparison to the TE introduced in Sect. 2.3 in order to make comparisons and then discuss the results in Sect. 2.4.

2.2 Correlations

A statistical process generates a temporal sequence of data: $\mathbf{X}_t = \{\ldots, x_{t-1}, x_t\}$, X_t is a random variable taking possible states S_X at time t, $x_t \in S_X$ and $\mathbf{X}_t^k = \{x_{t-k}, \ldots, x_{t-1}\} \in \{S_X\}^{k-1}$ is a random variable called the k-lagged history of X_t. The marginal probability is $p(X_t)$, the conditional probability of X_t given its k-lagged history is $p(X_t|\mathbf{X}_t^k)$ and further conditioned upon the second process $\mathbf{Y}_t^k$ is $p(X_t|\mathbf{X}_t^k, \mathbf{Y}_t^k)$. The Pearson correlation coefficient r between such time series is:

$$r_t^k = \frac{\mathrm{cov}(\mathbf{X}_t^k, \mathbf{Y}_t^k)}{\sigma_X \sigma_Y} \tag{2.1}$$

where $\mathrm{cov}(\cdot,\cdot)$ is the covariance, σ_X and σ_Y are standard deviations and r_t^k is calculated over a finite historical window of length k where in order to calculate the dynamics of r_t^k this window is allowed to slide over the data, updating r_t^k as t progresses. A key issue with data that arrives at irregular or stochastic time intervals and r_t^k is desired is what counts as a co-occurrence at time t of new data. The most common method is to bin the data into equally separated time intervals of length δ_t and if two observations x_t and y_t occur in the interval $[t - \delta_t, t]$ then x_t and y_t are said to co-occur at time t, this approach is used for the correlations calculated in this article. Throughout the change in the log price is the stochastic event of interest: if at time t the price is p_t and at time t' it changes to $p_{t'}$ then the stochastic observable is $x_{t'} = \log(p_{t'}) - \log(p_t)$ [5], the increment $t' - t$ may be fixed in which case it is labelled δt or may dynamically vary, more on this below.

2.3 Transfer Entropy

Transfer Entropy was developed by Schreiber [6] as a rigorous way of measuring the directed transfer of information from one stochastic process to another after accounting for the history of the primary process (see below) for arbitrary distributions. This is a natural extension of Granger Causality, based on covariances rather than information measures, first introduced by Granger [7] in econometrics and in the case of Gaussian processes Granger causality and Transfer Entropy are equivalent [8]. Specifically, the entropic measures we are interested in are:

$$\mathbf{H}(X_t) = -\mathbf{E}_{p(X_t)}[\log p(X_t)], \tag{2.2}$$

$$\mathbf{H}(X_t, Y_t) = -\mathbf{E}_{p(X_t,Y_t)}[\log p(X_t, Y_t)], \tag{2.3}$$

$$\mathbf{H}(X_t|\mathbf{X}_t^k) = -\mathbf{E}_{p(X_t)}[\log p(X_t|\mathbf{X}_t^k)], \tag{2.4}$$

$$\mathbf{H}(X_t|\mathbf{X}_t^k, \mathbf{Y}_t^k) = -\mathbf{E}_{p(X_t)}[\log p(X_t|\mathbf{X}_t^k, \mathbf{Y}_t^k)], \tag{2.5}$$

where $\mathbf{E}_{p(\cdot)}[\cdot]$ is the expectation with respect to distribution $p(\cdot)$. The mutual information between two stochastic time series $\mathbf{X}_t$ and $\mathbf{Y}_t$ is:

$$\mathbf{I}(\mathbf{X}_t; \mathbf{Y}_t) \equiv \mathbf{H}(\mathbf{X}_t) - \mathbf{H}(\mathbf{X}_t|\mathbf{Y}_t) = \mathbf{H}(\mathbf{Y}_t) - \mathbf{H}(\mathbf{Y}_t|\mathbf{X}_t) \tag{2.6}$$

with a finite data window of length k this is the information theoretical analogue of r_t^k and the k-lagged *transfer entropy* (TE) from the *source* $\mathbf{Y}$ to the *target* $\mathbf{X}$ is:

$$\mathbf{T}^k_{Y\to X} \equiv \mathbf{H}(X_t|\mathbf{X}_t^k) - \mathbf{H}(X_t|\mathbf{X}_t^k, \mathbf{Y}_t^k). \tag{2.7}$$

$\mathbf{T}^k_{Y\to X}$ measures the degree to which X_t is disambiguated by the k-lagged history of Y_t beyond that to which X_t is already disambiguated by its own k-lagged history. This work presents recent developments in TE [9], information theory and the 'critical phenomena' of markets [2], and adds new results for real systems to the recent success in using it as a predictive measure of the phase transition in the 2-D Ising model [10]. The implementation of TE used in this work was done in Matlab using [11].

2.3.1 Transfer Entropy Without Binning

The most common and direct method of calculating any of r_t^k, $\mathbf{I}(\mathbf{X}_t; \mathbf{Y}_t)$ or $\mathbf{T}^k_{Y\to X}$ is to use discrete time series data. This is made possible either by the nature of the study itself where discrete time steps are inherent or through post-processing of the data by binning it into a discrete ordered sequence. However, a lot of interesting data, including intra-day financial markets data, is inherently unstructured and binning the data loses some of the temporal resolution and obfuscates the relationship between past and future events making causal relationships difficult to establish, so an alternative is proposed that addresses these issues.

I define a modified form of $\mathbf{T}^k_{Y\to X}$ by first redefining the stochastic time series in order to capture the continuous nature of the price arrival process. With t and $t' \in \mathbb{R} > 0$ where 0 is taken as the start of trading on any given trading day and $\{t_i\}$ and $\{t'_j\}$ are the finite sequence of times at which the (log) price changes for two different equities during that day. Define the arrival indices of time series of length I and J as $\{i \le I\} \in \mathbb{N}$ and $\{j \le J\} \in \mathbb{N}$. Now there are two finite sequences of price changes on a single trading day d: $\{X^d(t_i)\}$ and $\{Y^d(t'_j)\}$. The entropy of $\{X^d(t_i)\}$ conditioned on its most recent past value is:

$$\mathbf{H}(X^d(t_i)|X^d(t_{i-1})) = -\mathbf{E}_{p(X^d)}\big[\log(p(X^d(t_i)|X^d(t_{i-1}))\big], \quad i > 1. \tag{2.8}$$

An equivalent definition for the entropy conditioned on the most recent past of both $\{X^d(t_i)\}$ and $\{Y^d(t'_j)\}$ is:

$$\mathbf{H}(X^d(t_i)|X^d(t_{i-1}), Y^d(t'_{j-1})) = -\mathbf{E}_{p(X^d)}\big[\log(p(X^d(t_i)|X^d(t_{i-1}), Y^d(t'_{j-1}))\big] \tag{2.9}$$

where $i, j > 1$ and t'_{j-1} is the minimum value such that, for a given t_i: $(t_i - t'_{j-1}) > 0$. This modified definition of the TE (for the rest of this article this is simply referred to as *the* TE) is:

$$\overline{\mathbf{T}}_{Y^d\to X^d} \equiv \mathbf{H}(X^d(t_i)|X^d(t_{i-1})) - \mathbf{H}(X^d(t_i)|X^d(t_{i-1}), Y^d(t'_{j-1})). \tag{2.10}$$

The relationship between this and other measures is illustrated in Fig. 2.1. The first row shows the log price changes for two equities (Alcoa and Boeing) as a stochastic time series with an irregular arrival rate. The black arrows indicate the direction and magnitude of the log price changes. The second row shows the changes in prices binned into time intervals of width δt so that changes that occur in the same time interval are considered co-occurring. In the third row is the lag-1 Pearson correlations or lag-1 mutual information, the causal direction of correlations is implicit in the time ordering of the bins, hence the arrows point forward in time. This does not account for the shared signal between x_{t-1} and y_{t-1}. The fourth row shows the lag-1 Granger causality or transfer entropy, the signal driving y_t is x_{t-1} after excluding the common driving factor of y's past: y_{t-1}. Red arrows indicate the measured signal from the source (Alcoa) to the target (Boeing) and blue arrows indicate y's signal that is being removed. Fifth row (fewer price changes shown for clarity): An alternative way to calculate the TE. Choose the target time series (in this case Boeing) and condition out the most recent previous price change in Boeing and then use only the most recent change in Alcoa as the source signal. Note that some Alcoa price signals are missed and some are used more than once and that price changes will rarely co-occur.

The definition of Eq. (2.10) has a number of appealing properties:

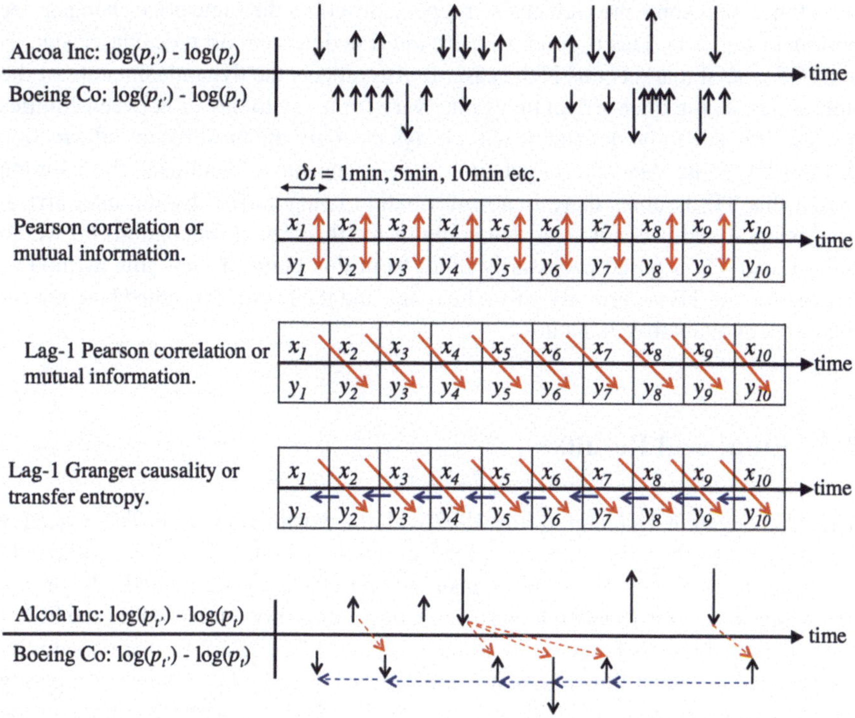

Fig. 2.1 A representation of different measures of 'instantaneous' and 'lagged' relationships between stochastic time series data

- Using a fixed interval in which the price at the beginning is compared with the price at the end of the interval conflates signals that may occur before or after another signal but arrives during the same binning interval, thereby mixing future and past events in the measured relationships between bins.
- Similarly, multiple price changes within δt may net to zero change and so some price signals are missed.
- As bin sizes get smaller they are less statistically reliable as fewer events occur within each bin, equally as bin sizes get larger there are fewer bins per day, thereby also reducing the statistical reliability.
- Over the period of a single day, for each bin size the number of total bins is: $\delta t = 30$ min: 13 bins/day, $\delta t = 1$ min: 390 bins/day, whereas the raw data may have 50–5000+ price changes in a day.

The proposed heuristic for the TE introduced above addresses some of these shortcomings but not without introducing some other issues. First, it will always condition out the most recent price change information in the target equity (Boeing in Fig. 2.1) and so uses every bit of relevant information in the target time series. It also uses the most recent price change from the source time series, however it will sometimes miss some price changes or repeatedly count the same price changes (see bottom of Fig. 2.1). This is good if we are interested in the most recent price signals and in financial markets this is the case. It also reflects the dynamical nature of the time series, as the inter-arrival times may vary from day to day or between equities no new δt needs to be defined, it will always use only the most recent information in both the source and the target time series. The most significant shortcoming is that this TE assumes there is no information being carried by the inter-arrival time interval and it is not clear that some of the theoretical foundations on which the original TE is based necessarily hold, from this point of view this method of calculating the TE is currently only a heuristic and the results presented here are for the moment qualitative in nature.

2.4 Empirical Results

The AFC began in Thailand in July 1997 with the devaluation of the Thai currency (the Bhat) and the crisis rapidly spread throughout South East Asia, ultimately resulting in the October 27 "mini-crash" of the DJIA, losing around 7 % on the day which was at the time the largest single day points drop on record for the DJIA, for a review of the crisis see [12] and the top plot of Fig. 2.2. Note that the entropy measurements shown illustrate that some care needs to be taken when comparing simple systems with data from real 'complex systems': the increase in the entropy of the DJIA on the 24th of June looks like what might be described as a 'first order' phase transition as studied in complex systems [13], but it is almost certainly caused by the rescaling of price increments on the New York Stock Exchange.[1]

This rescaling did have an interesting impact on the TE though, as can be seen in Fig. 2.3. Prior to the 24th of June there is considerable structure in the TE measure (warm colours denote high TE values, cooler colours denote lower TE values), however all signals drop off significantly immediately after this date although much of the structured signal eventually returns (not shown). The most notable signals are equities that act as targets of TE for multiple other equities, seen as yellow vertical strips indicating that many equities act as relatively strong sources of TE for a single equity: AT&T (equity 26), Wall Mart (equity 30) and McDonalds (equity 31) stand out in this respect. Notable single sources of TE are less obvious but Cocoa Cola and AT&T (equities 19 and 26) show some coherent signals indicated by multiple red points loosely forming a horizontal line. It is intriguing to note that the Pearson correlations showed no similar shift on the 24th of June (not shown) while conversely in Fig. 2.4 the mini-crash on the 27th October 1997 (day 64) there is a clear signal that the DJIA equities are significantly more correlated with no corresponding increase in the TE on that day (not shown) despite the general turmoil of the markets, as seen by significant fluctuations in the correlations on nearby days.

[1] For details see: http://www1.nyse.com/nysenotices/nyse/rule-changes/detail?memo_id=97-33.

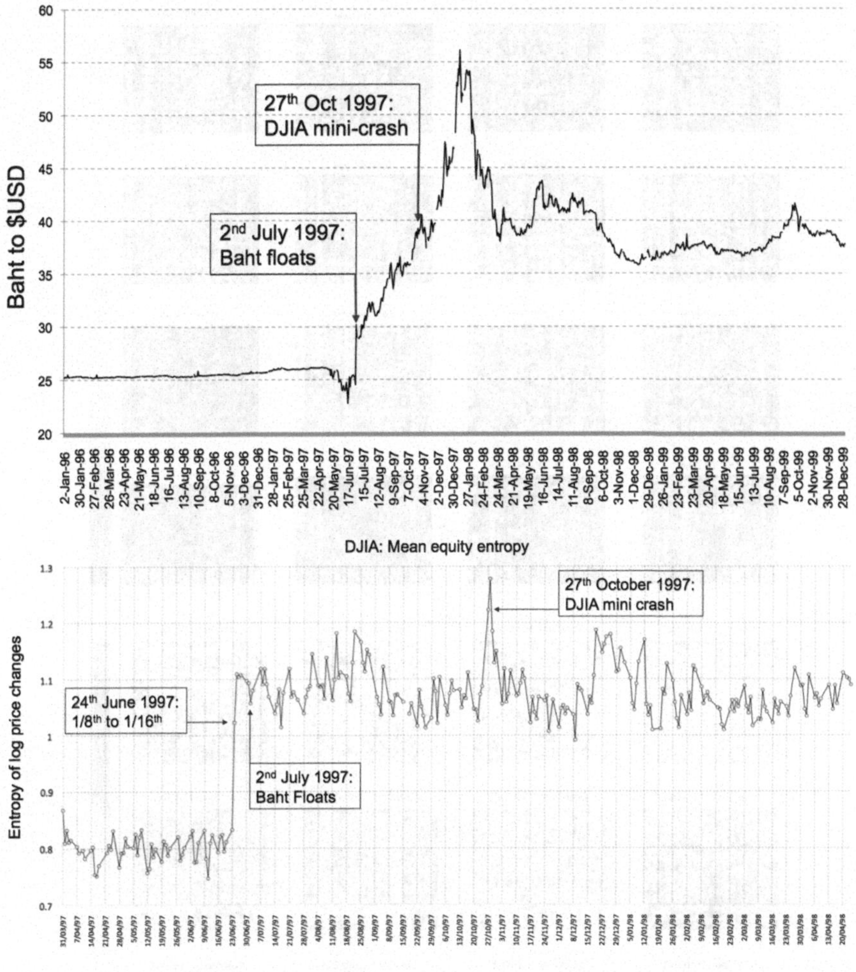

Fig. 2.2 The AFC and its key components. *Top plot*: the AFC is thought to have begun as the Baht was allowed to float against the US dollar on the 2nd of July 1997. The crisis contagion spread through the asian markets ultimately leading to the mini-crash of the DJIA on the 27th October 1997. *Bottom plot*: on the 24th of June 1997 the New York Stock Exchange changed its minimum incremental buy/sell price from 1/8th of a dollar to 1/16th of a dollar, causing the entropy of the price changes to shift suddenly and permanently, but not influencing the DJIA index itself. The crash on the 27th October 1997 is seen as the second largest peak in the entropy, the largest being the 28th of October

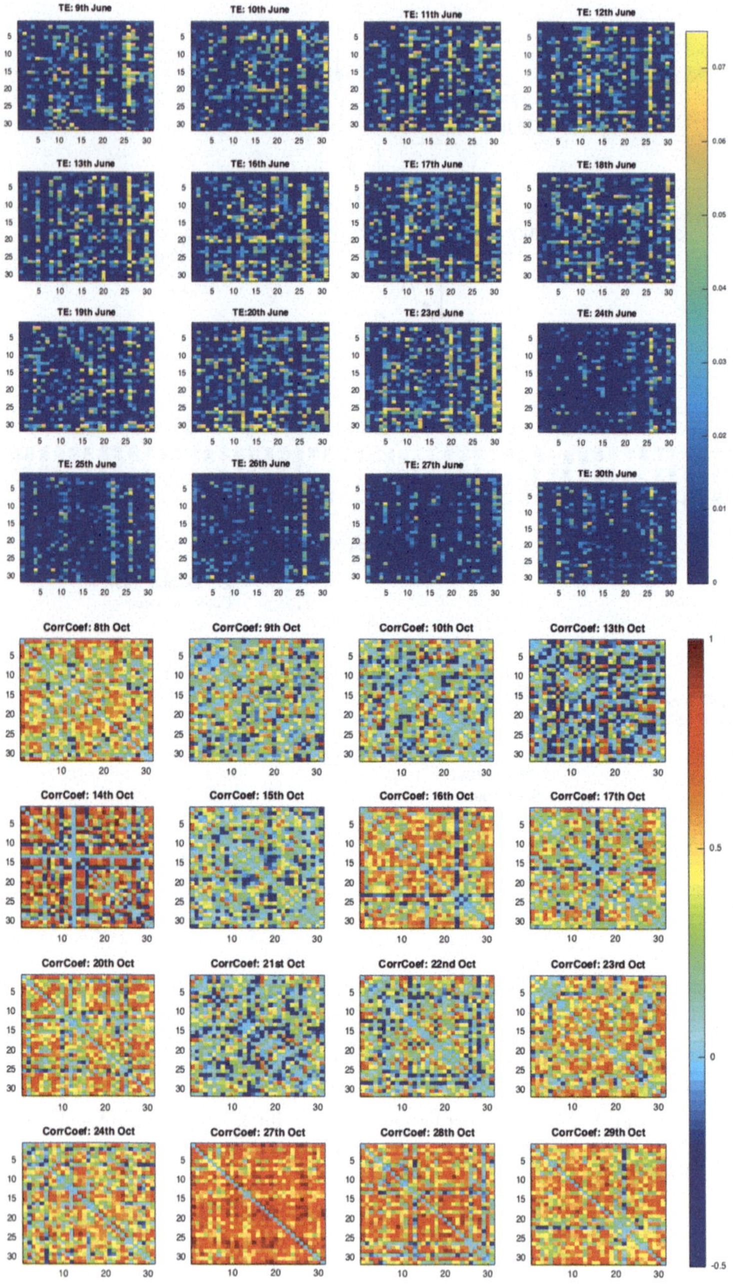

Fig. 2.3 *Top*: the TE from one DJIA equity to another equity indexed from 1 to 31. Index 1 = the DJIA, vertical axis is the *source* equity, horizontal axis is the *target* equity. The 24th of June 1997 clearly stands out as the first day of a substantive reduction in the TE between equities. *Bottom*: the Pearson correlation for the DJIA data binned using $\delta t = 30$ min. The market crash on the 27th October stands out during a turbulent time in the market's dynamics

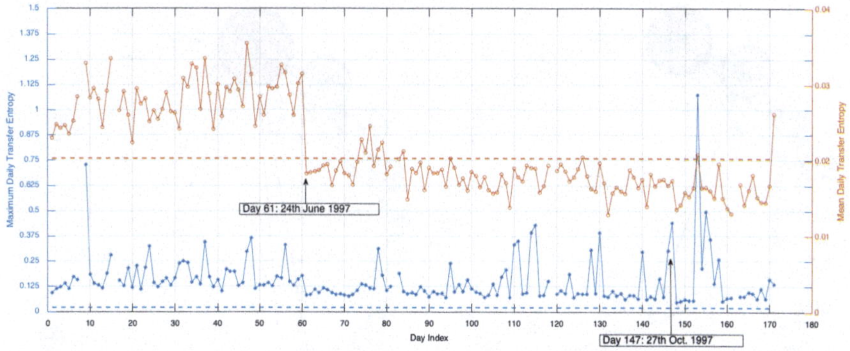

Fig. 2.4 Maximum (*blue*, *lower line* measured from the left axis) versus average (*orange*, *upper line* measured from the right axis) TE. Shuffled average of TE $\simeq$ 0.02 are *blue dashed* for left axis and *orange dashed* for right axis

In Fig. 2.4 it is shown that the average TE across all equities is quite stable except for the drop occurring at the time of the change in minimum price increments on the 24th June. A simple shuffling test [14] estimates the TE for unrelated data to be approximately 0.02 nits on average (see the dashed lines, randomly sampled before and after the drop on day 61) but note that numerical estimations of TE are difficult so the TE sometimes drop below zero. This suggests that *on average* the TE across the DJIA is close to negligible but that some equities clearly have TE values *significantly exceeding* the 0.02 nits level, as shown by the blue line values. The largest peak in the Maximum TE plot occurs 6 days after the Dow crashes and is from the Disney equity to the McDonalds equity.

Finally in Fig. 2.5 is plotted two networks of relationships between the equities based on Pearson correlations and TE. The Pearson correlation network is ordered counterclockwise according to the total link weight of each equity and a link was included if its correlation was greater than 0.4. The TE network is ordered counterclockwise by total link weight, the colour represents the total weight of incoming links and the node size represents the total weight of outgoing links and a link was included if its TE was greater than 0.05 nits. Thresholds were chosen such that 10 % of all links in each network are included. The most notable differences between these networks is the changes in the relative importance of the individual equities. The overall DJIA index (DJI) is significantly correlated with other equities whereas this index is the *least* significant node in the TE network. Similarly, Walmart (WMT) is very well connected in the TE network but it is the least relevant node in the Pearson correlation network.

These are preliminary results using the comparatively small dataset of the 30 equities that make up the DJIA and will need to be confirmed on other indices and other crashes. There is one very significant point that comes out of this study: The driver of correlations between equities in financial markets is not necessarily the changes in the prices of other equities. This is true in the sense that changes in

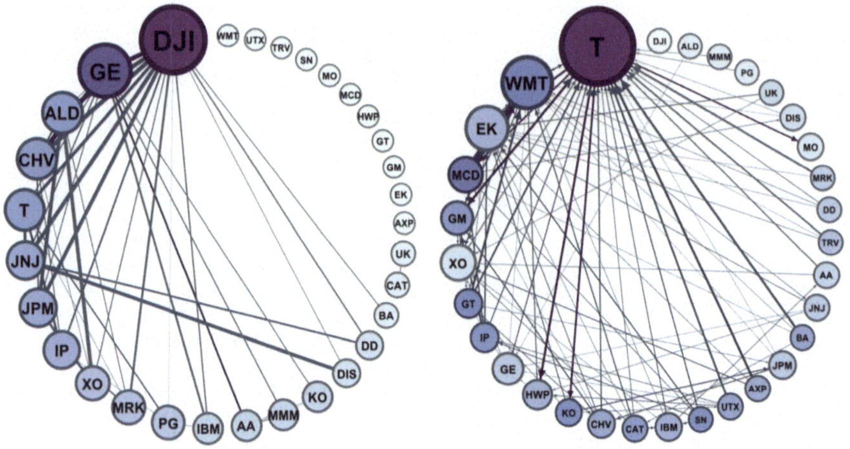

Fig. 2.5 The Pearson correlation (*left*, undirected links) compared to TE (*right*, directed links) network of relationships for a typical trading day (16th June 1997)

transfer entropy may leave correlations unchanged and changes in correlations are not necessarily driven by changes in transfer entropy. The former is a consequence of the top plot of Fig. 2.3 (the plots showing the lack of change in correlations is not shown due to space limitations), the latter is a consequence of the lower plot Fig. 2.3 for the Asian crisis crash (the plots showing the lack of change in transfer entropy are not shown). However, in the case of the Asian crash, the transfer entropy significantly peaked several days after the crisis but the significance of this is not clear from the data. This result is not peculiar to trading days in which known 'significant' events have occurred. Figure 2.5 shows an ordinary trading day in which the DJIA index plays a significant role in the correlation structure (left plot) but this relationship vanishes for the transfer entropy structure (right plot), compare for example the position of Walmart (WMT) in the two plots. In fact there appears to be very little relationship between strongly correlated equities and those that 'transfer' high values of entropy.

One of the goals of this work was to explore the analogy between phase transitions in statistical physics and market crashes in finance. Although recent work on precursors to phase transitions in physics has shown that it is a peak in a global measure of TE acts as precursor [10], it is interesting that peaks in Pearson correlations are not necessarily coincidental with peaks in TE for financial markets suggesting that it is not the transfer of entropy between equities within the DJIA that is driving the correlations but some signal external to the market. The results in [10] suggest that if the DJIA mini-crash was analogous to the second order phase-transition in the Ising model then peaks in the pairwise TE, mutual information and Pearson correlation [15] would be observed at the crash. However, in this and earlier studies only peaks in Pearson correlations and mutual information have so far been established during a market crash requiring verification and opening up a number of interesting questions for further work.

References

1. Mantegna RN (1999) Eur Phys J B Cond Matter Complex Syst 11(1):193
2. Harré M, Bossomaier T (2009) EPL (Europhys Lett) 87(1):18009
3. Vandewalle N, Boveroux P, Minguet A, Ausloos M (1998) Physica A: Stat Mech Appl 255(1):201
4. Plerou V, Gopikrishnan P, Stanley HE (2003) Nature 421(6919):130
5. Stanley HE, Mantegna RN (2000) An introduction to econophysics. Cambridge University Press, Cambridge
6. Schreiber T (2000) Phys Rev Lett 85(2):461
7. Granger CW (1969) Econometrica: J Econ Soc 37:424–438
8. Barnett L, Barrett AB, Seth AK (2009) Phys Rev Lett 103(23):238701
9. Barnett L, Bossomaier T (2012) Phys Rev Lett 109(13):138105
10. Barnett L, Lizier JT, Harré M, Seth AK, Bossomaier T (2013) Phys Rev Lett 111(17):177203
11. Lizier JT (2014) JIDT: an information-theoretic toolkit for studying the dynamics of complex systems. Front Robot AI 1:11. doi:10.3389/frobt.2014.00011
12. Radelet S, Sachs J (1998) The onset of the east asian financial crisis. Tech. rep., National bureau of economic research
13. Li W, Packard NH, Langton CG (1990) Physica D: Nonlinear Phenom 45(1):77
14. Kwon O, Yang JS (2008) EPL (Europhys Lett) 82(6):68003
15. Matsuda H, Kudo K, Nakamura R, Yamakawa O, Murata T (1996) Int J Theor Phys 35(4):839

Chapter 3
Execution and Cancellation Lifetimes in Foreign Currency Market

Jean-François Boilard, Hideki Takayasu, and Misako Takayasu

Abstract We analyze mechanisms of foreign currency market order's annihilation with a focus on the lifetime of these orders. Limit orders submitted in this market are approximately executed according to the random walk theory. In consequence, the distribution of execution lifetime can be approximated by a power law with exponent 1/2. Alternatively, limit orders submitted in foreign currency markets are roughly cancelled according to a mixed distribution; as a random walk with a tail following a power law. The cancellation lifetime distribution can be approximated by using a scaling relationship between the distance from mid-price and the random walk theory. In addition, we introduce the concept that market participants cancel orders depending on the market price's movement which is represented as the movement of the mid-price. Taking into consideration market conditions when orders have been injected, market participants do not have symmetric decision rules. This behavior could at least partially explain the shape of price change distribution.

J.-F. Boilard (✉) • M. Takayasu
Department of Computational Intelligence and Systems Science, Interdisciplinary Graduate School of Science and Engineering, Tokyo Institute of Technology, 4259-G3-52, Nagatsuta-cho, Midori-ku, Yokohama 226-8503, Japan
e-mail: boilard.j.aa@m.titech.ac.jp; takayasu@dis.titech.ac.jp

H. Takayasu
Sony Computer Science Laboratories, 3-14-13, Higashigotanda, Shinagawa-ku, Tokyo 141-0022, Japan

Meiji Institute of Advanced Study of Mathematical Sciences, Meiji University, 4-21-1 Nakano, Nakano-ku, Tokyo 164-8525, Japan

Department of Computational Intelligence and Systems Science, Interdisciplinary Graduate School of Science and Engineering, Tokyo Institute of Technology, 4259-G3-52, Nagatsuta-cho, Midori-ku, Yokohama 226-8503, Japan
e-mail: takayasu@csl.sony.co.jp

H. Takayasu et al. (eds.), *Proceedings of the International Conference on Social Modeling and Simulation, plus Econophysics Colloquium 2014*, Springer Proceedings in Complexity, DOI 10.1007/978-3-319-20591-5_3

3.1 Introduction

Financial market movements have been studied firstly by Bachelier who described price movements as a random walk [1]. Since then, many researchers studied the microstructure of markets [2–4] and more recently some of them have focused on order book fluctuations using physics methodology [5–7]. Availability of new detailed database help researchers precisely describe the impacts of some characteristics on the order book even if orders do not directly contribute to execution.

In addition, some researchers describe order injection deal and cancellation from a physics viewpoint. In a paper appeared in 2014, Yura et al. studied the correlation between layers in the order book and the market price movement based on the analogy with the colloidal motion in water molecules [8].

Recently, a high proportion of market transactions are done by automated traders (financial algorithms) which increase the reaction speed in the market as a whole [9]. Regulators focus on analysing lifetime of orders—the amount of time spent in the market before annihilation—to understand the impact of market speed on the stability of the financial market [10, 11]. An extraordinary event in May 2010, better known as the "flash crash", puts light on direct impact of cancellation orders on market volatility [12]. Actually, governmental organizations recognize that cancellation orders should be studied more carefully concerning cancellation lifetime [13].

The next section details the database used for this study. In Sect. 3.3 we describe lifetime statistical properties of annihilated limit orders with a focus on the relationship between executed and cancelled orders. Section 3.4 describes how market participants cancel their orders considering movement in market price. The final section contains a summary.

3.2 Description of the Database

We use a special database of Electronic Broking System (EBS) which contains identifications of every order. The database is from March 13th 21:00 (GMT) to March 18th 21:00 (GMT) 2011, and contains information about injected and annihilated orders with minimal tick time of 1 ms. This foreign exchange market is open 24 h per day during weekdays.

Traders must have a direct access to the market server via a private network and most of them are either banks or financial institutions. The majority of participants use financial algorithms for trading activity. In consequence, EBS market implements a minimum cancellation time rule of 250 ms. Then, it is not possible for participants to cancel a previous submitted order if the quote life is lower than 250 ms. As soon as the life quote is higher than this minimum, it is free for participants to do as they want.

Table 3.1 Details concerning limit orders

Currency	Submitted	Cancelled	Executed
USD/JPY	2,695,128	2,436,385 (90.4 %)	258,743 (9.6 %)
EUR/JPY	1,481,285	1,439,801 (97.2 %)	41,484 (2.8 %)
EUR/CAD	370,903	370,717 (99.9 %)	186 (0.1 %)

The order book is initialized at the beginning of the week and similarly all orders remaining at the end of the week are deleted. This market is described as over-the-counter (OTC). It means that market participants must already have a credit agreement with other participants before they transact with each other. If this kind of agreement does not already exist, it is not possible for them to make a transaction. In other words, it is possible to see a buying order at a higher price than a selling order (negative spread) if there is no credit agreement between these market participants.

The EBS database contains forty-eight currencies but our study focuses on two of them: USD/JPY and EUR/JPY. Both are very liquid currencies in EBS market. During our studied period, the minimal tick (pip) for USD/JPY and EUR/JPY is 0.001 yen.

Table 3.1 describes the occurrence of limit orders submitted during our studied period and the high proportion of cancelled orders is explained by the type of traders within this market. In fact, over 90 % of traders are automated softwares (algorithmic trading) and most of them are market makers. We can define market making strategy as a provider of limit orders in the electronic order book which in exchange is remunerated with the spread charged between bid and ask orders and possibly a rebate fee (if available). In consequence, they have a high incentive to protect themselves against execution risk (risk to see their limit orders executed at the worst time). As cancelling and submitting orders do not incur costs, they often update their orders depending on market conditions (volatility, news, etc.) which increase significantly the proportion of cancelled orders [14].

The most traded currency in EBS database is the USD/JPY in term of submitted limit orders and globally 90 % of all limit orders are ultimately cancelled. Another widely traded currency is EUR/JPY especially for triangular arbitrage. The percentage of cancellations increase when the currency is less liquid, for example, almost all submitted orders in EUR/CAD currency pair are cancelled which makes the proportion of cancelled orders extremely high (99.9 %).

In this current research, we use the notation $b(t)$ as the best bid price at time t and $a(t)$ as the best ask (offer) price at time t. The mid-price $m(t)$ is the average between the best bid and best ask price: $m(t) = [b(t)+a(t)]/2$. Figure 3.1 represents histogram of initial distance of orders in pips from the mid-price. The black histogram is the frequency of submitted orders at this initial distance and the grey histogram is the frequency of submitted orders at this initial distance knowing they will be executed.

We can observe that orders initially submitted around mid-price have higher chances to be executed than orders far from these prices. In addition, the proportion of executed orders decreases for orders far from the mid-price. In our research, we

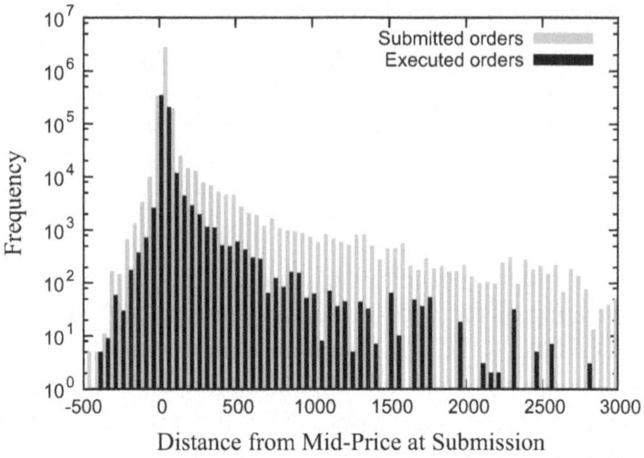

Fig. 3.1 Semi-log frequency plot of initial distance of limit orders from the mid-price of USD/JPY currency pair. *Grey* histogram represents all limit orders and *black* histogram represents limit orders to be executed

analyze both bid and ask orders symmetrically as they have similar properties in our market.

3.3 Execution and Cancellation Lifetimes

The life of an order starts when it is injected in the electronic order-book. Similarly, this order's life ends when it is finally executed or cancelled (annihilation process). The time between the beginning and ending of this order is defined as lifetime. In this section we focus on statistical properties of execution and cancellation lifetimes.

3.3.1 Execution Lifetime

An electronic order-book is filled by limit orders. A transaction occurs when a buying order (bid) is at a price equal or higher than a selling order price (ask or offer). In addition, both market participants must already have a credit agreement to fulfil the transaction. Figure 3.2 represents the cumulative distribution of lifetime of limit orders that are executed.

Execution lifetime (t_L) can be described as the time between the injection of the order and execution. Assuming the market price is approximated by a random walk and knowing a typical injection is close to the market price, the lifetime is estimated by a recurrence time of random walk [8]:

$$P(> t_L) \propto t_L^{-\frac{1}{2}} \tag{3.1}$$

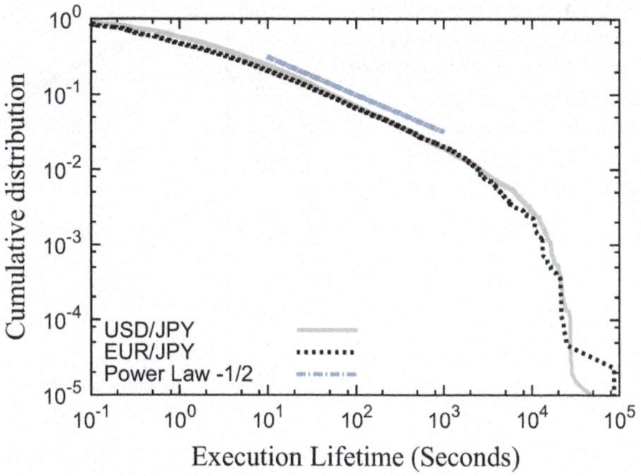

Fig. 3.2 Log-log cumulative distribution of execution orders lifetime. The *grey bold* and *black dotted lines* respectively represent USD/JPY and EUR/JPY currency pairs

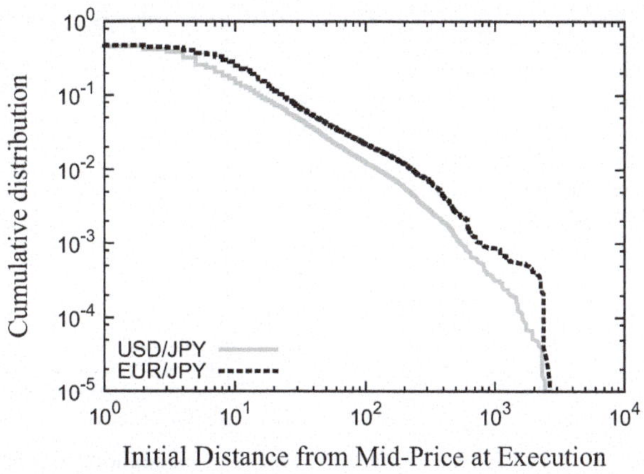

Fig. 3.3 Log-log cumulative distribution of the initial distance in pips from mid-price when orders are executed. The *grey bold* and *black dotted lines* respectively represent USD/JPY and EUR/JPY currency pairs

As confirmed in Fig. 3.2 lifetimes of execution orders in EBS foreign market are approximately following the random walk theory both for USD/JPY and EUR/JPY currency pairs.

Figure 3.3 represents the distance from mid-price at injection of orders later annihilated via execution. There are a lot of orders initially injected close from market price and ultimately executed.

3.3.2 Cancellation Distance and Lifetime

In this foreign currency market, participants can always decide to cancel their initial limit orders if their lifetime is longer or equal to 250 ms. Contrary to execution orders, cancellation orders are annihilated at a certain distance from the mid-price.

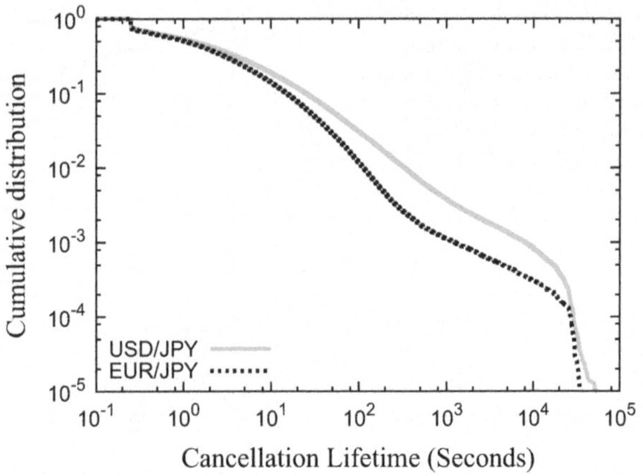

Fig. 3.4 Log-log cumulative distribution of cancellation orders lifetime. The *grey bold* and *black dotted lines* respectively represent USD/JPY and EUR/JPY currency pairs

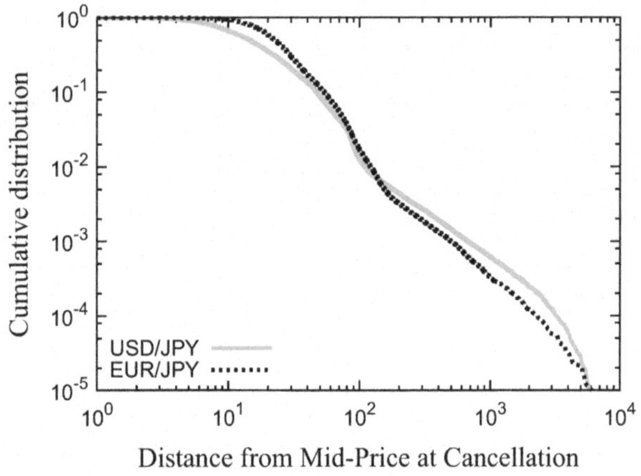

Fig. 3.5 Log-log cumulative distribution of distance in pips from mid-price when orders are cancelled. The *grey bold* and *black dotted lines* respectively represent USD/JPY and EUR/JPY currency pairs

Many reasons can push a market participant to cancel an order such as being more or less aggressive with the price of his limit order.

Figure 3.4 represents the cancellation lifetime (T_L) cumulative distribution. In both currency pairs, more than 20 % of orders are cancelled at the minimum allowed cancellation lifetime (250 ms). Furthermore, the cancellation lifetime distribution bends in comparison to execution lifetime. In other words, there is significantly more orders staying a long time in the electronic order book and ultimately cancelled.

In Fig. 3.5 we analyze the cumulative distribution of cancellation orders from the mid-price [$m(t)$]. For small distance from mid-price represented from 10^0 to 10^1, both currencies initially have a plateau because of the spread between bid and ask orders. USD/JPY and EUR/JPY generally have a spread of approximately ten pips. If the order is closer from the mid-price, the chance to be executed is greater. Inversely, there are lower chances to be executed if orders are far from the mid-price. The distribution of the distance from mid-price at cancellation (Fig. 3.5) is significantly different from the distribution of the initial distance from mid-price at execution (Fig. 3.3).

3.3.3 Power Law Relation Between Execution and Cancellation Lifetimes

In Sects. 3.3.1 and 3.3.2 we described distributions of execution and cancellation lifetimes. In this section, we show there is a relationship between both distributions.

In Fig. 3.6 cumulative distribution of lifetime of annihilated events are plotted for cancellation and execution in log-log scale. Difference between execution and cancellation lifetimes observed in this figure is mostly due to the non-constant proportion between execution and cancellation of orders. In consequence, orders far from the mid-price takes longer time before being annihilated via cancellation. Limit orders far from the mid-price are often referred as *extreme orders*. EBS market is open from Sunday 21:00 (GMT) to Friday 21:00 (GMT) which means the maximum cancellation and execution lifetime is $\sim 6 \times 10^5$ s. In other words, there is a natural cutoff in the lifetime of orders in our database. At the end of the week (Friday 21:00 GMT), all orders remaining in the order book are deleted and those deleted orders are not include in cancellation statistics. New injections of orders to fill the order book are only allowed on the next Sunday 21:00 GMT.

In the electronic order book, market participants initially submit their orders at a certain distance ($\bar{r}$) from the market price. If the order is closer from market price, market participants will assume there is a higher probably to see their orders executed in a shorter period of time (T). Inversely, orders far from the market price will take longer time before being executed. When the market participants inject their orders, they are aware of this fact and will cancel their orders if the initial scenario does not hold anymore. If this is the case, they might choose to cancel and inject their orders again at a different price. Considering this mechanism, it is

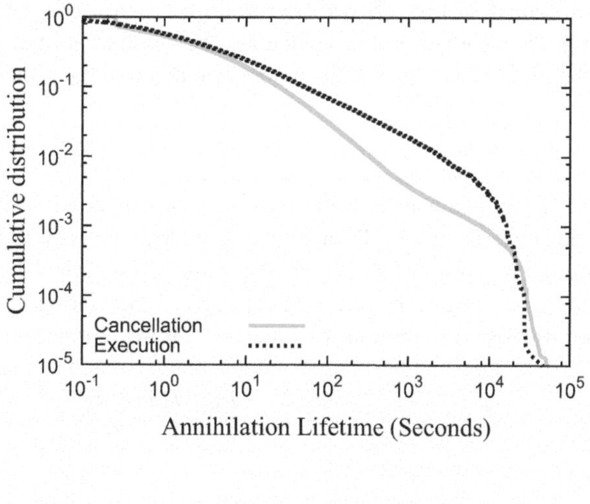

Fig. 3.6 Log-log cumulative distribution of annihilated orders lifetime of USD/JPY currency pair

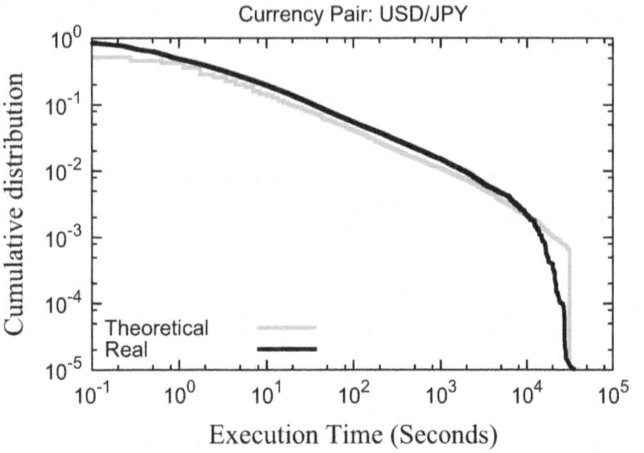

Fig. 3.7 Log-log cumulative distribution of execution orders lifetime. The *grey line* represents theoretical distribution estimated from Fig. 3.3 using Eq. (3.3) and the *black line* shows real data execution lifetime (Fig. 3.2)

reasonable to assume the random walk theory just like the case of derivation of Eq. (3.1). We can introduce the following scaling relation:

$$T \propto \bar{r}^2 \tag{3.2}$$

As demonstration of the scaling relation presenting in Eq. (3.2), we apply this relationship on the USD/JPY currency pair. Execution lifetime is approximated in using the probability distribution of the initial distance from mid-price of orders executed presented in Fig. 3.3 with the above scaling relation multiplied by a factor. Equation (3.3) is used to calculate the theoretical line from Fig. 3.3. Exempt the slower decay at the plateau on Fig. 3.7, the theoretical and real data lines fit well.

$$T = 0.07 \times \bar{r}^2 \tag{3.3}$$

To continue, we use the same methodology to approximate the cancellation lifetime of USD/JPY currency pair. Cancellation lifetime is approximated in using the probability distribution of the distance from mid-price of orders cancelled presented in Fig. 3.5 with the scaling relation (Eq. (3.2)) multiplied by a factor. Equation (3.4) is used to calculate the theoretical line. Exempt the slower decay at the plateau on Fig. 3.8, the theoretical and real data lines agree approximately.

$$T = 0.01 \times \bar{r}^2 \tag{3.4}$$

If the distance from the mid-price at cancellation follows a power-law such as the interval 10^2 to 10^4 in Fig. 3.8, we can directly link each other exponent. We define two power law exponents, α and δ respectively representing the cancellation lifetime and distance from mid-price at cancellation. For cancellation lifetime (T_L) cumulative distribution, we have the following relationship:

$$P(> T_L) \propto T_L^{-\alpha} \tag{3.5}$$

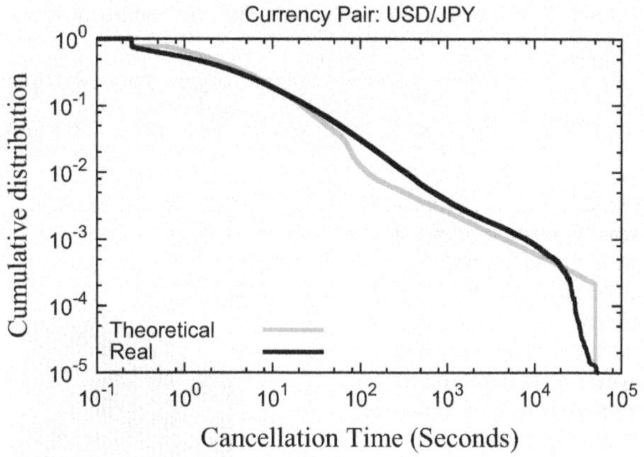

Fig. 3.8 Log-log cumulative distribution of cancellation orders lifetime. The *grey line* represents theoretical distribution estimated from Fig. 3.5 using Eq. (3.4) and the *black line* shows real data cancellation lifetime (Fig. 3.4)

In addition, cancellation orders from mid-price follows:

$$P(>\bar{r}) \propto \bar{r}^{-\delta} \tag{3.6}$$

By incorporating Eqs. (3.2), (3.5) and (3.6) we have the following relationship for cancellation lifetime cumulative distribution:

$$P(>T_L) \propto T_L^{-\frac{\delta}{2}} \tag{3.7}$$

Therefore, we find that δ can be linked with α :

$$\frac{\delta}{2} \approx \alpha \tag{3.8}$$

3.4 Cancellation Orders and Market Price Movement

One might suspect that market participants cancel their orders depending on the market conditions. To analyze the relative behavior of market participants toward market price movements, we introduce new quantities A_i and B_i as shown below. The numerator of both quantities represents the distance of an order from the mid-price at annihilation (time T). $m(t)$ is the mid-price at time t, $b_i(t)$ refers to the bid order *i-th* submitted at time t and $a_i(t)$ refers to the ask order *i-th* submitted at time t. The denominator represents the distance of the order from the mid-price at injection (time t).

In the case that $A_i = 0$ ($B_i = 0$) means that the mid-price did not move from the injection of the *i-th* ask (bid) to its annihilation moment. A_i (B_i) higher than zero means that the market movement moves further from the order price; it becomes more difficult for this order to be executed because the distance from the mid-price

Table 3.2 Frequency of cancellation orders filtered by the ratio value

Currency	Total	> 0	< 0	= 0	Neg. spread	Exclude
USD/JPY	2,436,385	1,063,715	877,185	457,397	7157	30,931
EUR/JPY	1,439,801	672,453	446,137	315,960	429	4822

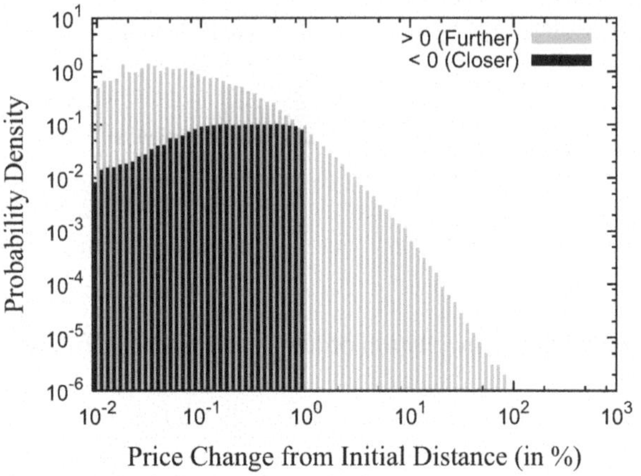

Fig. 3.9 Log-log probability density function (PDF) of price change from initial distance for USD/JPY currency pair. *Grey bars* represent situations where the market price is getting further from individual orders price. *Black bars* represent situations where the market price is getting closer from individual orders price

is getting higher. Inversely, A_i (B_i) lower than zero means that the market moves closer toward the order price.

$$A_i = \frac{[a_i(t) - m(T)]}{[a_i(t) - m(t)]} - 1 \tag{3.9}$$

$$B_i = \frac{[m(T) - b_i(t)]}{[m(t) - b_i(t)]} - 1 \tag{3.10}$$

Table 3.2 describes the frequency of cancellations filtering for different situations. The label *Neg. Spread* means that the bid-ask spread is negative at the moment of the cancellation order (time T). Again, it can happen because EBS market requires a credit agreement between every market participant before realizing transactions. If there is no agreement, even a buying price higher than a selling price will not trigger a transaction. In our probability density function analysis, we do not include this type of data. The label *Exclude* represents situations where the injection of order was initially done at a negative spread (time t). This situation has been excluded to the probability density function analysis.

To construct Fig. 3.9, we use bin size of $10^{-0.3}$. Bars with label > 0 *(Further)* represent cases where the ratio of Eqs. (3.9) and (3.10) is higher than zero. Inversely, bars with label < 0 *(Closer)* represent cases where the ratio of Eqs. (3.9) and (3.10) is lower than zero.

3.5 Discussion

Our study focuses on lifetime of annihilated orders either by execution or cancellation. Using the example of USD/JPY currency pair, we demonstrate the scaling relation between the distance from mid-price and the random walk theory and compare the result with real data. The non-trivial difference between cancellation and execution lifetime can be explained by a non-random cancellation process. The EBS database especially helps us to discover that cancellation probability depends on the distance of those orders from the mid-price.

In addition, we have seen that market participants have a non-symmetric cancellation behavior, which depends on the movement of the market price. Further research linking the impact of market movement (volatility) and cancellation behavior may contribute to quantify the impact of cancellation orders on market stability. It may represent promising future research subjects.

References

1. Bachelier L (1900) Annales Scientifiques de lÉcole Normale Supérieure 17:21
2. Eisler Z, Bouchaud JP, Kockelkoren J (2010) Quant Finan 12:1395
3. Zovko I, Farmer JD (2002) Quant Finan 2:387
4. Challet D, Stinchcombe R (2002) http://arxiv.org/abs/cond-mat/0208025
5. Takayasu M, Watanabe T, Takayasu H (2010) Approaches to large-scale business data and financial crisis. Springer, Tokyo
6. Hasbrouck J (2007) Empirical market microstructure. Oxford University Press, New York
7. Mantegna RN, Stanley HE (2000) An introduction to econophysics. Cambridge University Press, Cambridge
8. Yura Y, Takayasu H, Sornette D, Takayasu M (2014) Phys Rev Lett 112:098703
9. SEC (2013) The speed of the equity markets, 2013-05
10. SEC (2013) Quote lifetime distributions. Data Highlight 2013-04
11. SEC (2014) Equity market speed relative to order placement. Data Highlight 2014-02
12. Lauricella T, Patterson S (2010) SEC probes canceled trades. Wall Street Journal, Sept 1 2010
13. Menkveld AJ, Yueshen ZB (2013) http://ssrn.com/abstract=2243520
14. Menkveld AJ (2013) J Finan Mark 16:712–740

Chapter 4
Signs of Market Orders and Human Dynamics

Joshin Murai

Abstract A time series of signs of market orders was found to exhibit long memory. There are several proposed explanations for the origin of this phenomenon. A cogent one is that investors tend to strategically split their large hidden orders into small pieces before execution to prevent the increase in the trading costs. Several mathematical models have been proposed under this explanation.

In this paper, taking the bursty nature of the human activity patterns into account, we present a new mathematical model of order signs that have a long memory property. In addition, the power law exponent of distribution of a time interval between order executions is supposed to depend on the size of hidden order. More precisely, we introduce a discrete time stochastic process for polymer model, and show it's scaled process converges to a superposition of a Brownian motion and countably infinite number of fractional Brownian motions with Hurst exponents greater than one-half.

4.1 Introduction

Empirical studies [2, 6, 8, 11] on high frequency financial data of stock markets that employ the continuous double auction method have revealed a time series of signs of market orders has long memory property. In contrast, a time series of stock returns is known to have short memory property. A time series of *order signs* is defined by changing transactions at the best ask price into $+1$ and transactions at the best bid price into -1. The auto-correlation function of the order signs decays as a power law of the lag and the exponent of the decay is less than 1, which is equivalent to a *Hurst exponent* of the time series is greater than one-half.

In this paper, we propose a new mathematical model which takes account of origin of the long memory in order signs. As a first step, we define a discrete time stochastic process of cumulative order signs in accordance with some explanation

---

J. Murai (✉)
Graduate School of Humanities and Social Sciences, Okayama University, 3-1-1, Tsushima-Naka, Okayama 700-8530, Japan
e-mail: murai@e.okayama-u.ac.jp

H. Takayasu et al. (eds.), *Proceedings of the International Conference on Social Modeling and Simulation, plus Econophysics Colloquium 2014*, Springer Proceedings in Complexity, DOI 10.1007/978-3-319-20591-5_4

for the origin of the phenomenon. Subsequently, we verify increments of the process has the long memory property. In general, there are three ways to verify a discrete time process has some property. The first one runs computer simulations. The second calculates the distribution of the discrete time process directly. And the third, which we use in this paper, is to show that the scaled discrete time process converges to a continuous process which has that property.

There are various explanations for the origin of the long memory property of order sings [3]. A cogent one, which was proposed by Lillo et al. [9], is that investors tend to strategically split their large *hidden orders* into small pieces before execution to prevent the increase in the trading costs. Empirical findings partially support this explanation. A long memory phenomenon is found in a time series of order signs of transactions initiated by a single *member* of the stock market [3, 8]. Investors enter their orders into the market through one of its members.

Assuming the *size* of hidden orders distributes as a power law, Lillo et al. [9] considered a discrete time mathematical model with this explanation. Under an additional technical assumption that the number of hidden orders is fixed, they showed rigorously the model has a long memory property. However, this technical assumption does not seem natural.

Taking account of the bursty nature of human dynamics [1], Kuroda et al. [7] proposed another theoretical model with this explanation. They assumed that a time interval between order executions distributes as a power law, and that the power law exponent does not depend on the size of hidden order. Under an additional technical assumption that the size of hidden order is bounded above, they showed the scaled discrete time process converges to a superposition of a Brownian motion and a finite number of *fractional Brownian motions* with Hurst exponents greater than one-half. Moreover, the number of hidden orders is not fixed in their model, and it randomly varies. Although, the maximum Hurst exponent of obtained process depends on the largest hidden order.

The Hurst exponent of order signs expected by the theory of splitting large hidden order is smaller than the value of the empirical study [3]. About stocks with high liquidity, the fluctuation of Hurst exponents of order signs is small by a stock and a period [6]. These findings suggest that there might be some other cause about the long memory of order signs. We can pay attention not only to large hidden orders but also to small hidden orders. Vázquez et al. found two universality class in human dynamics [13]. On the other hand, Zhou et al. observed that in an online movie rating site, a power law exponent of the time interval between user's postings depends on user's activity [15].

In this paper, we propose a new mathematical model with an explanation for the origin of long memory of order signs that investors split their hidden order of any size into small pieces before execution. We assumed that the power law exponent of distribution of a time interval between order executions depends on the size of hidden order. We showed the scaled discrete time process converges to a superposition of a Brownian motion and countably infinite number of fractional Brownian motions with Hurst exponents greater than one-half. We note that the number of hidden orders randomly varies, that Hurst exponents are not bounded above and that the maximum Hurst exponent of obtained process depends on hidden order of medium size.

4.2 Model

In this section, we introduce a probability space (Ω_n, P_n), where n is a natural number. In the next section, we will define a discrete time stochastic process in time interval $\Lambda_n = \{1, 2, \ldots, n\}$ which describes *cumulative order signs* on the probability space. And we will show the increment of the process has a long memory property. We note that in this paper we only study the order sign and do not consider the stock price.

All essential assumptions our model requires for the market is as follows:

- Investors tend to split their hidden orders into small pieces before execution.
- The distribution of a time interval between order executions obeys a power law.
- The power law exponent of the inter-event distribution depends on the size of the hidden order.

A hidden order of one investor in a stock market and execution times of its small pieces is denoted by $\mathbf{p}$. Namely, $\mathbf{p}$ has two quantities: the order sign $s(\mathbf{p}) = s$ and the set of times of executions $b(\mathbf{p}) = \{u_1, \ldots, u_m\}$, where $m \geq 1$ is the number of small pieces of the hidden order split by the investor. We call $\mathbf{p}$ a *polymer* using the terminology of a mathematical method called the *cluster expansion*, which we will use to prove our main theorem. A method of the cluster expansion is developed in the study of the statistical physics and is applied for instance to convergence theorems of the phase separation line of the two dimensional Ising model [4, 12]. Since the cluster expansion is defined in an abstract setting [5], it can be applied to a financial model [7, 14].

It is known that a time series of trading volume in a stock market exhibits long memory [10]. However, we do not consider the memory of the trading volume in this paper; we suppose the volume of each piece is 1 just for the sake of simplicity, and we emphasize it is not technical assumption. Consequently, the number m of small pieces is equivalent to the size (or the total volume) of the hidden order. For any polymer $\mathbf{p}$, we can also regard m as the amount of *activity* of a investor in time period of her holding the polymer. Meanwhile, investors often do not split their own orders and submit it in a stock market at once. This situation is also included in our model as $m = 1$. Although the model proposed by Kuroda et al. [7] assumed that m is bounded above, that is, the maximum value of m is finite, our model does not require any restriction on the maximum value of m. More precisely, m has an upper bound $\log \log n$, and n tends to infinity in our main theorem.

The order sign $s(\mathbf{p})$ is assigned to $+1$ or -1 according to whether the hidden order is a buy order or a sell order. For any polymer $\mathbf{p}$, its order sign $s(\mathbf{p})$ is a single value. Obviously, different valued order signs are possibly assigned to different polymers possessed by one investor. In our model, the distribution of order signs is symmetry:

$$P_n(s(\mathbf{p}) = +1) = P_n(s(\mathbf{p}) = -1) = \frac{1}{2}. \tag{4.1}$$

Each element of the set of times of executions $b(\mathbf{p}) = \{u_1, \ldots, u_m\}$ is an integer. Their magnitude relation is given as $u_1 < u_2 < \cdots < u_m$, that is, the first piece of the hidden order is executed at u_1, and the last one is executed at u_m. For any distinct two polymers $\mathbf{p}_1$ and $\mathbf{p}_2$, their execution times do not overlap:

$$b(\mathbf{p}_1) \cap b(\mathbf{p}_2) = \emptyset. \tag{4.2}$$

Since we will observe the discrete time stochastic process of cumulative order signs in time interval Λ_n, it is enough to consider only polymers $\mathbf{p}$ which satisfy

$$b(\mathbf{p}) \cap \Lambda_n \neq \emptyset. \tag{4.3}$$

Taking the bursty nature of the human activity patterns into account, we assume that the distribution of a time interval between order executions obeys a power law and that its exponents $\alpha(m)$ depends on the size (or the activity) of the polymer:

$$P_n(u_i, u_{i-1} \in b(\mathbf{p}),\ \mathbf{p} \text{ is a polymer of size } m) \propto (u_i - u_{i-1})^{-\alpha(m)}. \tag{4.4}$$

According to an empirical study on human dynamics [15], power law exponents of the inter-event times are increasing in a parameter of activity. Hence, it suggests that the exponent $\alpha(m)$ is increasing in m. Our model does not require that the exponent is increasing, though it requires some condition on the exponent.

In the following, we define our model using the mathematical terminology. Let n be a natural number and $\Lambda_n = \{1, 2, \ldots, n\}$ be an observation time of a discrete time process. We describe a hidden order and its execution times of small pieces by a polymer.

Let $\mathscr{P}_{n,1}$ be the set of all polymers corresponds to hidden order of size 1:

$$\mathscr{P}_{n,1} = \{\mathbf{p} = (s, u)\,;\, s \in \{+1, -1\},\, u \in \Lambda_n\}. \tag{4.5}$$

For each polymer $\mathbf{p} = (s, u) \in \mathscr{P}_{n,1}$, we denote the order sign by $s(\mathbf{p}) = s$, the time of execution by $b(\mathbf{p}) = \{u\}$ and the size of hidden order by $|\mathbf{p}| = 1$.

For any m, $(2 \leq m \leq \log\log n)$, we define the set of all polymers corresponds to hidden order of size m by

$$\mathscr{P}_{n,m} = \{\mathbf{p} = (s, u_1, \ldots, u_m)\,;\, s \in \{+1, -1\},\, \{u_1, \ldots, u_m\} \cap \Lambda_n \neq \emptyset,$$
$$1 \leq u_i - u_{i-1} \leq n \quad (i = 2, \ldots, m)\}.$$

For each polymer $\mathbf{p} = (s, u_1, \ldots, u_m) \in \mathscr{P}_{n,m}$, we denote the order sign by $s(\mathbf{p}) = s$, the set of times of executions by $b(\mathbf{p}) = \{u_1, \ldots, u_m\}$ and the size of hidden order by $|\mathbf{p}| = m$. The set of all polymers is denoted by

$$\mathscr{P}_n = \bigcup_{m=1}^{\log\log n} \mathscr{P}_{n,m}. \tag{4.6}$$

The configuration space is denoted by

$$\Omega_n = \bigcup_{k=1}^{n} \{\omega = \{\mathbf{p}_1, \ldots, \mathbf{p}_k\} \subset \mathscr{P}_n\,;\, b(\mathbf{p}_i) \cap b(\mathbf{p}_j) = \emptyset,\ (1 \le i < j \le k)\}. \tag{4.7}$$

Example 1 We consider the case that $n = 10$ and $\omega = \{\mathbf{p}_1, \mathbf{p}_2, \mathbf{p}_3, \mathbf{p}_4\} \in \Omega_{10}$. Each polymer has order sings and the set of execution times as follows:

Polymer	Order sign	The set of execution times	Size
$\mathbf{p}_1$	$s(\mathbf{p}_1) = -1$	$b(\mathbf{p}_1) = \{-1, 3\}$	$m = 2$
$\mathbf{p}_2$	$s(\mathbf{p}_2) = +1$	$b(\mathbf{p}_2) = \{2, 5, 6\}$	$m = 3$
$\mathbf{p}_3$	$s(\mathbf{p}_3) = -1$	$b(\mathbf{p}_3) = \{4, 9, 11, 13\}$	$m = 4$
$\mathbf{p}_4$	$s(\mathbf{p}_4) = +1$	$b(\mathbf{p}_4) = \{8\}$	$m = 1$

We note that the sets of execution times intersect with Λ_{10} and do not intersect each other (see Fig. 4.1).

The power law exponent $\alpha(m)$ of distribution of a time interval between order executions depends on the size m of hidden order, and satisfies

$$1 - \frac{1}{m-1} < \alpha(m) < 1 - \frac{1}{m}\left(\frac{3}{4}\right)^m. \tag{4.8}$$

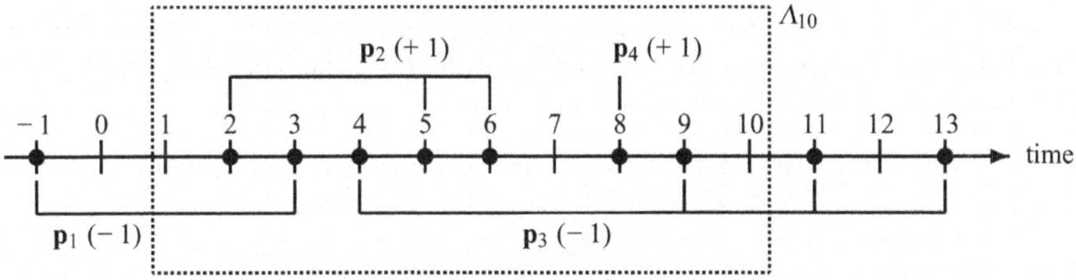

Fig. 4.1 Configuration of Example 1. The configuration consists of four polymers $\mathbf{p}_1$, $\mathbf{p}_2$, $\mathbf{p}_3$ and $\mathbf{p}_4$. Numbers $+1$ or -1 in parentheses are order signs of polymers. *Black circles* are times of executions. A discrete time stochastic process of cumulative order signs will be defined in time interval Λ_{10}

A probability intensity function of a polymer $\mathbf{p} \in \mathscr{P}_n$ is given by

$$\varphi(\mathbf{p}) = \begin{cases} d(n,1) & (\mathbf{p} = (s,u) \in \mathscr{P}_{n,1}) \\ d(n,m)\displaystyle\prod_{i=2}^{m}(u_i - u_{i-1})^{-\alpha(m)} & (\mathbf{p} = (s,u_1,\dots,u_m) \in \mathscr{P}_{n,m}, \\ & \quad 2 \le m \le \log\log n), \end{cases} \tag{4.9}$$

where scale factors $d(n,1)$ and $d(n,m)$ are given by

$$d(n,1) = c\,(\log n)^{-4},$$
$$d(n,m) = d(n,1)\left\{\frac{c\cdot(1-\alpha(m))}{\mathrm{e}\cdot n^{1-\alpha(m)}}\right\}^{m-1},$$

and c $(0 < c < 1)$ is a constant. We define a probability measure on Ω_n by

$$P_n(\omega) = \frac{1}{\Xi_n}\prod_{\mathbf{p}\in\omega}\varphi(\mathbf{p}), \quad (\omega \in \Omega_n),$$

where $\Xi_n = \sum_{\omega\in\Omega_n}\prod_{\mathbf{p}\in\omega}\varphi(\mathbf{p})$ is a normalization constant.

4.3 Main Theorem

We define a discrete time stochastic process of cumulative order signs by

$$S_u(\omega) = \sum_{\mathbf{p}\in\omega} s(\mathbf{p})\sum_{v=1}^{u} 1_{\{v\in b(\mathbf{p})\}}, \quad (u \in \Lambda_n,\ \omega \in \Omega_n). \tag{4.10}$$

We note that the increment $S_u(\omega) - S_{u-1}(\omega)$ of the process is order signs. In order to verify that the increment exhibits long memory, we show that the scaled process of the discrete time process converges to a continuous time stochastic process, the increment of which has a long memory property.

Example 2 Let $n = 10$ and the configuration $\omega = \{\mathbf{p}_1, \mathbf{p}_2, \mathbf{p}_3, \mathbf{p}_4\}$ be the same one given in Example 1. The discrete time stochastic process is as follows:

Time u	1	2	3	4	5	6	7	8	9	10
Order sign		+1	−1	−1	+1	+1		+1	−1	
$S_u(\omega)$	0	+1	0	−1	0	+1	+1	+2	+1	+1

A scaled process of $S_u(\omega)$ is given by

$$X_t^{(n)}(\omega) = \frac{1}{c(n)} \sum_{\mathbf{p}\in\omega} S_{[nt]}(\mathbf{p}), \quad (0 \le t \le 1,\ \omega \in \Omega_n), \tag{4.11}$$

where $c(n) = \sqrt{n \cdot d(n,1)} = \sqrt{c} n^{1/2} (\log n)^{-2}$ is a scale function, and $[nt]$ indicates the greatest integer less than or equal to nt.

Theorem 1 *The distribution of $X_t^{(n)}$ weakly converges to the distribution of*

$$X_t = \sqrt{c_1} B_t + \sum_{m=2}^{\infty} \sum_{\ell=1}^{m-1} \sqrt{c_2(m,\ell)} B_t^{H_{m,\ell}}, \qquad (0 \le t \le 1), \tag{4.12}$$

where

$$c_1 = 2 \sum_{m=1}^{\infty} m \left(\frac{c}{e}\right)^{m-1},$$

$$c_2(m,\ell) = \frac{4\,(1-\alpha(m))^{\ell-1}\,(m-\ell) B_\ell(\alpha(m))}{\ell\,\{(1-\alpha(m))\,\ell + 1\}} \left(\frac{c}{e}\right)^{m-1},$$

$$B_\ell(\alpha(m)) = \frac{\Gamma(1-\alpha(m))^\ell}{\Gamma(\ell(1-\alpha(m)))},$$

B_t is a standard Brownian motion, $B_t^{H_{m,\ell}}$ is a fractional Brownian motion with Hurst exponent

$$H_{m,\ell} = \frac{1}{2}\{(1-\alpha(m))\ell + 1\} \tag{4.13}$$

and $\left\{B_t,\ B_t^{H_{m,\ell}}\ ;\ m \ge 2, 1 \le \ell \le m-1\right\}$ are independent.

Remark 1 For any $m \ge 2$ and $1 \le \ell \le m-1$, it follows from the condition

$$1 - \frac{1}{m-1} < \alpha(m) \tag{4.14}$$

that $H_{m,\ell} < 1$. And it follows from the condition

$$\alpha(m) < 1 - \frac{1}{m}\left(\frac{3}{4}\right)^m \tag{4.15}$$

that

$$\sum_{m=2}^{\infty} \sum_{\ell=1}^{m-1} c_2(m,\ell) < \infty. \tag{4.16}$$

Remark 2 Let us consider a continuous time stochastic process Z_t of a superposition of two independent (fractional) Brownian motions B_t^H and $B_t^{\tilde{H}}$ with Hurst exponents $\frac{1}{2} \leq H < \tilde{H} < 1$:

$$Z_t = aB_t^H + \tilde{a}B_t^{\tilde{H}} \tag{4.17}$$

where $a, \tilde{a}$ are constants. We note that when $H = \frac{1}{2}$, since $B_t^{1/2}$ is a Brownian motion, the process Z_t is a superposition of a Brownian motion and a fractional Brownian motion. We define an increment of the process by $\Delta Z_t = Z_t - Z_{t-1}$. Since $E[\Delta Z_t] = 0$ and $Var(\Delta Z_t) = a^2 + \tilde{a}^2$, the auto-correlation function of the increment is

$$\begin{aligned}\rho_Z(\tau) &= \frac{E[\Delta Z_t\, \Delta Z_{t+\tau}]}{a^2 + \tilde{a}^2} = \frac{a^2}{a^2 + \tilde{a}^2} E\left[\Delta B_t^H \Delta B_{t+\tau}^H\right] + \frac{\tilde{a}^2}{a^2 + \tilde{a}^2} E\left[\Delta B_t^{\tilde{H}} \Delta B_{t+\tau}^{\tilde{H}}\right] \\ &\sim \frac{\tilde{a}^2}{a^2 + \tilde{a}^2} \tilde{H}(2\tilde{H} - 1)\tau^{2\tilde{H}-2} \quad (\tau \to \infty).\end{aligned} \tag{4.18}$$

Hence, we see that the Hurst exponent of Z_t is $\tilde{H} = \max\left\{H, \tilde{H}\right\}$. In a similar way, it can be verified that the Hurst exponent of the process X_t in Theorem 1 is

$$\begin{aligned}H_{\max} &= \max\{H_{m,\ell}\ ;\ m \geq 2, 1 \leq \ell \leq m-1\} \\ &= \max_{m \geq 2} \frac{1}{2}\{(1 - \alpha(m))(m-1) + 1\}.\end{aligned} \tag{4.19}$$

Remark 3 In the model of Kuroda et al. [7], since they assume that the exponent of inter-event time distribution is a constant $\alpha(m) = \alpha$, they need to put a limitation on the size of hidden order: $m \leq m_{\max}$ where $m_{\max}$ is a positive number. Then, they derive finite number of fractional Brownian motions. And the maximum Hurst exponents $H_{\max}$ is attained by the largest size $m_{\max}$ of hidden orders.

In our model, we set no limitation on the size m of hidden orders. As a result, we derive countably infinite number of fractional Brownian motions. On the other hand in the empirical study the size of the hidden order has some limitation, and the upper bound of the size possibly depends on markets or stocks. A finite number of fractional Brownian motions appears in the case.

If exponents $\alpha(m)$ is increasing in size m [15], then

$$\frac{1}{2}\{(1 - \alpha(m))(m-1) + 1\} \tag{4.20}$$

is not monotone function in m. Hence the maximum Hurst exponents in (4.19) is attained by middle size m^* of hidden orders.

4.4 Outline of the Proof of Theorem 1

In order to prove the main theorem, we show a convergence of a finite dimensional distribution, and show the tightness. In this section, we give an outline of the proof of Theorem 1. The detail of the proof is complicated. The interested reader is referred to Kuroda et al. [7].

For any $0 < t_1 < \ldots < t_k \le 1$ and any $\mathbf{z} = (z_1, \ldots, z_k) \in \mathbf{R}^k$, we define

$$Y^{(n)}(\omega) = Y^{(n)}_{\mathbf{t},\mathbf{z}}(\omega) = \prod_{\mathbf{p}\in\omega} \sum_{i=1}^{k} z_i X^{(n)}_{t_i}(\mathbf{p}), \quad (\omega \in \Omega_n). \tag{4.21}$$

Its characteristic function is denoted by

$$\phi^{(n)}_{\mathbf{t}}(\mathbf{z}) = E_n\left[e^{\sqrt{-1}Y^{(n)}}\right] \tag{4.22}$$

Using the method of the cluster expansion [5, 12], we have

$$\log \phi^{(n)}_{\mathbf{t}}(\mathbf{z}) = \sum_{A\in\mathscr{A}_n} \left(e^{\sqrt{-1}Y^{(n)}(A)} - 1\right) \varphi(A) \frac{\alpha^T(A)}{A!} \tag{4.23}$$

where $\mathscr{A}_n = \{A : \mathscr{P}_n \to \{0, 1, 2, \ldots\}\}$, and for any $A \in \mathscr{A}_n$, $A! = \prod_{\mathbf{p}\in\mathscr{P}_n} A(\mathbf{p})!$,

$$Y^{(n)}(A) = \sum_{\mathbf{p}\in\mathscr{P}_n} Y^{(n)}(\mathbf{p})A(\mathbf{p}) = \sum_{\mathbf{p}\in\mathscr{P}_n} \sum_{i=1}^{k} z_i X^{(n)}_{t_i}(\mathbf{p})A(\mathbf{p}),$$

$$\varphi(A) = \prod_{\mathbf{p}\in\mathscr{P}_n} \varphi(\mathbf{p})^{A(\mathbf{p})},$$

$$\alpha(A) = \begin{cases} 1 & A! = 1 \text{ and } \operatorname{supp}(A) \in \Omega_n \\ 0 & \text{o.w.,} \end{cases}$$

$\operatorname{supp}(A) = \{\mathbf{p}\}$ and $\alpha^T(A) = \operatorname{Log}\alpha(A)$. Applying the Taylor's expansion, we obtain

$$\log \phi^{(n)}_{\mathbf{t}}(\mathbf{z}) = \sqrt{-1}I_1(n) - \frac{1}{2}\left\{I_2(n) + \hat{I}_2(n)\right\} - \frac{\sqrt{-1}}{3!}\left\{I_3(n) + \hat{I}_3(n)\right\} \tag{4.24}$$

where

$$I_1(n) = \sum_{A\in\mathscr{A}_n} Y^{(n)}(A)\varphi(A)\frac{\alpha^T(A)}{A!},$$

$$I_2(n) = \sum_{\mathbf{p}\in\mathscr{P}_n} \left\{Y^{(n)}(\mathbf{p})\right\}^2 \varphi(\mathbf{p}), \quad \hat{I}_2(n) = \sum_{A\in\mathscr{A}_n, |A|\geq 2} \left\{Y^{(n)}(A)\right\}^2 \varphi(A)\frac{\alpha^T(A)}{A!},$$

$$I_3(n) = \sum_{\mathbf{p}\in\mathscr{P}_n} \left\{Y^{(n)}(\mathbf{p})\right\}^3 e^{\sqrt{-1}\theta Y^{(n)}(\mathbf{p})}\varphi(\mathbf{p}),$$

$$\hat{I}_3(n) = \sum_{A\in\mathscr{A}_n, |A|\geq 2} \left\{Y^{(n)}(A)\right\}^3 e^{\sqrt{-1}\theta Y^{(n)}(A)}\varphi(A)\frac{\alpha^T(A)}{A!}$$

for some $\theta \in (0, 1)$ and $|A| = \sum_{\mathbf{p}\in\mathscr{P}_n} A(\mathbf{p})$ for any $A \in \mathscr{A}_n$. From the symmetry property of the model, we have

$$I_1(n) = 0. \tag{4.25}$$

It is easy to see that

$$\lim_{n\to\infty} I_3(n) = 0. \tag{4.26}$$

It follows from the theory of Kotecký and Preiss [5] and the Cauchy formula that

$$\lim_{n\to\infty} \hat{I}_2(n) = 0, \quad \lim_{n\to\infty} \hat{I}_3(n) = 0. \tag{4.27}$$

It can be seen that

$$\lim_{n\to\infty} I_2(n) = c_1 \sum_{i=1}^{k}\sum_{j=1}^{k} z_i z_j \min\{t_i, t_j\}$$
$$+ \sum_{i=1}^{k}\sum_{j=1}^{k} z_i z_j \sum_{m=2}^{\infty}\sum_{\ell=1}^{m-1} c_2(m,\ell)\frac{1}{2}\left\{t_i^{H_{m,\ell}} + t_j^{H_{m,\ell}} - \left|t_i - t_j\right|^{H_{m,\ell}}\right\}.$$

Hence we have shown a convergence of the finite dimensional distribution.

Using Pfister's lemma (Lemma 3.5 in [12]), it can be shown that there are a positive constant $c_3 > 0$ and a positive number $n_0 \in \mathbf{N}$ such that for any $n \geq n_0$ and any $0 \leq r \leq s \leq t \leq 1$,

$$E_n\left[\left(X_s^{(n)} - X_r^{(n)}\right)^2 \left(X_t^{(n)} - X_s^{(n)}\right)^2\right] \leq c_3(t-r)^2. \tag{4.28}$$

Hence we have shown the tightness condition. Therefore we complete the proof of Theorem 1.

4.5 Conclusion

Using a method of the cluster expansion developed in the study of the statistical physics, we introduced a new mathematical model with the explanation for the origin of long memory of order signs that investors split their hidden order of any size into small pieces before execution. The power law exponent of distribution of a time interval between order executions was supposed to depend on the size of hidden order. The limit process of the scaled discrete time process was found to be a superposition of a Brownian motion and countably infinite number of fractional Brownian motions with Hurst exponents greater than one-half. Namely, increments of the limit process have a long memory property. The maximum Hurst exponent of obtained process was described as

$$H_{\max} = \max_{m \geq 2} \frac{1}{2} \{(1 - \alpha(m))(m - 1) + 1\}. \tag{4.29}$$

The power law exponent $\alpha(m)$ of distribution of a time interval between order executions was supposed to be increasing, cf. [15]. Thus, investors having a hidden order of medium size m^*, which attains the maximum in (4.29), have an influence on the Hurst exponent of order signs. It should be noted that in the empirical study, the power law exponent of the auto-correlation function $\rho(\ell)$ of order signs is determined by middle region of lag ℓ. Hence, investors who have an influence on the Hurst exponent of order signs depends on size m of their hidden order, the power law exponent $\alpha(m)$ and the distribution of investors.

Acknowledgements The author wish to thank Prof. Y. Higuchi, Prof. K. Kuroda, and Prof. J. Maskawa for fruitful discussion. The author thanks the anonymous referee for valuable comments and useful suggestions.

References

1. Barabási A (2005) The origin of bursts and heavy tails in human dynamics. Nature 435:207–211
2. Bouchaud JP, Gefen Y, Potters M, Wyart M (2004) Fluctuations and response in financial markets: the subtle nature of 'random' price changes. Quant Finan 4:176–190
3. Bouchaud JP, Farmer JD, Lillo F (2009) How markets slowly digest changes in supply and demand. In: Hens T, Schenkhoppe KR (ed) Handbook of financial markets: dynamics and evolution. Academic Press, London
4. Dobrushin R, Hryniv O (1997) Fluctuations of the phase boundary in the 2D Ising ferromagnet. Commun Math Phys 189(2):395–445

5. Kotecký R, Preiss D (1986) Cluster expansion for abstract polymer models. Commun Math Phys 103:491–498
6. Kuroda K, Murai J (2008) A probabilistic model on the long memory property in stock market. In: International conference 2008 in Okayama, rising economies and regional cooperation in the East Asia and Europe, 1–20
7. Kuroda K, Maskawa J, Murai J (2013) Application of the cluster expansion to a mathematical model of the long memory phenomenon in a financial market. J Stat Phys 4:706–723
8. Lillo F, Farmer J (2004) The long memory of the efficient market. Stud Nonlinear Dyn E 8:1–33
9. Lillo F, Mike S, Farmer JD (2005) Theory for long memory in supply and demand. Phys Rev E 7106(6 pt 2):287–297
10. Lobato IN, Velasco C (2000) Long memory in stock-market trading volume. J Bus Econ Stat 18(4):410–427
11. Maskawa Ji, Mizuno T, Murai J, Yoon H (2011) Kabuka no keizai butsurigaku (in Japanese). Baifukan Co., Ltd, Tokyo
12. Pfister CE (1991) Large deviations and phase separation in the two-dimensional Ising model. Helv Phys Acta 64(7):953–1054
13. Vázquez A, Oliveira J, Dezsö Z, Goh K, Kondor I, Barabási A (2006) Modeling bursts and heavy tails in human dynamics. Phys Rev E 73(3), 036127
14. Wang J, Deng S (2008) Fluctuations of interface statistical physics models applied to a stock market model. Nonlinear Anal Real 9(2):718–723
15. Zhou T, Kiet HAT, Kim BJ, Wang BH, Holme P (2008) Role of activity in human dynamics. Europhys Lett 82(2):1–5

Chapter 5
Damped Oscillatory Behaviors in the Ratios of Stock Market Indices

Ming-Chya Wu

Abstract This article reviews a recent finding on the properties of stock market indices (Wu, Europhys Lett 97:48009, 2012). A stock market index is an average of a group of stock prices with equal or unequal weights. Different stock market indices derived from various combinations of stocks are not expected to have fixed relations among them. From analyzing the daily index ratios of Dow Jones Industry Average (DJIA), NASDAQ, and S&P500 from 1971/02/05 to 2011/06/30 using the empirical mode decomposition, we found that the ratios NASDAQ/DJIA and S&P500/DJIA, normalized to 1971/02/05, approached and then retained the values of 2 and 1, respectively. The temporal variations of the ratios consist of global trends and oscillatory components including a damped oscillation in 8-year cycle and damping factors of 7183 days (NASDAQ/DJIA) and 138,471 days (S&P500/DJIA). Anomalies in the ratios, corresponding to significant increases and decreases of indices, are local events appearing only in the time scale less than 8-year cycle. The converge of the dominant damped oscillatory component implies that representative stocks in the pair-markets become more coherent as time evolves.

5.1 Introduction

The study of financial systems using the concepts and theories developed in physics has attracted much attention in recent years [1–22]. Such study have revealed interesting properties in financial data, which facilitate deeper understanding of the underlying mechanisms of the systems and are essential for sequent modelling. These include financial stylized facts [4, 7–9, 11, 23, 24], such as fat tails in asset return distributions, absence of autocorrelations of asset returns, aggregational normality, asymmetry between rises and falls, volatility clustering [10], phase clustering [13–15], and damped oscillation in ratios of stock market indices [22].

M.-C. Wu (✉)
Research Center for Adaptive Data Analysis, National Central University, Chungli 32001, Taiwan

Institute of Physics, Academia Sinica, Taipei 11529, Taiwan
e-mail: mcwu@ncu.edu.tw

H. Takayasu et al. (eds.), *Proceedings of the International Conference on Social Modeling and Simulation, plus Econophysics Colloquium 2014*, Springer Proceedings in Complexity, DOI 10.1007/978-3-319-20591-5_5

Successful empirical analysis and modelling of financial criticality have suggested possible physical pictures for financial crashes and stock market instabilities [5, 6, 12, 18–21].

In this article, we briefly review our recent study on the damped oscillations in daily stock market indices of Dow Jones Industry Average (DJIA), NASDAQ, and S&P500, from 1971/02/05 to 2011/06/30 [22]. The daily data were downloaded from Yahoo Finance (http://finance.yahoo.com/), and were preprocessed to have the three indices aligned with the same length by removing three data points in DJIA and S&P500 (1973/9/26, 1974/10/7, and 1975/10/16) which do not exist in NASDAQ. There are finally 10,197 data points involved in the study. Figure 5.1a shows the daily index data of the three stock markets. It is interesting that by keeping DJIA as a reference and multiplying the NASDAQ index by a factor of 5.2 and S&P500 by 8.5, the curves of the rescaled indices coincide very well in several epoches, except large deviations in NASDAQ for the periods 1999–2001 and 2009–2011, as shown in Fig. 5.1b. In year 2011, DJIA is the average price of 30 companies (http://www.djaverages.com/), NASDAQ consists of 1197 companies (http://www.nasdaq.com/), and S&P500 index is an average result of 500 companies (http://www.standardandpoors.com). Some companies, such as Intel and Microsoft, are included in all the three markets, but most of their compositions are different. The relations among the indices are not crucially determined by the common companies. The coincidence of the three indices via scaling is apparently not trivial, but may result from some kine of coherence among respective representative stocks in the markets. The properties of the relations among them deserve further study.

5.2 Data Analysis and Discussions

Let us first consider two indices x_i and x_j. The ratio between them $R_{ij}(t_n) = x_i(t_n)/x_j(t_n)$ at time t_n can be alternatively formulated as

$$R_{ij}(t_n) = R_{ij}(t_{n-1}) \frac{1 + g_i(t_{n-1})}{1 + g_j(t_{n-1})}, \tag{5.1}$$

where

$$g_i(t_n) = \frac{x_i(t_{n+1}) - x_i(t_n)}{x_i(t_n)}, \tag{5.2}$$

is the gain of the index x_i. The gain time series of the three indices are shown in Fig. 5.1c. After normalizing the ratio to the initial value of R_{ij} at t_0, we have

$$NR_{ij}(t_n) \equiv \frac{R_{ij}(t_n)}{R_{ij}(t_0)} = \prod_{m=1}^{n} \frac{1 + g_i(t_{m-1})}{1 + g_j(t_{m-1})}. \tag{5.3}$$

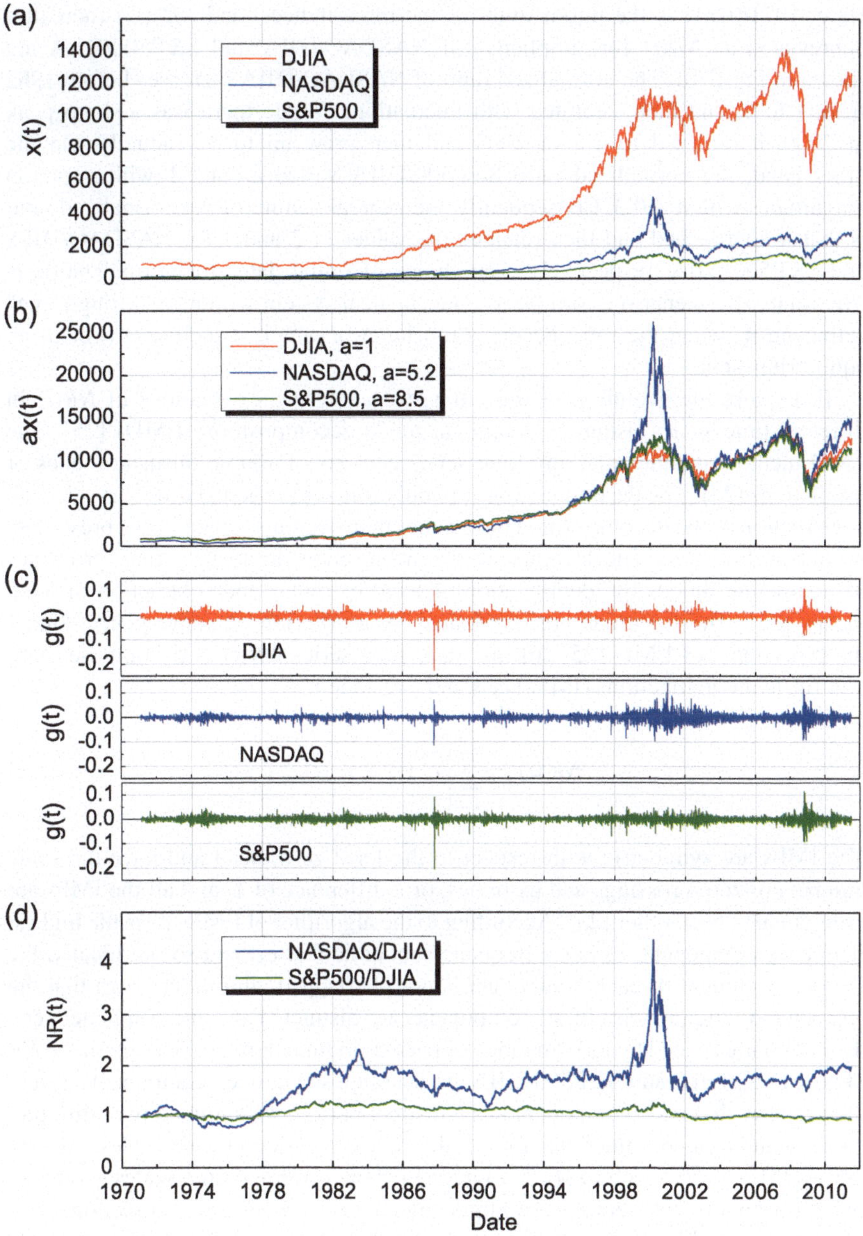

Fig. 5.1 (**a**) Daily indices of Dow Jones Industry Average (DJIA), NASDAQ, and S&P500. (**b**) Rescaled indices $ax(t)$. (**c**) The gains of the three indices. (**d**) Ratios of paired indices, normalized to the values on 1971/02/05. (Reproduced from Fig. 1 of [22])

Here, 1971/02/05 is the initial time for the three indices, and $NR_{ij}(t_n)$ (hereafter abbreviated as $NR(t)$ for simplicity) of NASDAQ/DJIA and S&P500/DJIA are shown in Fig. 5.1d. The normalized ratio of NASDAQ/DJIA increases before 1982 from 1 to 2, and then saturated with fluctuations. While there was a sharp peak in 2000, it returned to 1.5 in 2003 and then grew up to 2 gradually. On the other hand, the normalized ratio S&P500/DJIA varied around 1 with variation magnitudes within ± 0.3. Consequently, the general feature of the normalized ratio is that it approached and then retained the values of 2 and 1 for NASDAQ/DJIA and S&P500/DJIA, respectively. The choice of initial dates for normalization is irrelevant. The scenario is similar to a mechanical system with a "restoring force" acting on it: when the ratio becomes too large or small, it inclines to retain an equilibrium state.

To explore the evolution of the ratios, we analyze the variations of $NR(t)$ in different time scales using the empirical mode decomposition (EMD) [25]. The EMD method assumes that any time series consists of simple intrinsic modes of oscillations [25]. The decomposition explicitly utilizes the actual time series for the construction of the decomposition base rather than decomposing it into a prescribed set of base functions. The decomposition is achieved by iterative "sifting" processes for extracting modes by identification of local extremes and subtraction of local means [25]. The iterations are terminated by a criterion of convergence. Under the procedures of EMD [25, 26], the ratio time series $NR(t)$ is decomposed into n intrinsic mode functions (IMFs) c_k's and a residue r_n,

$$NR\,(t) = \sum_{k=1}^{n} c_k\,(t) + r_n\,(t)\,. \tag{5.4}$$

The IMFs are symmetric with respect to the local zero mean and have the same numbers of zero crossings and extremes, or a difference of 1, and all the IMFs are orthogonal to each other [25]. According to the algorithm of EMD, c_1 is the highest frequency component, c_2 has a frequency about half of c_1, and so on. Ideally, the frequency content of each component is not overlapped with others such that the characteristic frequencies of all components are distinct. Thus one component can then be characterized by its own range of periods in time domain. Here, both the NR of NASDAQ/DJIA and S&P500/DJIA are decomposed into ten components. Using the property that each component has a distinct period, we summed over different components to assess the behaviors of the ratios in different time scales. Among others, IMFs c_6 to c_9 are of special interest for their average time scales estimated by zero-crossing calculations are larger than 1 year (about 250 transaction days, which is a suitable time scale to analyze the behaviors in Fig. 5.1d. Figures 5.2a and 5.3a show the comparisons of NR, residue r_9 and combinations of the residue and IMFs, $c_9 + r_9$ and $c_{(6-9)} + r_9$ (here $c_6 + c_7 + c_8 + c_9$ has been abbreviated as $c_{(6-9)}$ for simplicity). For the current case, we are more interested in the residue r_9 and IMFs c_8 and c_9, shown in Figs. 5.2b and 5.3b. The residue r_9 is the trend of the ratio NASDAQ/DJIA which approaches 2 gradually from 1.2 (Fig. 5.2b), while r_9 of

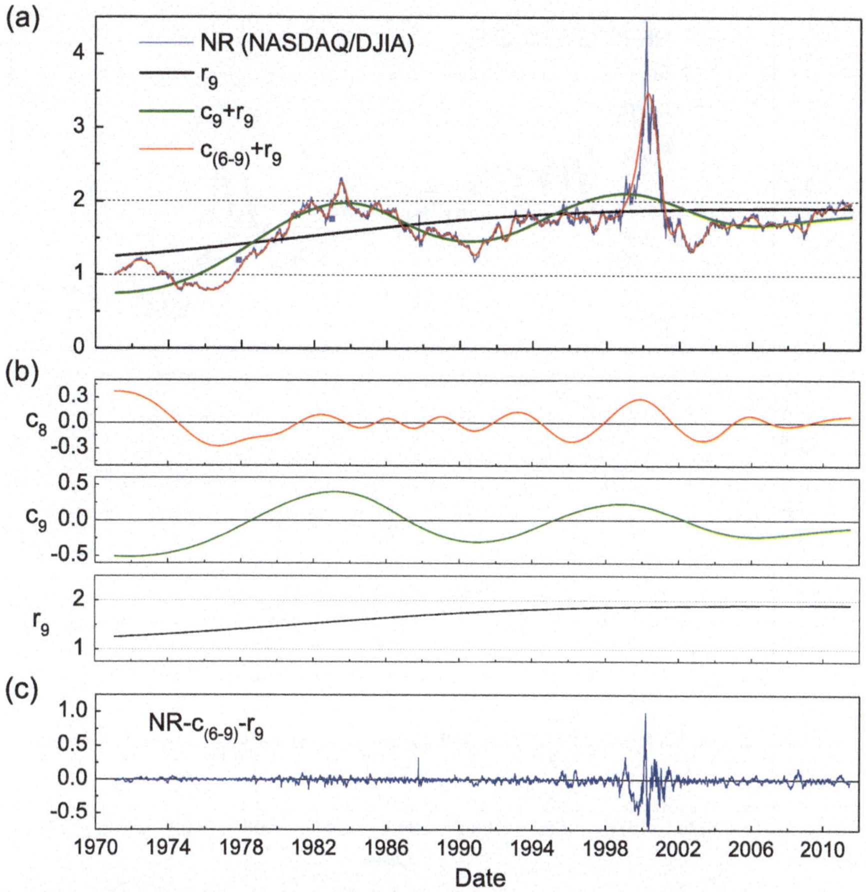

Fig. 5.2 Empirical mode decomposition (EMD) of the ratio *NR* for NASDAQ/DJIA. (**a**) Comparisons of *NR*, IMFs and IMF combinations. (**b**) IMFs c_8, c_9, and residue r_9. (**c**) The data of $NR - c_6 - c_7 - c_8 - c_9 - r_9$. (Reproduced from Fig. 2 of [22])

S&P500/DJIA grows up from 1 to 1.2 and then decreases back to 1 (Fig. 5.3b). With the aid of zero-crossing calculations and fitting, the IMF c_9 reveals that the variations of the ratios in the scale of 8-year cycle behave as a damped oscillation in the form of $\exp[-(t_n - t_0)/\gamma]$ with damping factors $\gamma \approx 7183$ days (NASDAQ/DJIA) and 138,471 days (S&P500/DJIA) determined from the local minima of IMF c_9. Thus, the combination of c_9 and r_9 shows the converge of oscillations to values 2 and 1 for NASDAQ/DJIA and S&P500/DJIA, respectively. Meanwhile, the IMF c_8 corresponding to (2–4)-year cycle is accompanied with frequency modulation in late of 1990s, implying the trigger of the anomaly in amplitude change and its recovery to regular situation lasts 1.5 oscillatory cycles, about 4–6 years. Since this anomaly does not appear in IMF c_9, it is a local event in time with time scale less than 8-year cycle. Here we should remark that the nature of the EMD method is adaptive. It

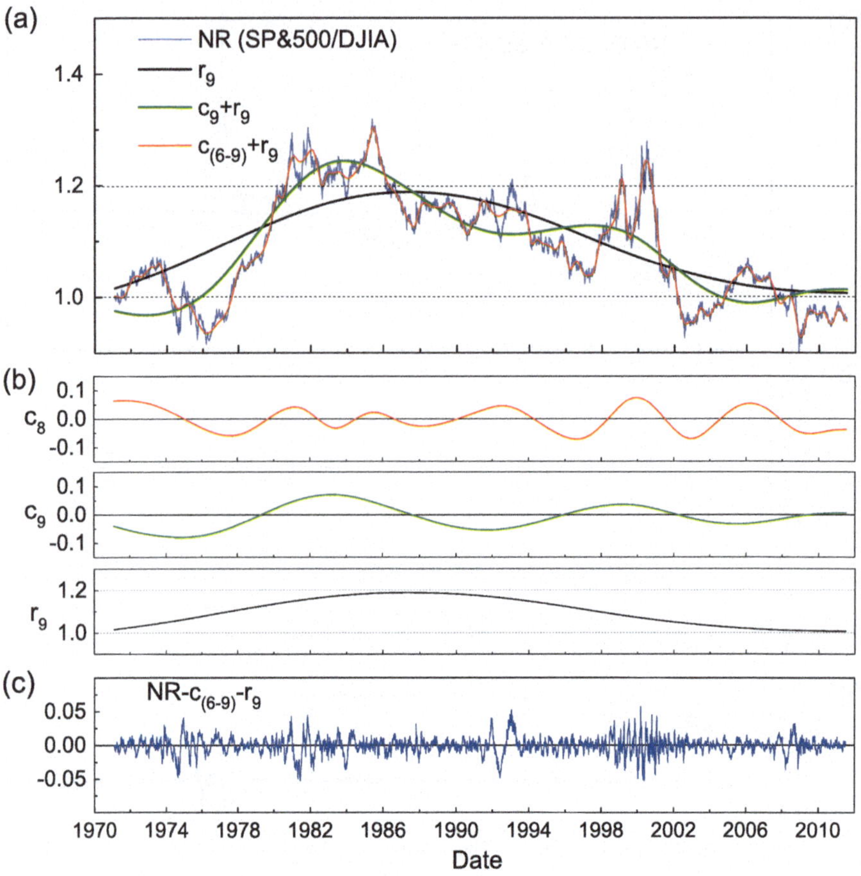

Fig. 5.3 Empirical mode decomposition (EMD) of the ratio *NR* for S&P500/DJIA. (**a**) Comparisons of *NR*, IMFs and IMF combinations. (**b**) IMFs c_8, c_9, and residue r_9. (**c**) The data of $NR - c_6 - c_7 - c_8 - c_9 - r_9$. (Reproduced from Fig. 3 of [22])

catches intrinsic oscillations in a time series, such that the number of IMFs depends on the properties of the data itself (i.e., the index of IMFs may change). The above-discussed behaviors can be observed no matter the data used here is considered as a whole or is split into two or more segments (if long enough to see components with particular time scales) for the same analysis.

The components of the ratios in the cycle less than half year (about 125 days) are derived by subtracting $c_{(6-9)} + r_9$ from *NR*. The data of $\delta NR = NR - c_{(6-9)} - r_9$ are shown in Figs. 5.2c and 5.3c for NASDAQ/DJIA and S&P500, respectively. Within cycles less than 1 year, there are no explicit repeat patterns in the data. Thus, we analyze their statistical properties by the detrended fluctuation analysis (DFA) [27–29] and the multiscale entropy (MSE) [30] analysis, and the results are presented in Fig. 5.4. The DFA analysis measures the fluctuation $F(n)$ of δNR with respect to a

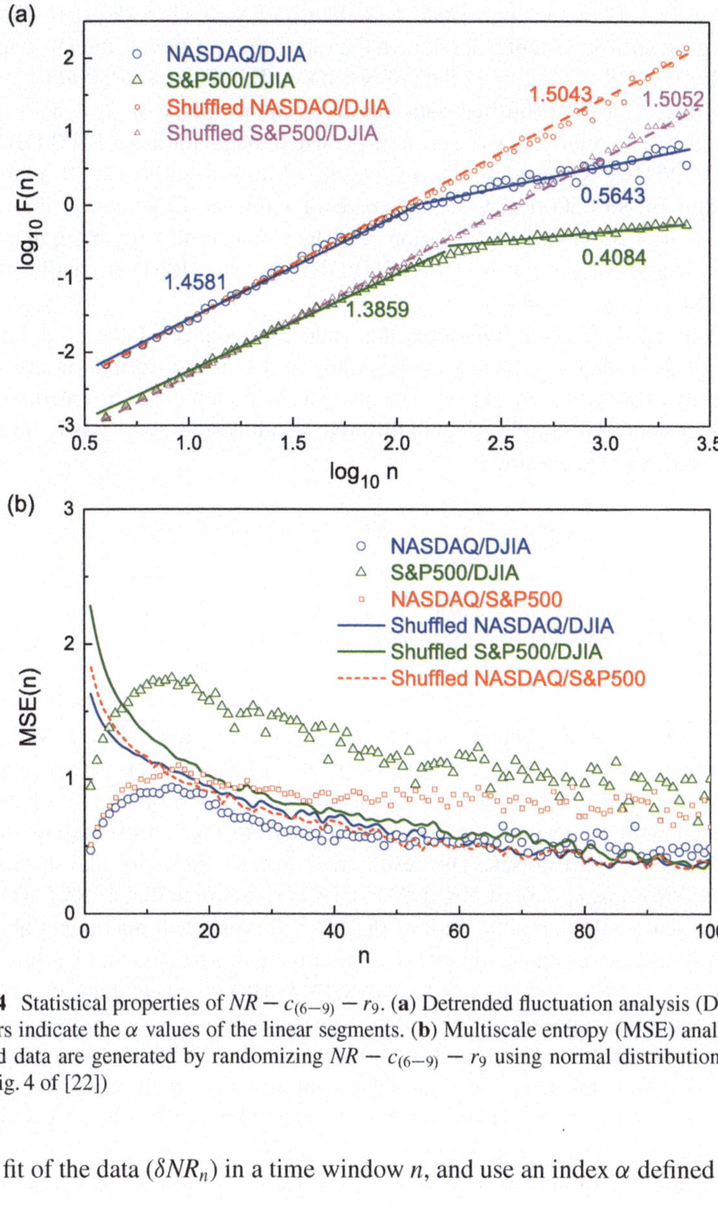

Fig. 5.4 Statistical properties of $NR - c_{(6-9)} - r_9$. (**a**) Detrended fluctuation analysis (DFA). The numbers indicate the α values of the linear segments. (**b**) Multiscale entropy (MSE) analysis. The shuffled data are generated by randomizing $NR - c_{(6-9)} - r_9$ using normal distribution. (Edited from Fig. 4 of [22])

linear fit of the data (δNR_n) in a time window n, and use an index α defined from

$$F(n) = \sqrt{\frac{1}{T}\sum_t [\delta NR(t) - \delta NR_n(t)]^2} \sim n^{\alpha} \tag{5.5}$$

to describe the correlation property of the data [27–29]. The results of $\alpha = 1.4851$ for NASDAQ/DJIA and $\alpha = 1.3859$ for S&P500/DJIA in Fig. 5.4a suggest that the property of $NR - c_{(6-9)} - r_9$ is similar to a Brownian motion with more negative correlation (<1.5) in the time scale less than half year (125 days) in Fig. 5.4a, indicating the anti-persistent behaviors in the ratios. For reference, the DFA analysis for the shuffled $NR - c_{(6-9)} - r_9$ data of NASDAQ/DJIA and S&P500/DJIA is also shown in Fig. 5.4a. The shuffled data manifests the property of Brownian motion with $\alpha = 1.5$. The relatively stronger anti-persistent behavior in S&P500/DJIA than in NASDAQ/DJIA is considered as a signature of more significant self-adjustment in the ratio of S&P500/DJIA. The change of slope at 125 days is due to the removal of high-order IMFs ($c_{(6-9)}$ and r_9). The slopes in this regime indicate that effective changes of the ratios in S&P500/DJIA ($\alpha = 0.4084$) is smaller than in NASDAQ/DJIA ($\alpha = 0.5643$).

Next, the MSE analysis measures the scale dependence of the complexity in the data [30]. Higher complexity corresponds to a higher information content or a superiority of system control [31]. The analysis is implemented by calculating the entropies of a set of resampled data in different window sizes, which is to be a scale factor in MSE plot, according to

$$s_n(t) = \sum_{\overline{\delta NR_n}(t)} P(\overline{\delta NR_n}(t)) \log[P(\overline{\delta NR_n}(t))], \tag{5.6}$$

$$\overline{\delta NR_n}(t) = \frac{1}{n} \sum_{i=1}^{n-1} \delta NR(t+i), 1 \le i \le \frac{T}{n}, \tag{5.7}$$

where $P(\overline{\delta NR_n}(t))$ is the occurrence probability of the value $\overline{\delta NR_n}(t)$. MSE is an average of successive difference of $s_n(t)$ over time. The analysis is finally presented by the curve of MSE as a function of n. Here, the relative complexity of the data is evaluated with respect to a reference defined from the corresponding shuffled data or some standard noises. The results in Fig. 5.4b show that the information content of $NR - c_{(6-9)} - r_9$ of NASDAQ/DJIA is richer than that of S&P500/DJIA in all time scales. Remarkably, both of the MSE curves reach maxima at about 14 days, implying reassessments on ratios are relatively more active in this time scale. The entropy of $NR - c_{(6-9)} - r_9$ for NASDAQ/DJIA is lower than the shuffled data, generated by randomizing the time series of $NR - c_{(6-9)} - r_9$ using normal distribution, in the scale less than 60 days, and that for S&P500/DJIA is less than the shuffled data in the scale less than 7 days. Interestingly, the information content in $NR - c_{(6-9)} - r_9$ for NASDAQ/DJIA is relatively lower than the corresponding shuffled data resembling to a white noise. There is a weaker correlation between NASDAQ and DJIA than between S&P500 and DJIA. As a result, larger deviations of the rescaled indices in Fig. 5.1b for DJIA and NASDAQ than DJIA and S&P500 can be observed in the period from 1999 to 2002. Here for reference, the same analysis applied to $NR - c_{(6-9)} - r_9$ of NASDAQ/S&P500 is also presented in Fig. 5.4b, which shows that the data for NASDAQ/S&P500 also reaches maximum at about 14 days and is less than its shuffled data in the scale less than 12 days.

We further calculate the dynamical cross-correlations for pairs of the stock market indices using logarithmic return, $lr_i(t_n) = \log[x_i(t_{n+1})/x_i(t_n)]$. The dynamical cross-correlation between returns of two indices is defined as

$$\rho_{ij}(t_n) = \frac{(lr_i - \langle lr_i \rangle)\left(lr_j - \langle lr_j \rangle\right)}{\sigma_i \sigma_j}, \tag{5.8}$$

with $\sigma_i^2 = \langle lr_i^2 - \langle lr_i \rangle^2 \rangle$ the variance of the index, and $\langle \cdots \rangle$ indicates an average over a time window T. Despite of phase differences in short time scale, the variations of the indices in large time scale are generally positive correlated (more in phase). Figure 5.5a shows the window size dependence of the average correlation of the stock indices. The average correlation between S&P500 and DJIA is stronger than NASDAQ and DJIA for all window sizes, consistent with inference from the MSE analysis in Fig. 5.4b that the information content in &P500/DJIA in the cycle less than half-year is richer than NASDAQ/DJIA. DJIA and NASDAQ have the strongest correlation at $T = 60$ days, while the correlation strength between DJIA and S&P500 grows gradually with time and saturates at $T > 1000$ days.

Note that the normalized ratio $NR_{ij}(t_n)$ in Eq. (5.3) can be rewritten as

$$NR_{ij}(t_n) = \frac{1 + \sum_{k=1}^{n} G_i^{(k)}}{1 + \sum_{k=1}^{n} G_j^{(k)}}, \tag{5.9}$$

with

$$G_i^{(k)} = \frac{1}{k!} \sum_{t_{m_1} \neq t_{m_2} \cdots} \prod_{l=1}^{k} g_i(t_{m_l}). \tag{5.10}$$

The term $G_i^{(1)}$ is a sum of all the gains. The means of the grains are 0.00032095, 0.00040822, and 0.00031729 for DJIA, NASDAQ, and S&P500, respectively, and the value of $G_i^{(1)}$ is in the order of 1. The term $G_i^{(2)}$ is proportional to the autocorrelation function of the gain, defined as $C(\tau) = \int_{t_0}^{t_N - \tau} \delta g(t) \delta g(t + \tau) dt / \eta^2$, with variance $\eta^2 = \langle g^2 - \langle g \rangle^2 \rangle$, and its value is also in the order of 1. The $G_i^{(k)}$'s with $k \geq 3$ are combinations of the sum of gains and autocorrelation functions. Further calculations of $G_i^{(k)}$ show that the values of all $G_i^{(k)}$'s of Eq. (5.10) are in the order of 1. Consequently, all $G_i^{(k)}$'s have equal contributions to the ratios. We then calculate the autocorrelation of the absolute gain and the results are shown in Fig. 5.5b. Using exponential decay model to fit the autocorrelation function, the correlation length is determined to be 194 days for DJIA, 766 days for NASDAQ, and 238 days for S&P500, which are less than 4 years. Consequently, from above analysis, we confirmed that the damped oscillation in 8-year cycle is not a consequence of cross-correlation and autocorrelation of the indices.

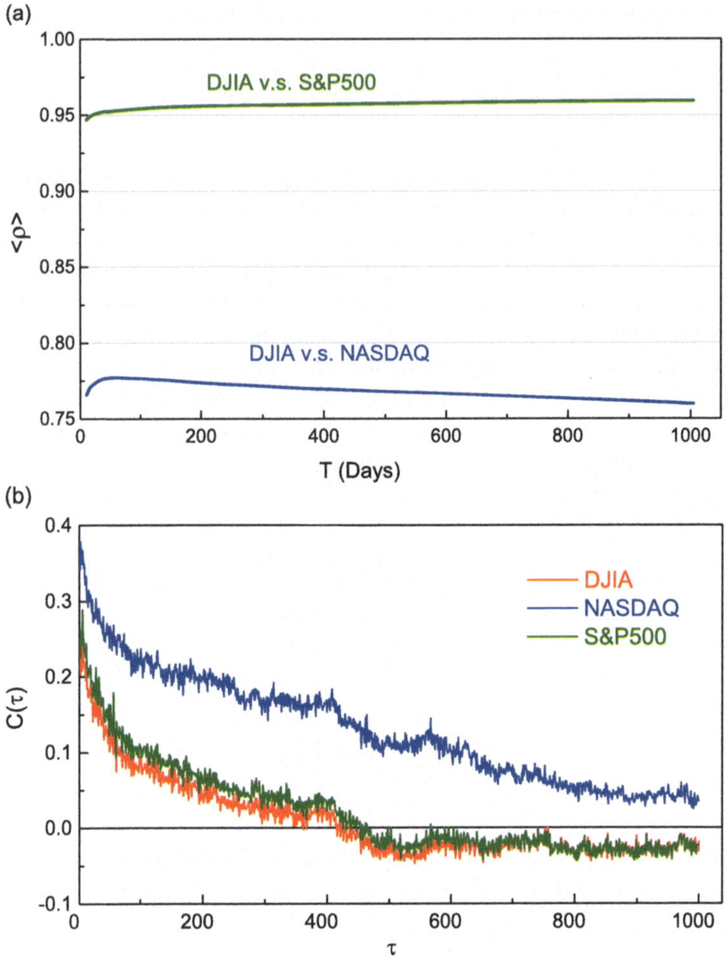

Fig. 5.5 (**a**) Window size dependence of the average dynamic cross-correlation stock market indices. DJIA and NASDAQ has the strongest correlation at $T = 60$ days, and the correlation between DJIA and S&P500 grows gradually with time and saturates at $T > 1000$ days. (Edited from Fig. 5c of [22]) (**b**) Autocorrelation functions of the absolute gains. The correlation length is 194 days for DJIA, 766 days for NASDAQ, and 238 days for S&P500. (Edited from Fig. 6b of [22])

5.3 Conclusion

In conclusion, from analyzing the ratios of the daily index data of DJIA, NASDAQ, and S&P500 from 1971/02/05 to 2011/06/30, it can be shown that though three indices are distinct from one another, using suitable scaling factors, the indices can be made coincidence very well in several epoches, except NASDAQ in the periods 1999–2001 and 2009–2011. Sophisticated time series analysis based on

EMD method further shows that the ratios NASDAQ/DJIA and S&P500/DJIA, normalized to 1971/02/05, approached and then retained 2 and 1, respectively, from 1971 to 2011, through damped oscillatory components in 8-year cycle and damping factors of about 29 years (7183 days for NASDAQ/DJIA) and 554 years (138,471 days for S&P500/DJIA). Note that the damped oscillation of 8-year cycle is not associated with the characteristic time scales in the auto-correlation of the gains and cross-correlation of the returns of the indices. Furthermore, the peak of NASDAQ/DJIA in the period from 1998 to 2002, which is considered as an anomaly in the ratio, is a local event that does not appear in the 8-year cycle. The converge of the damped oscillatory component implies that representative stocks in the pair-markets become more coherent as time evolves. For the components with cycles less than half-year, behaviors of self-adjustments are observed in the ratios, and there is a relatively active reassessment on the ratio in the time scale of 14-days according to the results of MSE analysis. The behavior of self-adjustment in the ratio for S&P500/DJIA is more significant than in NASDAQ/DJIA.

Finally we would like to remark that the damped components found in the study set reasonable bounds to the variations of the indices. It may be informative for risk evaluation of the markets. This requires further investigations.

Acknowledgements This work was supported by the Ministry of Science and Technology (Taiwan) under Grant No. MOST 103-2112-M-008-008-MY3.

References

1. Laloux L, Cizeau P, Bouchaud JP, Potters M (1999) Phys Rev Lett 83:1467
2. Plerou V, Gopikrishnan P, Rosenow B, Amaral LAN, Stanley HE (1999) Phys Rev Lett 83:1471
3. Mantegna RN (1999) Eur Phys J B11:193
4. Mantegna RN, Stanley HE (2000) An introduction to econophysics, correlations and complexity in finance. Cambridge University Press, Cambridge
5. Sornette D, Johansen A (2001) Quant Finan 1:452
6. Sornette D (2003) Phys Rep 378:1
7. Bouchaud JP, Potters M (2003) Theory of financial risk and derivative pricing: from statistical physics to risk management. Cambridge University Press, Cambridge
8. Gabaix X, Gopikrishnan P, Plerou V, Stanley HE (2003) Nature 423:267
9. McCauley JL (2004) Dynamics of markets: econophysics and finance. Cambridge University Press, Cambridge
10. Yamasaki K, Muchnik L, Havlin S, Bunde A, Stanley HE (2005) Proc Natl Acad Sci USA 102:9424
11. Lax M, Cai W, Xu M (2006) Random processes in physics and finance. Oxford University Press, Oxford
12. Kiyono K, Struzik ZR, Yamamoto Y (2006) Phys Rev Lett 96:068701
13. Wu M-C, Huang M-C, Yu H-C, Chiang TC (2006) Phys Rev E 73:016118

14. Wu M-C (2007) Phys A 375:633
15. Wu M-C (2007) J Korean Phys Soc 50:304
16. Bouchaud JP (2008) Nature 455:1181
17. Podobnik B, Stanley HE (2008) Phys Rev Lett 100:084102
18. Sornette D, Woodard R, Zhou W-X (2009) Phys A 388:1571
19. Preis T, Reith D, Stanley HE (2010) Phil Trans R Soc A 368:5707
20. Preis T (2011) Eur Phys J Special Topics 194:5
21. Preis T, Schneider JJ, Stanley HE (2011) Proc Natl Acad Sci USA 108:674
22. Wu M-C (2012) Europhys Lett 97:48009
23. Cont R (2001) Quant Finan 1:223
24. Engle RF, Patton AJ (2001) Quant Finan 1:237
25. Huang NE, Shen Z, Long SR, Wu MC, Shih HH, Zheng Q, Yen N-C, Tung C-C, Liu HH (1998) Proc R Soc Lond A 454:903
26. Wu M-C, Hu C-K (2006) Phys Rev E 73:051917
27. Peng C-K, Buldyrev SV, Havlin S, Simons M, Stanley HE, Goldberger AL (1994) Phys Rev E 49:1685
28. Chen Z, Ivanov PC, Hu K, Stanley HE (2002) Phys Rev E 65:041107
29. Chen Z, Hu K, Carpena P, Bernaola-Galvan P, Stanley HE, Ivanov PC (2005) Phys Rev E 71:011104
30. Costa M, Goldberger AL, Peng C-K (2002) Phys Rev Lett 89:062102
31. Costa M, Goldberger AL, Peng C-K (2005) Phys Rev E 71:021906

Chapter 6
Exploring Market Making Strategy for High Frequency Trading: An Agent-Based Approach

Yibing Xiong, Takashi Yamada, and Takao Terano

Abstract This paper utilizes agent-based simulation to explore market making strategy for high frequency traders (HFTs) and tests its performance under competition environments. After proposing a model representing HFTs' activities in financial market when they act as market makers, we carry out simulations to explore how order price and order quantity affect HFTs' profits and risks. As the result, we find that offering prices around last trading price, as well as taking advantage of order imbalance, increases HFTs' returns. On the other hand, our results show utilizing adaptive order size based on previous order execution rate and setting a net threshold based on average trading volume helps to control the risks of end-of-day inventory. In addition, we introduce the competition environments of increased competitors and decreased latency, so as to see how these factors affect the performance of market making strategy.

6.1 Introduction

On March 11th, 2014, Virtu Financial Inc., the high frequency market maker, who had just one day of trading losses in 1238 days, filed for an initial public offering. Many people were astonished by their near-perfect record as a market maker, while others argued that the profit is becoming unsustainable due to competitions. In this paper we address two questions: what kind of market making strategy helps to increase HFTs' profit, and how this strategy performances under competition environments?

The Securities and Exchange Commission (SEC) generalized four types of trading strategies that often utilized by HFTs [1]. Among them, market making is the most transparent one and constitutes more than 60 % of HFT volume [2]. Menkveld carefully studied the profits and net position of a large HFT who acts as a modern

Y. Xiong (✉) • T. Yamada • T. Terano
Tokyo Institute of Technology, Tokyo, Japan
e-mail: ybxiong@trn.dis.titech.ac.jp; tyamada@trn.dis.titech.ac.jp; terano@dis.titech.ac.jp

H. Takayasu et al. (eds.), *Proceedings of the International Conference on Social Modeling and Simulation, plus Econophysics Colloquium 2014*, Springer Proceedings in Complexity, DOI 10.1007/978-3-319-20591-5_6

market maker [3]. But the strategies under the performance of this HFT, as well as the relationship between strategy and market condition, remains unrevealed.

The formidable challenge for better revealing HFT activities, is obtaining comprehensive and detailed data. Agent-based simulation provides an effective way to solve this problem, and has already been used to design trading strategies. Kendall and Su use an agent-based model to evolve successful trading strategies by integrating individual learning and social learning [4]. Nevmyvaka et al. use a simple class of non-predictive trading strategies to test electronic market making, and examine the impact of various parameters on the market maker's performance [5]. Wang et al. implement a learning algorithm for market makers to search the optimal trading frequency, and they study how different trading frequencies of market makers affect the market [6]. Wah and Wellman employ simulation based methods to evaluate heuristic strategies for market makers and find the presence of the market maker is benefit to both impatient investors and overall market [7]. Comparing with their approaches, our agent-based simulation focuses on testing the combinations of different classes of market making strategies for HFTs, and further examining the performance in competitive environments. Our main contributions are as follows:

- We build an artificial transaction system to represent HFTs' activities in stock market when they act as market makers. This system fits with main statistical properties of financial markets and is used to compare the performance of different market making strategies.
- We find one market making strategy which increases daily return and decreases end-of-day inventory, is offering prices around last trading price, as well as take advantage of order imbalance, together using adaptive order size based on previous order execution rate and a net threshold based on average trading volume.
- We further introducing the environment of increased competitors and decreased latency, in order to test the strategy under different market conditions.

6.2 Modelling of High Frequency Trading

In this section, we propose an artificial stock market in which agents trade through a limit-order book (LOB). Agents are classified into two categories according to their goals and strategies. The one is Low Frequency Traders (LFTs), who concern the value of the asset and try to earn the profit using an integrated strategy of fundamentalist and chartist. The other is High Frequency Traders (HFTs), who ignore the value of the asset but only pay attention to the trading environment itself, and they mainly try to accumulate the profit on the spread using the market making strategy. We stimulate the intra-day transaction scenario where both agents trade on one single asset. The framework of the model are presented first, following by details.

6.2.1 Framework

The model is an extension of the one presented by Leal et al. [8]. There are totally T sessions in each of which both HFTs and LFTs trade. The trading procedure is as follows:

1. Active LFTs decide whether to enter the market according to their expected returns. If enter, they submit either a sell or a buy order with size and price based on their expectations.
2. Knowing the orders submitted by LFTs, HFTs decide whether to enter the market. If enter, they usually submit both a sell and a buy order with size and price in order to absorb the orders of LFTs and earn the profit on the spread.
3. LFTs and HFTs' orders are matched and executed according to their price and arrival time. the last trading price is determined then and unexecuted orders rest in the LOB for the next trading session.
4. After each session, LFTs and HFTs decide whether to update their trading parameters according to their performances.

6.2.2 Low Frequency Traders Activity

For each trading session, LFT i acts as following:

1. Decides whether to be active according to its active possibility LF^i_ap. LF^i_ap is drawn from a uniform distribution with support $[\alpha^L_{min}, \alpha^L_{max}]$ and may be changed according to individual profit.
2. If being active, LFT i first calculates the expected price of the asset LF^i_EP based on its expected return LF^i_ER, then generates the ask price $LF^i\_AP_t$ and bid price $LF^i\_BP_t$ at time t based on last trading price p_t.
   The return at time t are defined as

$$R_t = log(\frac{p_t}{p_{t-1}}) \tag{6.1}$$

   Utilizing the idea of LeBaron and Yamamoto [9]. LFTs form their weighted forecasts on the future returns by combining fundamental-, chart-, and noise-based forecasts as follows:

$$LF^i\_ER = n_1^i \times log(\frac{p^f}{p_t}) + n_2^i \times \frac{1}{l_i}\sum_{j=1}^{l_i} log(\frac{p_{t-1}}{p_{t-j-1}}) + n_3^i \times N(0,1) \tag{6.2}$$

   and the expected price LF^i_EP is calculated as $p_t \times e^{LF^i\_ER}$.
   l_i represents the memory length of LFT i, and $l_i \sim U(1, l_{max})$. n_1^i, n_2^i, n_3^i are weights for fundamentalist, chartist, and noise-induced components for LFT i,

respectively, and randomly assigned according to normal distributions: $n_1^i \sim |N(0, \sigma_1)|$, $n_2^i \sim N(0, \sigma_2)$, and $n_3^i \sim N(0, \sigma_3)$.

And price for sell/buy orders at time t are formed as follows:

$$LF^i\_AP_t = p_{t-1} \times (1 - \kappa_i^L)$$
$$LF^i\_BP_t = p_{t-1} \times (1 + \kappa_i^L) \quad (6.3)$$

where $\kappa_i \sim U(\kappa_{min}, \kappa_{max})$ represents the price fluctuation parameter.

3. If the ask price of LFT i is higher than its expected future price, LFT i will submit a sell order at price $LF^i\_AP_t$ with size LF^i_AS; if the bid price of LFT i is lower than its expected future price, it will submit a buy order at price $LF^i\_BP_t$ with size LF^i_BS. The valid time of the order is γ^L.
The size of the orders are proportional to expected returns and are formed as:

$$LF^i\_AS = LF^i\_ER \times \eta_i^L$$
$$LF^i\_BS = -LF^i\_ER \times \eta_i^L \quad (6.4)$$

where η_i^L represents the size fluctuation parameter of LFTs and $\eta_i^L \sim U(\eta_{min}^L, \eta_{max}^L)$, it can be changed according to individual profit. And if the order has not been fully executed within γ^L, the rest of it will be removed from the LOB.

4. After τ sessions, LFT i decide whether to update its trading parameters based on its profit $LF^i\_P_t$.
if $LF^i\_P_t > 0$, LFT i will update some of its parameters as:
$LF^i\_ap \sim U(LF^i\_ap, \alpha_{max})$ $\eta_i^L \sim U(\eta_i^L, \eta_{max}^L)$
if $LF^i\_P_t \leqslant 0$, it becomes:
$LF^i\_ap \sim U(\alpha_{min}, LF^i\_ap)$ $\eta_i^L \sim U(\eta_{min}^L, \eta_i^L)$
if $LF^i\_P_t \leqslant 0$ and a random number $\sim U[0, 1] < \lambda$, then the component-weighted parameters and memory length will be renewed based on the distributions:
$n_1^i \sim |N(0, \sigma_1)|$, $n_2^i \sim N(0, \sigma_2)$, $n_3^i \sim N(0, \sigma_3)$, and $l_i \sim U(1, l_{max})$.

6.2.3 High Frequency Traders Activity

For each trading session, HFT j acts as follows:

1. HFT j decides whether to be active based on the price fluctuation P_t^{flu} (bps) at time t and its action threshold HF^j_at.
Since evidence suggests HFT activities prefer higher volatility. We calculate

$$P_t^{flu} = |\frac{p_{t-1} - p_{t-2}}{p_{t-1}}| \times 10,000 \quad (6.5)$$

and $HF^j\_at \sim U(\alpha_{min}^H, \alpha_{max}^H)$. If $P_t^{flu} > HF^j\_at$, then HFT j becomes active.

2. If being active, it submits both a sell order at price HF^j_AP with size HF^j_AS and a buy order at price HF^j_BP with size HF^j_BS. All orders from HFTs are submitted in a random order, and the valid time of the orders is γ^H.
Under the defaulted setting, $HF^j\_AP = p_t + \kappa_j^H$, $HF^j\_BP = p_t - \kappa_j^H$, κ^H refers to price fluctuation. While HFTs decide the order quantity based on the quotes in the LOB. $HF^j\_AS = HF^j\_BS = 0.5 \times (q_b + q_s) \times \eta_j^H$. Where q_b (q_s) refers to the total size of buy (sell) orders in the LOB at this session, and η^H refers to order absorption rate.
3. Like LFTs, after τ sessions, HFT j decide whether to update η_j^H based on its performance.

6.2.4 Model Validation

Table 6.1 lists the defaulted value of all the parameters. The number of sessions is set as 400 for intra-day trading. There are 441 traders, and 2 % (9) of them are HFTs [2]. Price and volume related parameters are calibrated to fit with market volatility and liquidity condition respectively, while keeping the diversity of agents.

Table 6.1 Parameters in initial simulation

Description	Symbol	Defaulted
Number of trading sessions	T	400
Number of traders	N	441
Fundamental value	p_f	50
Tick size	t_s	0.01
LFT initial active possibility	$[\alpha^L_{min}, \alpha^L_{max}]$	[0.01, 0.1]
LFT max memory length	l_{max}	30
LFT order price fluctuation	$[\kappa^L_{min}, \kappa^L_{max}]$	[−0.002, 0.01]
LFT order size fluctuation	$[\eta^L_{min}, \eta^L_{max}]$	[200, 1000]
LFT order life	γ^L	10
LFT parameter evolution circle	τ	30
LFT parameter evolution rate	λ	0.3
Std of fundamental component	σ_1	0.3
Std of chartists component	σ_2	0.6
Std of noise-trader component	σ_3	0.1
HFT percentage	HFT_{per}	2 %
HFT active threshold	$[\alpha^H_{min}, \alpha^H_{max}]$	[5, 5]
HFT order price fluctuation	κ^H	0.01
HFT order absorption rate	$[\eta^H_{min}, \eta^H_{max}]$	[0.1, 0.5]
HFT order life	γ^H	1

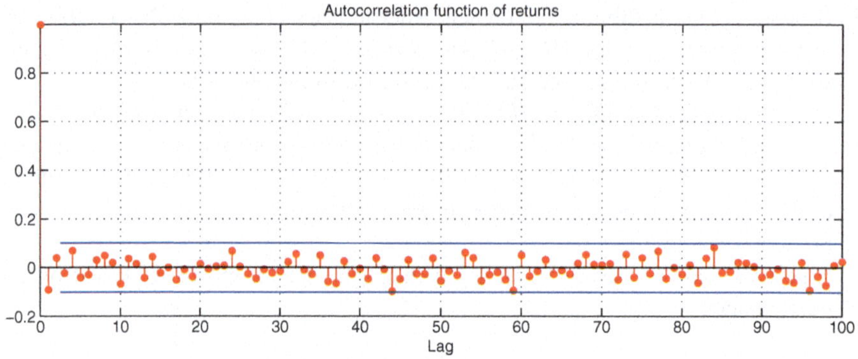

Fig. 6.1 Autocorrelation of return in HFT simulations

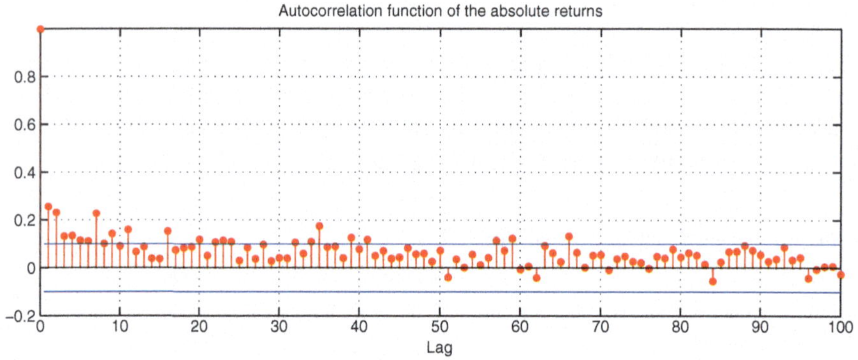

Fig. 6.2 Volatility clustering in HFT simulations

We check whether the model is able to account for the main stylized facts of financial markets. The price movement generated by the model is in line with the empirical evidence as absence of autocorrelation, see as Fig. 6.1. In contrast, the autocorrelation function of absolute return display a slow decaying pattern, see as Fig. 6.2. In addition, the existence of fat tails in the distribution of price is shown in Fig. 6.3.

6.3 Exploration of Market Making Strategy

In this section, we design experiments concerning HFT order price, order quantity respectively, trying to find what kind of order price and order quantity help HFTs increasing profits and decreasing risks.

Based on [3], our model assumes HFTs usually utilize passive market making and try to earn the spread. But when they speculated order imbalance by detecting

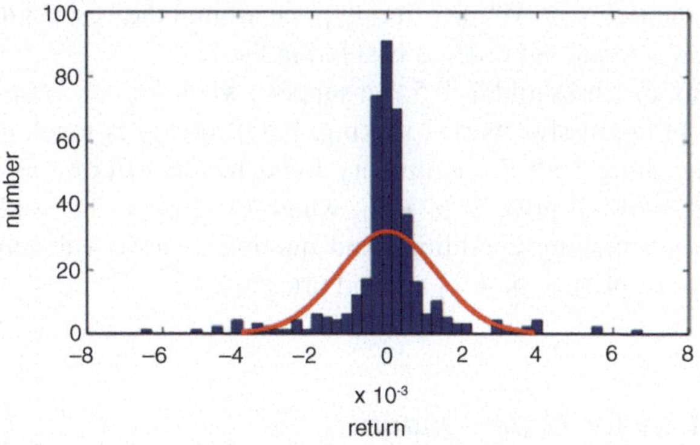

Fig. 6.3 Return distribution in HFT simulations

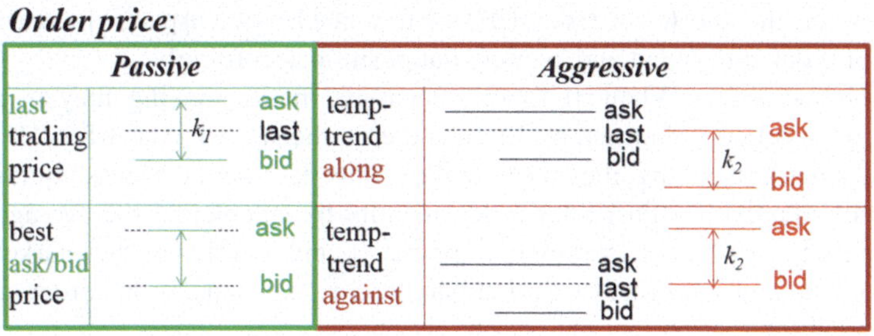

Fig. 6.4 Passive and aggressive market making

the transaction trends of LFTs, they may utilize aggressive market making. In this case they trade on quickly to get the profit from price movement, or close their position. In the model, we consider HFT's profit as the daily return, and the risk as the end-of-day inventory. All transaction fees are ignored for simplicity.

6.3.1 Strategies for Order Price

HFTs switch between passive and aggressive market making. In usual case, HFT use passive market making as a liquidity maker, and they quote either based on last trading price or best ask/bid price to earn the spread. But when the volume difference between ask and bid quoting becomes significant, i.e. when $|q_b - q_s|/(q_b + q_s)$ is over a threshold, they will adopt aggressive market making. In this case HFTs will either quoting along with temporary trend to earn the profit based on the price movement, or against it to take liquidity in order to adjust their position. Shown as Fig. 6.4.

When HFTs adopt passive market making and offer price around the last trading price, the ask order is offered at $p_t + 0.5 \times k_1$ and bid order at $p_t - 0.5 \times k_1$, where k_1 refers to the ask-bid spread in passive market making condition and its default value is twice the tick size. While offering price around the best ask/bid, they offer ask order at best ask and bid order at best bid in the LOB.

While setting the threshold as 0.5, we suppose when $|q_b - q_s|/(q_b + q_s) > 0.5$, HFTs will adopt aggressive market making. If sell orders is much more than buy orders, quoting along with the temporary trend means offering ask price at last trading price p_t and bid price at $p_t - k_2$, where k_2 refers to the ask/bid spread in aggressive market making condition. And quoting against with temporary trend means offering ask price at $p_t + k_2$ and bid price at p_t.

6.3.2 Strategy for Order Quantity

Another question need to be discussed for the strategy is order quantity. In order to gain more profits and fewer risks, HFTs will consider two aspects, increasing the chance of order fulfillment and keeping flat position accordingly.

For the first aspect, When HFTs utilizes passive market making, they are aiming to absorb the orders submitted by LFTs and earn the spread. And when HFTs use aggressive market making, they try to make profits base on the excess liquidity. All HFTs will adaptively adjust their order quantity by calculating the average order execution rate in past few sessions. Suppose r_m and r_h refer to the market order execution rate and self order execution rate in past τ sessions, so the order quantity Q is decide by:

$$\begin{aligned} Passive : Q &= min(q_b, q_s) \times 0.5 \times (r_m + r_h) \\ Aggressive : Q &= |q_b - q_s| \times 0.5 \times (r_m + r_h) \end{aligned} \tag{6.6}$$

For the second aspect, based on the previous work by Menkveld and Hendershott [3, 10], we introduce a parameter named net threshold, denoted as NT to supervise the net position of HFTs, denoted as np. Suppose V_i represents the trading volume in session i and NT is proportional to average trading volume in first τ periods, so after τ periods, NT is calculated and works as follows:

$$NT = \frac{1}{\tau} \sum_{i=1}^{\tau} V_i \tag{6.7}$$

- When $|np| < 0.5 \times NT$, it trades as usual.
- $0.5 \times NT \leqslant |np| < NT$, it applies price pressure and adjusts its quotes by one tick size.
- When $|np| \geqslant NT$, it stops one side (buy or sell) trading.
- a HFT's max order size equals NT.

6.3.3 Comparison of Strategies

In last two sections, we mentioned the strategies used for passive and aggressive market making, as well as the method to control end-of-day inventory. We suppose HFTs can use either single or combination of market making, so totally eight types of quoting methods are list in Table 6.2.

Supposing 2 % of the traders are HFTs, all of them use a same quoting strategy, selected from the list, together with adaptive order quantity and net threshold for controlling inventory, and their orders are submitted in a random order. We test the performance of these strategies focused on the daily return and end-of-day inventory, denoted as *EDI*. After testing each strategy for 250 simulations, Fig. 6.5 shows the average return and inventory of these strategies.

$$EDI\% = \frac{|np|}{NT} \times 100\,\% \tag{6.8}$$

Table 6.2 List of quoting strategies

	Market conditions					
	$(q_b - q_s)/(q_b + q_s) > 0.5$		$(q_s - q_b)/(q_b + q_s) > 0.5$		Other	
Strategy	ask price	bid price	ask price	bid price	ask price	bid price
last trading	$p_t + 0.5 \times k_1$	$p_t - 0.5 \times k_1$	$p_t + 0.5 \times k_1$	$p_t - 0.5 \times k_1$	$p_t + 0.5 \times k_1$	$p_t - 0.5 \times k_1$
ask/bid	best ask	best bid	best ask	best bid	best ask	best bid
trend-along	$p_t + k_2$	p_t	p_t	$p_t - k_2$	*NA*	*NA*
trend-against	p_t	$p_t - k_2$	$p_t + k_2$	p_t	*NA*	*NA*
last+along	$p_t + k_2$	p_t	p_t	$p_t - k_2$	$p_t + 0.5 \times k_1$	$p_t - 0.5 \times k_1$
last+against	p_t	$p_t - k_2$	$p_t + k_2$	p_t	$p_t + 0.5 \times k_1$	$p_t - 0.5 \times k_1$
ask/bid+along	$p_t + k_2$	p_t	p_t	$p_t - k_2$	best ask	best bid
ask/bid+against	p_t	$p_t - k_2$	$p_t + k_2$	p_t	best ask	best bid

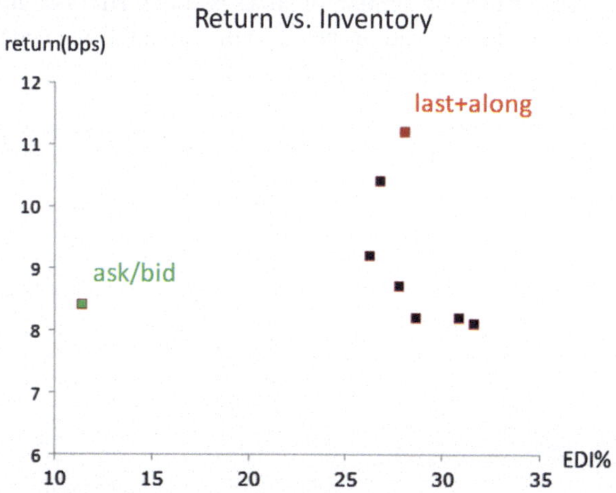

Fig. 6.5 Comparison of different market making strategies

According to the graph, we see that quoting based on best ask/bid price has its advantage on inventory, which indicate this strategy has the lowest risk. While quoting based on last trading price together using a temporary trend along strategy, lead to the highest return for HFTs. Here, we choose the last+along strategy as the benchmark for further experiments.

6.4 Experiments on Competition

In following simulations, we change the percentage of HFTs (2 % in previous experiments) to see its influence. On the other hand, all HFTs submit their orders in a random order in past experiments, which means they have similar latencies. Considering HFTs are pursuing lower latency in order to run in front of their competitors nowadays, we arrange different latencies for HFTs. In this case HFTs submit their orders one after another in a fixed order, thus a HFT with lower latency submits its orders earlier and is likely to have higher order execution probabilities.

6.4.1 Total Return of HFTs

We first consider the total return of HFTs. We concern how HFTs' profit will be affected by increased competitors and decreased latency. While all HFTs using the last+along strategy, we adjust the number of HFTs, and run the simulation 250 times for each different HFT percentage (0.5 %, 1 %, 1.5 %, ...5 %) first in similar then in different latency settings. The total return of HFTs is calculated and shown in Fig. 6.6:

There are two things interesting according to this result. On one hand, in both latency conditions, the total return of HFTs first went up and then down. This may because when the number of HFTs is small, HFTs do not fully absorb LFTs' orders and there are still surplus profit on the spread. But when this number becomes larger,

Fig. 6.6 Total return of HFTs

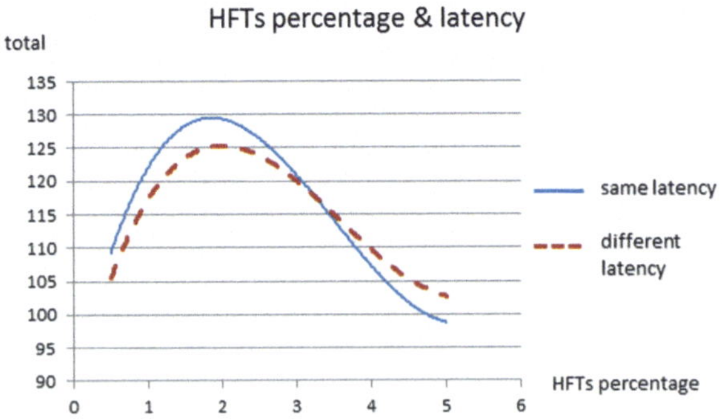

HFTs suffer from position imbalance and need to pay the price to trade out of their positions, thus causes a decline in their total return. On the other hand, the red curve shows the condition that HFTs compete with each other on speed and have different latencies. Comparing to the blue one, it indicates this competition actually decreases HFTs' total return when the number of HFTs is small but increases their total return when the number is larger.

6.4.2 *Individual Return of HFTs*

We then turn to the individual return of HFT and concern the value of low-latency. Supposing HFTs have different latencies, Fig. 6.7 depicts the return difference among LFTs, normal HFTs and the fastest HFT.

In the result, the red curve shows the average individual return of HFTs, it decreases with the increase of HFTs percentage and can be seen as the return of a normal HFT. Since LFTs' population is much more than HFTs' and their average return can be seen as zero, the red curve also illustrates the return difference between a normal HFT and a LFT approximately, and can be taken as a reference for a LFT to decide whether it is worth taking part in HFT. The green curve, on the other hand, is calculated as the difference between the average and the highest return (return of the HFT with the lowest latency). It can be interpreted as how much potential profit can be earned for a normal HFT to become the fastest one by renewing its devices or using co-location method. This chart may suggest that although the profit for becoming a HFT decreases with the increase of HFTs, it is always profitable for a HFT to pursue lower latency.

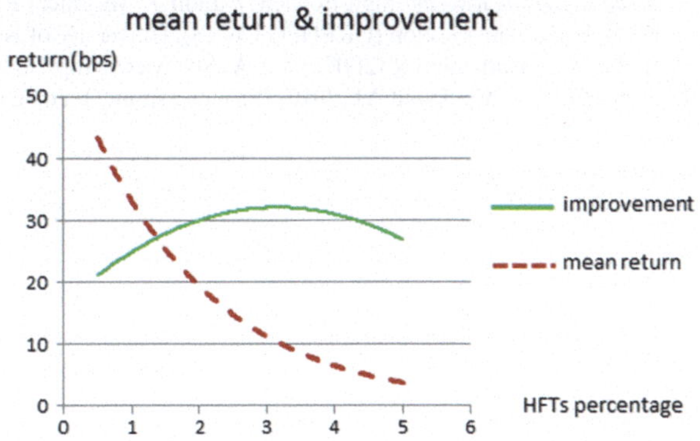

Fig. 6.7 Individual return of HFTs

6.5 Conclusions

This paper focuses on exploring market making strategy for High Frequency Trading and consists of three parts. First we combine previous work and build an intra-day transaction model based on limit order book to simulate the trading activities of HFTs and LFTs. In addition, by analyzing both passive and aggressive market making, we try to figure out what kind of order price and order quantity help HFTs increasing their profits and decreasing their risks. Finally, we test the strategy in competition environments including increased competitors and decreased latency, in order to see its performance.

References

1. Securities and Exchange Commission (2010) Fed Reg 75(13):3594
2. Hagströmer B, Nordén L (2013) J Financ Mark 16(4):741
3. Menkveld AJ (2013) J Financ Mark 16(4):712
4. Kendall G, Su Y (2003) Proceedings of IEEE, pp 2298–2305
5. Nevmyvaka Y, Sycara K, Seppi D (2003) In: Proceedings of the third international workshop on computational intelligence in economics and finance
6. Wang C, Izumi K, Mizuta T, Yoshimura S (2013) SICE J Control Meas Syst Integr 6(3):216
7. Wah E, Wellman MP (2015) In: Fourteenth international conference on autonomous agents and multi-agent systems
8. Jacob Leal S, Napoletano M, Roventini A, Fagiolo G (2014) Rock around the clock: an agent-based model of low-and high-frequency trading. Technical Report, Groupe de REcherche en Droit, Économie, Gestion (GREDEG CNRS), University of Nice Sophia Antipolis
9. LeBaron B, Yamamoto R (2007) Phys A: Stat Mech Appl 383(1):85
10. Hendershott T, Menkveld AJ (2010) Price pressures. Technical Report, CFS Working Paper

Chapter 7
Effect of Cancel Order on Simple Stochastic Order-Book Model

Shingo Ichiki and Katsuhiro Nishinari

Abstract We investigate the effect of the order canceling rule in the trading model of financial exchanges. This study employs a stochastic order-book model. Such models are widely used to study the relation between price fluctuation and price formation in continuous double auction. The model herein incorporates simple mechanisms such as limit order and trading rules without considering investors' strategies. It captures the transaction structure used in financial exchanges. Using three simple stochastic order-book models, we indicate the comparative analysis of the effectiveness of the cancel order.

7.1 Introduction

Major financial exchanges employ continuous double auctions wherein sellers and buyers simultaneously present respective prices. To determine the relation between price formation and price fluctuation, scholars use stochastic order-book models, which replicate transactions occurring under mechanisms customarily used in financial exchanges. In particular, Maslov's proposal is a good example of its pioneering model [1]. In this model, limit and market order were chosen with equal probability. Bid and ask orders were chosen with equal probability as well. The limit order price was selected by a uniform random number within a specified range from the transaction price. By doing so, he captured the power law in the distribution of price differences gathered via simulations. Since the publication of this model, various stochastic order-book models have been proposed [2, 3].

S. Ichiki (✉)
Department of Advanced Interdisciplinary Studies, Graduate School of Engineering, The University of Tokyo, 4-6-1 Komaba, Meguro-ku, Tokyo 153-8904, Japan
e-mail: ichiki@jamology.rcast.u-tokyo.ac.jp

K. Nishinari
Research Center for Advanced Science and Technology, The University of Tokyo, 4-6-1 Komaba, Meguro-ku, Tokyo 153-8904, Japan
e-mail: tknishi@mail.ecc.u-tokyo.ac.jp

H. Takayasu et al. (eds.), *Proceedings of the International Conference on Social Modeling and Simulation, plus Econophysics Colloquium 2014*, Springer Proceedings in Complexity, DOI 10.1007/978-3-319-20591-5_7

In previous studies, the order-book models which have mechanism such as random cancel and automatic cancel due to passing specific period are introduced. T. Preis *et al.* incorporated the mechanism that a limit order is deleted with a probability per time unit in his model [2]. Considering the cancel order is important in the construction of a simple mechanical simulation of a continuous double auction. Using examples from Maslov's model, we suggest three simple stochastic order-book models that incorporate the cancel order mechanism which exclude investors' strategies. Then we focus on the cancel order and compare the effectiveness of the cancel orders in the three models.

7.2 Model

This section explains the structure of our three models, which incorporates financial exchanges' basic trading rules.

7.2.1 Basic Trading Rules of Financial Exchanges

Major financial exchanges operate with continuous double auction that uses an electronic board (an order book) on which buyers (sellers) enter bid prices (ask prices) and transactions are matched. Investors place limit and market orders. Limit orders specify prices at which investors will execute trades, whereas market orders do not. When an exchange receives sellers' limit orders, it matches them with buyers' bids that equal or exceed sellers' ask prices. Conversely, exchanges match buyers' limit bids with sellers' asking identical or lower prices. A market order is immediately matched with any existing order on the order book. When there are ask (bid) orders on the order book, any incoming bid (ask) market order will be matched with the lowest ask (highest bid) order on the order book. For matching the orders, the price priority rule is used. The highest bid (lowest ask) order on the order book will be given priority over all other bid (ask) orders. When the order book contains multiple orders at identical price, the oldest is executed first. This rule is called the time priority rule. Price and time priorities are standard practices at exchanges worldwide.

7.2.2 Simulation Models

This subsection explains how orders are selected for execution in our simulation models.

7.2.2.1 No-Cancel Model

We explain about a model without a order canceling rule. The model assumes an equal probability that a new order is a bid or an ask, and the price of the order is selected randomly within a specified range from the most recent transaction price. We specify a range of ± 15 from the most recent transaction price. For example, if the most recent transaction price is 0, the bid or ask price is randomly selected within the range $[-15, 15]$, and one unit will be placed on the order book.

Moreover, in this study, we employ the price priority rule that is used in the trading mechanism of the major financial exchanges. The time priority rule is not meaningful in our simulation because we do not distinguish agents who send orders. Trading takes place whenever best ask $\leq$ best bid, where "best ask" is the lowest ask price and "best bid" is the highest bid price on the order book. The transaction price is either the price of the bid or ask order, whichever is on the order book first.

Figure 7.1 depicts a transaction. At State 1, the order book holds an order to sell three units at an ask price of 101 and an order to buy two units at a bid price of 99. The most recent transaction price is 100. At State 2, one unit of bid order at a price of 102 is entered. Because of this new order, best ask $\leq$ best bid; therefore, at State 3, the transaction occurs between the one unit ask order at a price of 101 and the new bid order at a price of 102. The transaction price is 101.

This model does not incorporate market orders. However, because an order is always placed in terms of one unit only, when an order is immediately executed, it could be interpreted as being a market order because it has having the same effect as a market order. This model with the above rules is called the "no-cancel model."

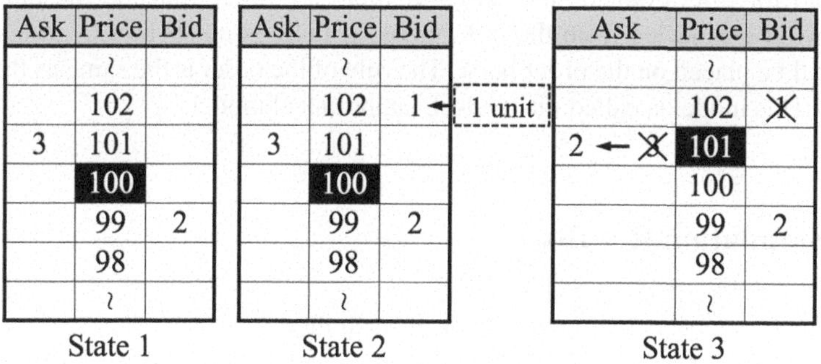

Fig. 7.1 Drawing exemplifying an order book transaction

7.2.2.2 Random Cancel Model

We explain about a model with a certain kind of order canceling rule.

First, we assort the order situations of the order book in the following cases.

1. There are both ask and bid orders in the order book.
2. There are orders only in the ask side in the order book.
3. There are orders only in the bid side in the order book.
4. There is no order in the order book.

In Case 1, the model selects an order or a cancel order in an equal probability. If the order is selected, the rule of the order is the same as the no-cancel model. If the cancel order is selected, either the ask or bid is selected in an equal probability, and an ask order or a bid order on the order book is canceled randomly. In Case 2, similar to Case 1, the order or the cancel order is selected in an equal probability. If the order is selected, the rule of the order is the same as the no-cancel model. If the cancel order is selected, a ask order on the order book is canceled randomly. Case 3 mirrors Case 2 but switches the ask and the bid. In Case 4, the order is always selected. And the rule of the order is the same as the no-cancel model. This model including the above order canceling rule is called the "random cancel model."

7.2.2.3 Out-of-Range Cancel Model

We explain about a model with a order canceling rule that are outside the specified range of ± 15 from the most recent transaction price. First, existing orders are examined for prices outside the range. If there are orders outside the range, the orders will be canceled from the order book. If there is no such order, then a new order will be placed on the order book. The rule of the order is the same as the other models. This model is called the "out-of-range cancel model."

7.3 Simulation Results

This section compares price movements in each model.

First, we performed simulations for the no-cancel model. Figure 7.2 shows 1,000 tick of price data.

Next, we performed simulations for the random cancel and out-of-range cancel models. One million simulations were performed 10 times for each model. Transactions occurred with the ratios $10.91\,\% \pm 0.04\,\%$ for the number of simulations using the random cancel model and $29.03\,\% \pm 0.04\,\%$ for the number of simulations using the out-of-range cancel model. Figure 7.3 shows 10,000 tick of price data.

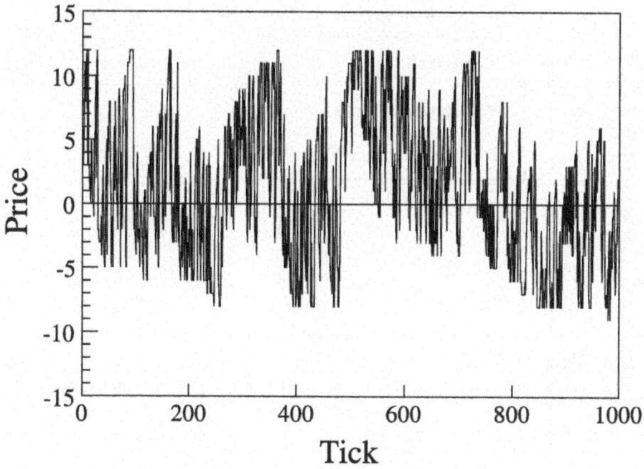

Fig. 7.2 Price movements for 1,000 ticks. Fluctuations are obtained by simulation using the no-cancel model

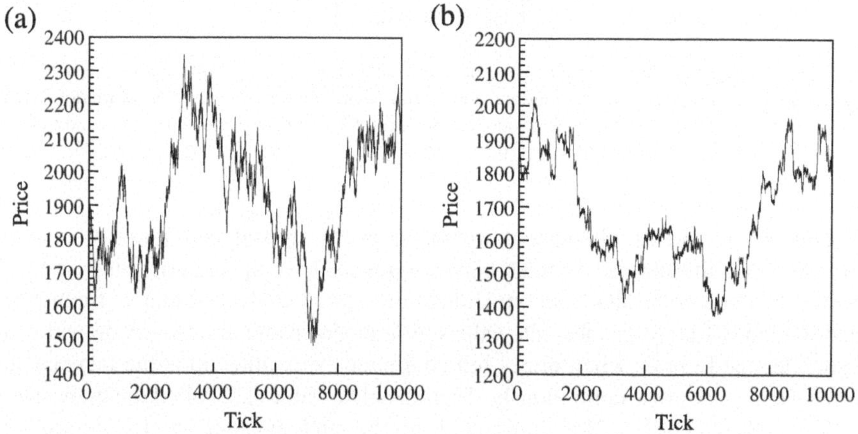

Fig. 7.3 Price movements for 10,000 ticks. Fluctuations are obtained by simulations using (**a**) the random cancel model and (**b**) the out-of-range cancel model

Illustrating how prices diffuse with time, Fig. 7.4 shows the relation between the standard deviation of the price gap and the time scale (tick) on a double-logarithmic graph. Also, we here estimate Hurst exponent by linearizing the points plotted on the double-logarithmic graph. Each Hurst exponent of the random cancel model and the out-of-range cancel model is 0.499 and 0.478. The dotted line is the one-half power of the time scale. In fact, the relationship $\sigma(\tau)$ between the standard deviation of the price gap and the time scale is as follows, where τ is the time scale, and H is Hurst exponent.

$$\sigma(\tau) \propto \tau^H, \quad H = 0.5, \quad \tau \geq 1.$$

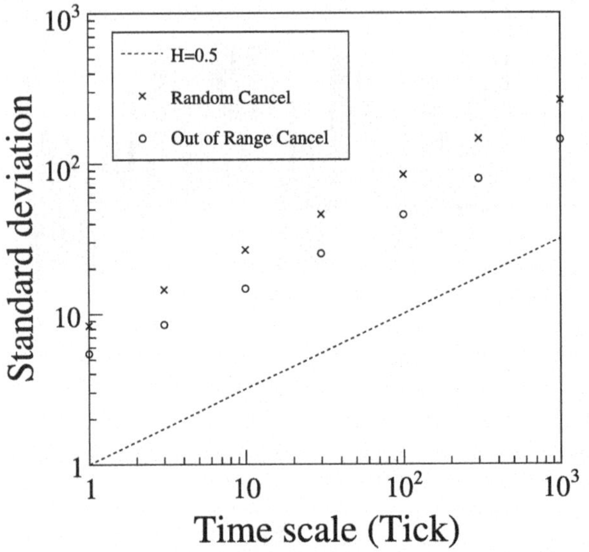

Fig. 7.4 Double-logarithmic graph of the standard deviations of price gaps with respect to the time scale (tick) derived from the random cancel model and the out-of-range cancel model. Each Hurst exponent of the random cancel model and the out-of-range cancel model is 0.499 and 0.478. The *dotted line* is the Hurst exponent of 0.5

Next, we analyzed the extent to which prices moved continuously in one direction. We did not differentiate between downside or upside price movements. So we obtained price data from our simulations and plotted a cumulative frequency distribution (CFD) of the absolute values of draw down and draw up. A draw down is the decline in price when prices fell continuously. A draw up is the increase in price when prices rose continuously. Their absolute values are called the draw size.

We analyzed CFDs of the draw size from the random cancel model. Figure 7.5 illustrates the CFDs that compare the draw size from the random cancel model and the draw size from shuffled price gap data from the random cancel model. The solid line depicts a linearization of CFD for draw size of 16 and larger from the random cancel model (slope = −0.040). The dotted line depicts a linearization of CFD for draw size of 16 and larger from the shuffled price gap data from the random cancel model (slope = −0.036).

Next, we analyzed the CFDs of the draw size from the out-of-range cancel model. Figure 7.6 illustrates the CFDs that compare the draw size from the out-of-range cancel model and the draw size from shuffled price gap data from the out-of-range cancel model. The chained line depicts a linearization of CFD for draw size of 16 and larger from the out-of-range cancel model (slope = −0.044). The double-dotted line depicts a linearization of CFD for draw size of 16 and larger from the shuffled price gap data from the out-of-range cancel model (slope = −0.061).

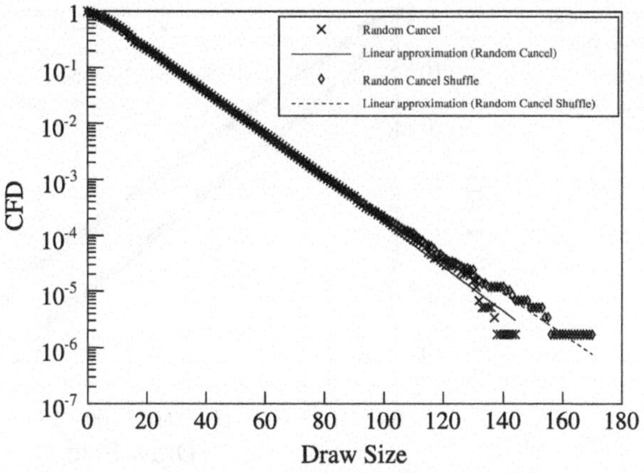

Fig. 7.5 Semilogarithmic graph of CFDs for a draw size from the random cancel model (Random Cancel) and shuffled price gap data from the random cancel model (Random Cancel Shuffle). The *solid line* depicts a linearization of CFD for draw size of 16 and larger from the random cancel model (slope = −0.040). The *dotted line* depicts a linearization of CFD for draw size of 16 and larger from the shuffled price gap data from the random cancel model (slope = −0.036)

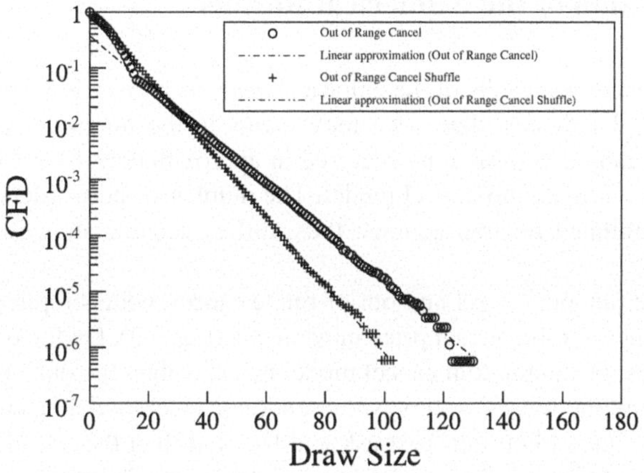

Fig. 7.6 Semilogarithmic graph of CFDs for a draw size from the out-of-range cancel model (Out-of-Range Cancel) and shuffled price gap data from the out-of-range cancel model (Out-of-Range Cancel Shuffle). The *chained line* depicts a linearization of CFD for draw size of 16 and larger from the out-of-range cancel model (slope = −0.044). The *double-dotted line* depicts a linearization of CFD for draw size of 16 and larger from the shuffled price gap data from the out-of-range cancel model (slope = −0.061)

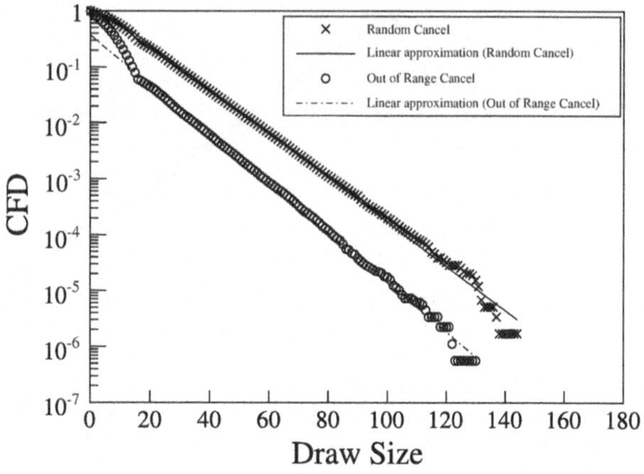

Fig. 7.7 Semilogarithmic graph of CFDs from the random cancel and out-of-range cancel models

Additionally, we compared the CFD of the draw size from the random cancel model and the out-of-range cancel model (Fig. 7.7).

7.4 Discussion of the Numerical Results

This section examines results of the empirical analysis in Sect. 7.3.

First, Fig. 7.2 indicates that price movements in the no-cancel model vibrate within a fixed range. Transactions occurred in approximately 40 % for the number of simulations using the no-cancel model. The number of new orders exceeds that of orders annihilated by transactions; thus, orders accumulate, restraining price movements.

Second, the random cancel and out-of-range cancel models replicated the price movements that resemble actual price movements (Fig. 7.3). On the other hand, the price movement of the random cancel model is larger than the out-of-range cancel model. We think the reason of this is the random cancel model has the possibility to have bigger spread between best ask and best bid than the out-of-range cancel model because of the difference of order canceling rule. The standard deviation of the price gap for the time scale (tick) is proportional to about one-half power of the time scale for each model (Fig. 7.4). This finding indicates that price data from each model diffuse at a speed characteristic of a random walk.

Third, the CFD shape in the random cancel model deviates slightly around a draw size of 16 (Fig. 7.5). The specified range of ± 15 possibly explains this finding. A draw size of 16 and larger reflects only the effects of consecutive unidirectional price movements. Moreover, this CFD can be approximated exponentially. Additionally, the CFD of the draw size from the random cancel model shares features with that from the shuffled price gap data from the random cancel model.

Fourth, Fig. 7.6 shows that as with Fig. 7.5, the CFD shape in the out-of-range cancel model deviates around a draw size of 16. Slopes of the linearized data differ, but for draw sizes of 16 and larger, the CFD from the out-of-range cancel model resembles that of the draw size from shuffled price gap data from the out-of-range cancel model.

Fifth, we examine CFDs of draw size from the random cancel and the out-of-range cancel models. The change in CFD shape of the former exceeds that of the latter (Fig. 7.7). This difference arises from differing methods of order cancelation. For draw sizes of 16 and larger, the slope of the linearized data is nearly identical, suggesting that draw size occur less frequently as it increase with a constant probability. This finding suggests that there is no strong serial correlation among some parts.

Finally, we conclude this section by discussing the application potentiality for those models. The out-of-range canceling rule is more convenient than the random canceling rule in reality. Because investing information is abundantly and readily available to investors; therefore, it is unlikely that their orders would be left on the order book when the transaction price has moved sufficiently away from their order price. In addition, in markets led by professional traders, traders are constantly calculating the theoretical price of product; therefore, the entire trading community has similar ideas regarding appropriate pricing. Therefore, it is more realistic to remove an order whose price is placed outside the established range from the most recent transaction price [4].

7.5 Conclusion

This study compared the effectiveness of the cancel order in three simple stochastic order-book models. Using a simple stochastic order-book model, it showed that the method of order cancelation is important in replicating actual price movements. Also, both random cancel and out-of-range cancel models replicated the price movements that resemble actual price movements. Price movements obtained from these models closely resemble a random walk. On the other hand, because of investors' aspect, the out-of-range canceling rule is more convenient than the random canceling rule in reality. Therefore, a comparative analysis that employs this base model with models that incorporate investors' strategies captures how investors' trading strategies affect price movements [4]. Future simulation analyses using these base models will deepen the understanding of investors' trading strategies.

Acknowledgements This work was supported by JSPS KAKENHI Grant Number 25287026.

References

1. Maslov S (2000) Phys A 278:571
2. Preis T, Golke S, Paul W, Schneider JJ (2007) Phys Rev E 76:016108
3. Slanina F (2008) Eur Phys J B 61:225
4. Ichiki S, Nishinari K (2015) Phys A 420:304

Part II
Robustness and Fragility

Chapter 8
Cascading Failures in Interdependent Economic Networks

Shlomo Havlin and Dror Y. Kenett

Abstract Throughout the past decade, there has been a significant advance in understanding the structure and function of networks, and mathematical models of networks are now widely used to describe a broad range of complex systems, such as socio-economic systems. However, the significant majority of methods have dealt almost exclusively with individual networks treated as isolated systems. In reality an individual network is often just one component in a much larger complex multi-level network (network of networks, NON). The NON framework provides critical new insights into the structure and function of real-world complex systems. One such insight is that NON system is significantly more vulnerable to shocks and damages, which has lead to the development of the theory of cascading failures in interdependent networks. Here we provide an overview of this theory, and one example of its application to economic systems.

8.1 Introduction

The growth of technology, globalization, and urbanization has caused world-wide human social and economic activities to become increasingly interdependent [1–13]. From the recent financial crisis it is clear that components of this complex system have become increasingly susceptible to collapse. Current models have been unable to predict instability, provide scenarios for future stability, or control or even mitigate systemic failure. Thus, there is a need of new ways of quantifying complex system vulnerabilities as well as new strategies for mitigating systemic damage and increasing system resiliency [14, 15]. Achieving this would also provide new insight into such key issues as financial contagion [16, 17] and systemic risk [18–20] and would provide a way of maintaining economic and financial stability in the future.

S. Havlin (✉)
Department of Physics, Bar-Ilan University, Ramar Gan, Israel
e-mail: havlins@gmail.com

D.Y. Kenett (✉)
Center for Polymer Studies and Department of Physics, Boston University, MA, USA
e-mail: drorkenett@gmail.com

H. Takayasu et al. (eds.), *Proceedings of the International Conference on Social Modeling and Simulation, plus Econophysics Colloquium 2014*, Springer Proceedings in Complexity, DOI 10.1007/978-3-319-20591-5_8

Throughout the past decade, there has been a significant advance in understanding the structure and function of networks, and mathematical models of networks are now widely used to describe a broad range of complex systems, from techno-social systems to interactions amongst proteins [21–32]. However, the significant majority of methods have dealt almost exclusively with individual networks treated as isolated systems. In reality an individual network is often just one component in a much larger complex multi-level network (network of networks). As technology has advanced, the coupling between networks is becoming stronger and stronger. For example, there is a strong coupling between human mobility (which can be tracked by mobile networks) and transport networks. In these interdependent networks, the failures of nodes in one network will cause failures of dependent nodes in other networks, and vice-versa [33–41]. This process happens recursively, and leads to a cascade of failures in the network of networks system. As in physics, when only the individual particles were studied it was made possible to understand the properties of gas; however, when the transition was made to study the interactions between these particles, it was finally made possible to understand and describe liquids and solids, as well as the concept of phase transitions. Such a development in network science has led to a significant paradigm shift, which has opened the door to the understanding of a multitude of new features and phenomena (see schematic overview in Fig. 8.1). Here we will review the theory of cascading failures in interdependent networks, and present one application in economic networks.

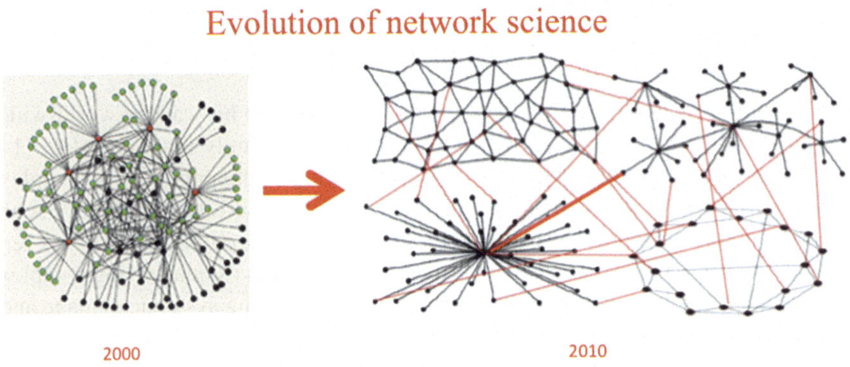

Fig. 8.1 Schematic representation of the scope of network science research from the beginning of the twenty first century, from focusing on the case of a single network, to the case of interconnected and interdependent networks. The *black links* represent connectivity links while the *red links* are dependency links. The concept of dependency links and the generalization of percolation theory to include such links was first introduced in Buldyrev et al. [33]

8.2 Overview of Cascading Failure Processes in Interdependent Networks

The theory for cascading failures in interdependent networks was introduced in [33, 34, 36, 37, 42, 43], and we review it shortly in this section. In order to model interdependent networks, consider two networks, A and B, in which the functionality of a node in network A is dependent upon the functionality of one or more nodes in network B (see Fig. 8.2), and vice-versa: the functionality of a node in network B is dependent upon the functionality of one or more nodes in network A. The networks can be interconnected in several ways. In the most general case we specify a number of links that arbitrarily connect pairs of nodes across networks A and B. The direction of a link specifies the dependency of the nodes it connects, i.e., link $A_i \rightarrow B_j$ provides a critical resource from node A_i to node B_j. If node A_i stops functioning due to attack or failure, node B_j stops functioning as well but not vice-versa. Analogously, link $B_i \rightarrow A_j$ provides a critical resource from node B_i to node A_j.

To study the robustness of interdependent networks systems, we begin by removing a fraction $1-p$ of network A nodes and all the A-edges connected to these nodes. As an outcome, all the nodes in network B that are connected to the removed A-nodes by $A \rightarrow B$ links are also removed since they depend on the removed nodes in network A. Their B edges are also removed. Further, the removed B nodes will cause the removal of additional nodes in network A which are connected to the removed B-nodes by $B \rightarrow A$ links. As a result, a cascade of failures that eliminates virtually all nodes in both networks can occur. As nodes and edges are removed, each network breaks up into connected components (clusters). The clusters in network A (connected by A-edges) and the clusters in network B (connected by B-edges) are different since the networks are each connected differently. If one assumes that small clusters (whose size is below certain threshold) become non-functional, this may invoke a recursive process of failures that we now formally describe.

The insight based on percolation theory is that when the network is fragmented the nodes belonging to the giant component connecting a finite fraction of the network are still functional, but the nodes that are part of the remaining small

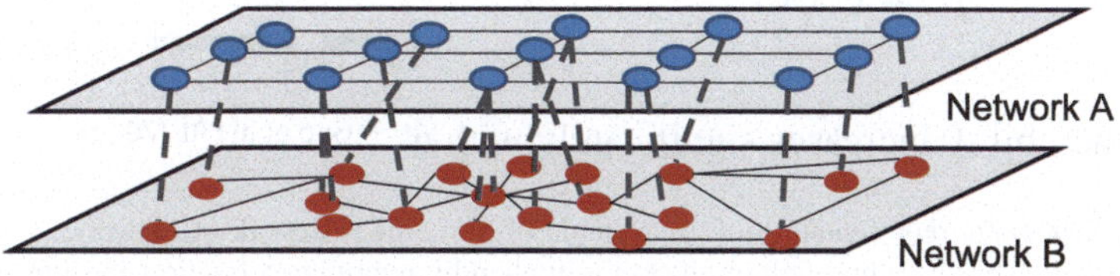

Fig. 8.2 Example of two interdependent networks. Nodes in network B (e.g. communications network) are dependent on nodes in network A (e.g. power grid) for power; nodes in network A are dependent on network B for control information

clusters become non-functional. Thus in interdependent networks only the giant mutually-connected cluster is of interest. Unlike clusters in regular percolation whose size distribution is a power law with a p-dependent cutoff, at the final stage of the cascading failure process just described only a large number of small mutual clusters and one giant mutual cluster are evident. This is the case because the probability that two nodes that are connected by an A-link and their corresponding two nodes are also connected by a B-link scales as $1/N_{\mathrm{B}}$, where N_{B} is the number of nodes in network B. So the centrality of the giant mutually-connected cluster emerges naturally and the mutual giant component plays a prominent role in the functioning of interdependent networks. When it exists, the networks preserve their functionality, and when it does not exist, the networks split into fragments so small they cannot function on their own. In Fig. 8.3 we present a schematic representation of an example of a tree-like network of networks, composed of five networks. The cascading failure process is applied by removing $1-p$ nodes, and calculating the size of the mutual giant component, P_∞. We present (Fig. 8.2) a comparison between P_∞ of $n = 1, 2, 5$ networks, and show that the network of networks system is more vulnerable to cascading failures. Finally, we show (Fig. 8.2) the analytical

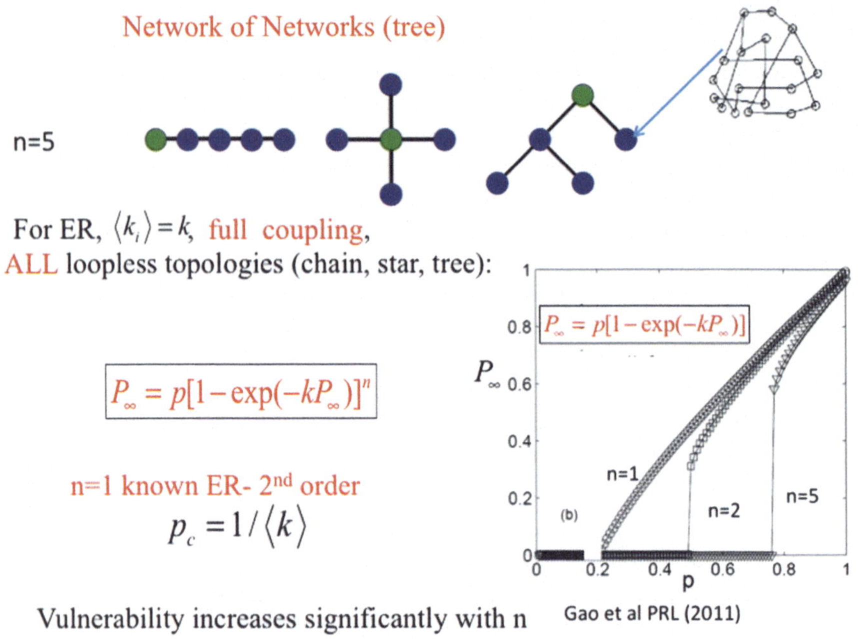

Fig. 8.3 Schematic representation of an example of a tree-like network of networks, composed of 5 networks. The cascading failure process is applied by removing a fraction $1-p$ nodes, and calculating the size of the mutual giant component, P_∞. We present a comparison between P_∞ of $n = 1, 2, 5$ networks, and show that the network of networks system is more vulnerable to cascading failures. Finally, we show the analytical relationship between P_∞, n, k and p, which for the case of one network collapses to the well known ER formalism

relationship between P_∞, n, k and p, which for the case of a single network ($n = 1$) collapses to the well known ER formalism [24].

8.3 Cascading Failures in Economic Networks

Network science has greatly evolved in the twenty first century, and is currently a leading scientific field in the description of complex systems, which affects every aspect of our daily life [2, 22–25]. Network theory provides the means to model the functional structure of different spheres of interest, and thus, understanding more accurately the functioning of the network of relationships between the actors of the system, its dynamics and the scope or degree of influence. In addition, it measures systemic qualities, e.g., the robustness of the system to specific scenarios, or the impact of policy on system actions. The advantage offered by the network science approach is that instead of assuming the behavior of the agents of the system, it rises empirically from the relationships that they really hold; hence, the resulting structures are not biased by theoretical perspectives or normative approaches imposed 'by the eye of the researcher'. On the contrary, the modeling by network theory could validate behavioral assumptions by economic theories. Network theory can be of interest to various edges of the financial world: the description of systemic structure, analysis and evaluation of contagion effects, resilience of the financial system, flow of information, and the study of different policy and regulation scenarios, to name a few [44–57]. Once the network structure and topology is uncovered, it is possible to test many features of the economic system. One critical issue is the resilience of economic and financial systems to shock scenarios, which is commonly investigated using stress tests [58–61]. Cascading failure processes can be applied to study the stability of economic and financial systems, and uncover global and local vulnerabilities to the system. Here, we review a recent application of the theory of cascading failures in interdependent economic systems to quantify and rank the economic influence of specific industries and countries, which was recently introduced by Li et al. [51].

Li et al [51] have examined the interdependent nature of economies between and within 14 countries and the rest of the world (ROW), using input-output table [62] during the period 1995–2011. The economic activity in each country is divided into 35 industrial classifications. Each cell in the table shows the output composition of each industry to all other 525 industries and its final demand and export to the rest of the world (see [63]). From the IO table, an output network is constructed using the 525 industries as nodes and the output product values as weighted links based on the input-output table. The goal of this work is to introduce a methodology for quantifying the importance of a given industry in a given country to global economic stability with respect to other industries in countries that are related to this industry. The authors use the theory of cascading failures in interdependent networks to gain valuable information on the local and global influence on global stability of different economic industries.

In order to identify and rank the influence of industries in the stability of this global network, the authors perform a cascading failure tolerance analysis [33, 51]. The model can be described as follows. Suppose industry A fails, other industries can no longer sell their products to industry A and thus they lose that revenue. The revenue of each industry is reduced by a fraction p', which for each industry is defined as the revenue reduction caused by the failure of industry A divided by that industry's total revenue. The tolerance fraction ϕ is the threshold above which an industry fails. This occurs when reduced revenue fraction p' is larger than tolerance fraction ϕ. Here we assume that (i) ϕ is the same for all industries and that every industry fails when its $p' > \phi$ and (ii) the failure of an industry in country A does not reduce the revenue of the other industries in the same country A because they are able to quickly adjust to the change. The methodology can be schematically illustrated as follows (see Fig. 8.4): In step 1, industry A in country i fails. This causes other industries in other countries to fail if their $p' > \phi$. Assume that in step 2 industries B, C, and D fail. The failure of these industries in step 2 will reduce other industries' revenue and cause more industries including those in country i to have a reduced fraction p'. Thus in step 3 there is an increased number of industries whose $p' > \phi$. Eventually the system reaches a steady state in which no more industries

Fig. 8.4 Schematic representation of each step in the cascading failure propagation in the world economic network (**a**⟶**b**⟶**c**⟶**d**). We present an example of two countries, where *circle* nodes represent country 1, and *triangles* represent country 2. Both countries have the same industries, and the *arrow* between two nodes points in the direction of money flow. The different subpanels demonstrate the cascade of the damage, after an initial failure in electrical equipment industry in Country 1 (*circle*) which causes a failure of electrical equipment industry in Country 2, which cascades into other industries. After [51]

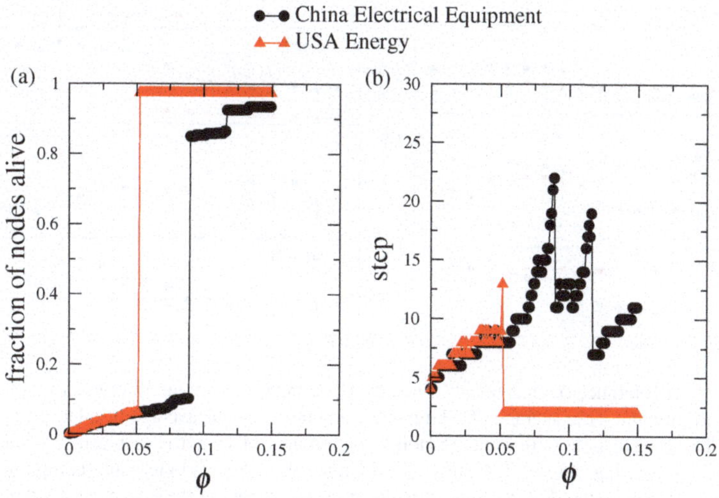

Fig. 8.5 Typical examples of industry tolerance threshold ϕ_c. (**a**)(*left*) the *black curve* shows the fraction of surviving industries as a function of tolerance threshold for the case when the electrical equipment industry in China fails in year 2009 and the *red curve* represents the case of the failure of the energy industry in 2009 in the USA. (**b**)(*right*) Number of failure steps as a function of p. The total number of steps is the number of cascades it takes for the network to reach a steady state after certain initial failure. After [51]

fail. The surviving industries will all have a reduced revenue fraction that is smaller than the tolerance fraction, i.e., $p' \leq \phi$.

Figure 8.5 shows an example of the failures of electric equipment industry in China and the energy industry in the US for the 2009 WIOT and shows the fraction of the largest cluster of connected industries as a function of the tolerance fraction ϕ after the Chinese electric equipment industry becomes malfunction and is removed from the network due to a large shock to the industry. The shock could result from different causes, such as natural environmental disaster, government policy changes, insufficient financial capability. The removal of China electric equipment industry will cause revenue reduction in other industries because China electric equipment industry is not able to buy products and provide money to other industries. When ϕ is small, the industries are fragile and sensitive to the revenue reduction, causing most of the industries fail, and the number of the surviving industries is very small. When ϕ is large, the industries can tolerate large revenue reduction and are more robust when revenue decreases. The number of the surviving industries tends to increase abruptly at a certain $\phi = \phi_c$ value as ϕ increases. Figure 8.5b shows the number of steps that elapse before a stable state is reached as a function of tolerance fraction ϕ after removing the Chinese electric equipment industry or the US energy industry. The number of steps reaches a peak when ϕ approaches criticality ϕ_c [64].

Finally, Li et al. [51] use the cascading failure methodology to rank the economic importance of individual countries, and track how it evolves in time. Figure 8.6 (left)

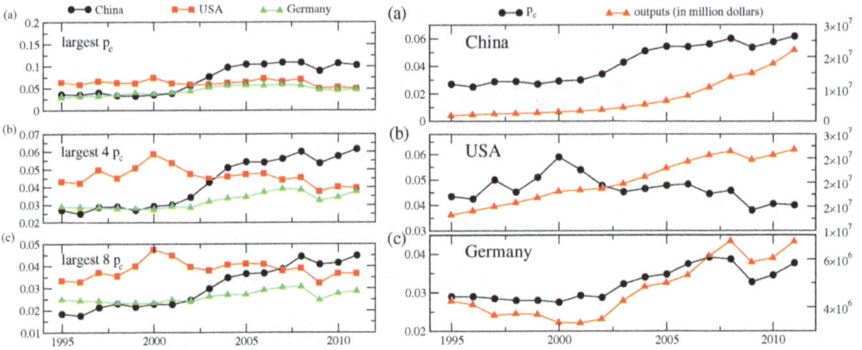

Fig. 8.6 (*left*) Tolerance ϕ_c changes of China, the USA and Germany for 17 years: *top*—the largest tolerance ϕ_c; *middle*—the average of 4 largest ϕ_c; and *bottom*—the average of 8 largest p_c in each country. These results show that the economic importance of China is increasing, while that of the USA is decreasing. (*right*) Tolerance ϕ_c of China, the USA and Germany comparing to the total product output value. For each country, the ϕ_c is an average of the largest four industry ϕ_c of this country (*black circles*). The product output (*red triangle*) value is the money flow a country supplies to the rest of the countries, which also indicates its impact to foreign countries. After [51]

shows the average of ϕ_c of country for the 17-year period investigated, for the case of China, USA and Germany: top—the largest tolerance ϕ_c; middle—the average of four largest industries ϕ_c; and bottom—the average of 8 largest ϕ_c in each country. The results of Li et al [51] present how the economic importance of China relative to that of the USA shows a consistent increase from year to year, illustrating how the economic power structure in the world's economy has been changing during time. Finally, to further validate these results, the total product output (see Fig. 8.6 (right), red triangles) and average tolerance ϕ_c (see Fig. 8.6 (right), black circles) for China, USA, and Germany, as a function of time. The product output value is the total money flow a country supplies to the other countries plus value added in the products, which also indicates its total trade impact on foreign countries.

8.4 Summary

In summary, this paper presents a review of the recently-introduced mathematical framework of for cascading failures in a Network of Networks (NON), particularly in economic NON. In interacting networks, when a node in one network fails it usually causes dependent nodes in other networks to fail which, in turn, may cause further damage in the first network and result in a cascade of failures with catastrophic consequences. This analytical framework enables to follow the dynamic process of the cascading failures step-by-step and to derive steady state solutions [65–67]. This formalism provides critical new information on the resilience and vulnerabilities of real world complex systems, such as economic and financial systems. In economics, some key applications include new stress test tools, such as those presented by Li et al. [51] and Levy-Carciente et al. [61]. Furthermore, these developed tools can be used to introduce intervention strategies in order to manage and mitigate once a cascade of failures is set off in the system (see for example [68]).

Acknowledgements We acknowledge financial support from Office of Naval Research (ONR), DTRA, BSF, the LINC (No. 289447) and the Multiplex (No. 317532) EU projects, the DFG, and the Israel Science Foundation.

References

1. Helbing D, Balietti S (2010) arXiv preprint arXiv:1012.4446
2. Havlin S, Kenett D, Ben-Jacob E, Bunde A, Cohen R, Hermann H, Kantelhardt J, Kertész J, Kirkpatrick S, Kurths J, et al (2012) Eur Phys J Spec Top 214(1):273
3. San Miguel M, Johnson JH, Kertesz J, Kaski K, Díaz-Guilera A, MacKay RS, Loreto V, Érdi P, Helbing D (2012) Eur Phys J Spec Top 214(1):245
4. Helbing D (ed) (2012) Social self-organization. Springer, Berlin, pp 261–284
5. Lazer D, Pentland AS, Adamic L, Aral S, Barabasi AL, Brewer D, Christakis N, Contractor N, Fowler J, Gutmann M, et al (2009) Science 323(5915):721. (New York, NY)
6. King G (2011) Science(Washington) 331(6018):719
7. Lorenz J, Rauhut H, Schweitzer F, Helbing D (2011) Proc Natl Acad Sci 108(22):9020
8. Yamasaki K, Fujiwara T, Yoshizawa K, Miyake S, Zheng Z, Gao X, Sakurai N (2012) Proceedings of international conference on business management & IS, pp 130–140
9. Meng B, Inomata S (2009) Production networks and spatial economic interdependence: An international input-output analysis of the asia-pacific region. Technical report. Institute of Developing Economies, Japan External Trade Organization (JETRO)
10. Rinaldi SM, Peerenboom JP, Kelly TK (2001) IEEE Control Syst 21(6):11
11. Solomon S, Levy M (2003) Quant Finan 3(1):c12
12. Levy M (2010) Phys A Stat Mech Appl 389(21):4913
13. Klimek P, Hausmann R, Thurner S (2012) PloS one 7(6):e38924
14. Farmer JD, Foley D (2009) Nature 460(7256):685
15. Lux T, Westerhoff F (2009) Nat Phys **5**(1):2
16. Forbes K, Rigobon R (2001) International financial contagion. Springer, New York, pp 43–66
17. Forbes KJ, Rigobon R (2002) J Financ 57(5):2223
18. Bodie Z, Kane A, Marcus AJ (2002) Investments (Tang Kinh Cac, 2002)
19. Billio M, Getmansky M, Lo AW, Pelizzon L (2010) Econometric measures of systemic risk in the finance and insurance sectors. Techniacl report, National Bureau of Economic Research
20. Bisias D, Flood M, Lo AW, Valavanis S (2012) Annu Rev Financ Econ 4(1):255
21. Albert R, Barabási AL (2002) Rev Mod Phys 74(1):47
22. Newman MEJ Networks: an introduction (OUP, 2009)
23. Jackson MO (2010) Social and economic networks. Princeton University Press, Princeton
24. Boccaletti S, Latora V, Moreno Y, Chavez M, Hwang DU (2006) Phys Rep 424(4):175

25. Cohen R, Havlin S 2010 Complex networks: structure, robustness and function. Cambridge University Press, Cambridge
26. May RM (2013) Philos Trans R Soc A Math Phys Eng Sci 371(1987):20120376
27. Bashan A, Bartsch RP, Kantelhardt JW, Havlin S, Ivanov PC (2012) Nat Commun 3:702
28. Barrat A, Barthelemy M, Vespignani A (2008) Dynamical processes on complex networks. Cambridge University Press, Cambridge
29. Vespignani A (2010) Nature 464(7291):984
30. Lux T (2011) Nature 469(7330):303
31. Li D, Fu B, Wang Y, Lu G, Berezin Y, Stanley HE, Havlin S (2015) Proc Natl Acad Sci 112(3):669
32. Ludescher J, Gozolchiani A, Bogachev MI, Bunde A, Havlin S, Schellnhuber HJ (2014) Proc Natl Acad Sci 111(6):2064
33. Buldyrev SV, Parshani R, Paul G, Stanley HE, Havlin S (2010) Nature 464(7291):1025
34. Parshani R, Buldyrev SV, Havlin S (2010) Phys Rev Lett 105(4): 048701
35. Bashan A, Berezin Y, Buldyrev SV, Havlin S (2013) Nat Phys 9(10):667
36. Gao J, Buldyrev SV, Stanley HE, Havlin S (2012) Nat Phys 8(1):40
37. Kenett DY, Gao J, Huang X, Shao S, Vodenska I, Buldyrev SV, Paul G, Stanley HE, Havlin S (2014) Networks of networks: the last frontier of complexity Springer, New York, pp 3–36
38. Boccaletti S, Bianconi G, Criado R, Del Genio C, Gómez-Gardeñes J Romance M, Sendina-Nadal I, Wang Z, Zanin M (2014) Phys Rep 544(1):1
39. Lee KM, Kim JY, Cho Wk, Goh K, Kim I (2012) New J Phys 14(3):033027
40. Baxter G, Dorogovtsev S, Goltsev A, Mendes J (2012) Phys Rev Lett 109(24):248701
41. Peixoto TP, Bornholdt S (2012) Phys Rev Lett 109(11):118703
42. Havlin S, Kenett D, Bashan A, Gao J, Stanley H (2014) Eur Phys J Spec Top 223(11):2087
43. Havlin S, Stanley HE, Bashan A, Gao J, Kenett DY (2014) Chaos, Solitons Fractals
44. Lillo F (2010) Encyclopedia of life support systems (EOLSS), Developed under the auspices of the UNESCO
45. Summer M (2013) Annu Rev Financ Econ 5(1):277
46. Tumminello M, Lillo F, Mantegna RN (2010) J Econ Behav Organ 75(1):40
47. Kenett DY, Tumminello M, Madi A, Gur-Gershgoren G, Mantegna R, Ben-Jacob E (2010) PloS one 5(12):e15032
48. Kenett DY, Raddant M, Lux T, Ben-Jacob E (2012) PloS one 7(2):e31144
49. Cont R (2013) Handbook on systemic risk. Cambridge University Press, Cambridge, p 285
50. Glasserman P, Young HP (2015) How likely is contagion in financial networks? J Bank Financ 50:383–399
51. Li W, Kenett DY, Yamasaki K, Stanley HE, Havlin S (2014) arXiv preprint arXiv:1408.0443
52. Garas A, Argyrakis P, Rozenblat C, Tomassini M, Havlin S (2010) New J Phys 12(11):113043
53. Haldane AG, May RM (2011) Nature 469(7330):351
54. Haldane AG, et al. (2009) Speech delivered at the financial student association, Amsterdam, April
55. Cont R, Moussa A, Bastos E (2010) e Santos. Available at SSRN:1733528
56. Amini H, Cont R, Minca A (2012) Stress testing the resilience of financial networks. Int J Theor Appl Financ 15(01):1250006
57. Chan-Lau J, Espinosa M, Giesecke K, Solé J (2009) IMF global financial stability report, vol. 2
58. Boss M, Elsinger H, Summer M, Thurner S (2004) Quant Finan 4(6):677
59. Langfield S, Liu Z, Ota T (2012) Financial sevices authority
60. Martínez Jaramillo S, Kabadjova B, Bravo Benítez B, Solórzano J (2012) Margain, Systemic risk analysis by means of network theory: An empirical study of the mexican banking system. Technical report Banco de México working papers
61. Levy Carciente S, Kenett DY, Avakian A, Stanley HE, Havlin S (2014) Available at SSRN 2482742
62. Timmer M, Erumban A, Gouma R, Los B, Temurshoev U, de Vries G, Arto I (2012) WIOD Background document available at www.wiod.org

63. World Input Output Database (2012) www.wiod.org
64. Parshani R, Buldyrev SV, Havlin S (2011) Proc Natl Acad Sci 108(3):1007
65. Gao J, Buldyrev S, Havlin S, Stanley H (2012) Phys Rev E 85(6):066134
66. Zhou D, Bashan A, Cohen R, Berezin Y, Shnerb N, Havlin S (2014) Phys Rev E 90(1):012803
67. Dong G, Gao J, Du R, Tian L, Stanley H.E, Havlin S (2013) Phys Rev E 87(5):052804
68. Majdandzic A, Podobnik B, Buldyrev SV, Kenett DY, Havlin S, Stanley HE Nat Phys 10(1):34 (2014)

Chapter 9
Do Connections Make Systems Robust? A New Scenario for the Complexity-Stability Relation

Takashi Shimada, Yohsuke Murase, and Nobuyasu Ito

Abstract Whether interactions among the elements make the system robust or fragile has been a central issue in broad range of field. Here we introduce a novel type of mechanism which governs the robustness of open and dynamical systems such as social and economical systems, based on a very simple mathematical model. This mechanism suggest a moderate number ($\sim$ 10) of interactions per element is optimal to make the system against successive and unpredictable disturbances. The relation between this very simple model and more detailed nonlinear dynamical models is discussed, to emphasize the relevance of this newly reported mechanism to the real phenomena.

9.1 Introduction

Most real complex systems of our interest are *ecosystem-like*. Good examples are reaction networks and gene regulatory networks in living organisms in evolutionary time scale, brain and immune system in developmental timescale, engineering systems with decentralized control scheme, ecosystems of companies or products, and

T. Shimada (✉)
Department of Applied Physics, Graduate School of Engineering, The University of Tokyo, 7-3-1 Hongo, Bunkyo-ku, Tokyo 113-8656, Japan

JST CREST, 4-1-8 Honcho, Kawaguchi, Saitama 332-0012, Japan
e-mail: shimada@ap.t.u-tokyo.ac.jp

Y. Murase
JST CREST, 4-1-8 Honcho, Kawaguchi, Saitama 332-0012, Japan

RIKEN AICS, 7-1-26, Minatojima-minami-machi, Chuo-ku, Kobe, Hyogo 650-0047, Japan
e-mail: yohsuke.murase@riken.jp

N. Ito
Department of Applied Physics, Graduate School of Engineering, The University of Tokyo, 7-3-1 Hongo, Bunkyo-ku, Tokyo 113-8656, Japan

JST CREST, 4-1-8 Honcho, Kawaguchi, Saitama 332-0012, Japan

RIKEN AICS, Minatojima-minami-machi, Chuo-ku, Kobe, Hyogo 650-0047, Japan
e-mail: ito@ap.t.u-tokyo.ac.jp

H. Takayasu et al. (eds.), *Proceedings of the International Conference on Social Modeling and Simulation, plus Econophysics Colloquium 2014*, Springer Proceedings in Complexity, DOI 10.1007/978-3-319-20591-5_9

social communities. In those ecosystem-like systems, there is no top-down or centralized mechanism for the system's growth and maintenance. And their complexity emerges as a result of successive introductions of new elements. In the following, we focus on the universal aspects of robustness of such ecosystem-like systems.

The robustness (or stability, fragility, resilience, etc.) of complex systems itself is indeed a classical problem [1]. Essential theoretical findings those have been found on this issue include the general instability of large and densely interacting systems [2], the self-organized criticality [3], and the relation between the robustness and the network structure of the systems [4, 5]. However, the key and universal feature of the real complex systems, openness, has not been well considered. Meanwhile, theoretical studies on ecosystems using various different models have indicated that they share universal behaviors independent of the detail of the dynamics [6–10]. Therefore it is natural to ask how can such ecosystem-like system grow to more complex structure by adding new elements to it, using a simpler model. In the following, we first introduce a minimal model for this problem and show that it yields a novel type of transitions, together with its underlying mechanism [11]. Then we show an example of direct relation between the minimal model and the more detailed nonlinear dynamical models, which corroborates the relevance of the newly found mechanism to the real phenomena.

9.2 A Universal Relation Between Robustness and Connection

9.2.1 A Minimal Model of Evolving Open Systems

We here introduce a minimal model of evolving open systems [11]. In this model, the entire system is structured as a collection of nodes connected by directed and weighted links (Fig. 9.1). The nodes may represent various kinds of *species* (e.g. chemical species, different genes and proteins, neurons, animal species, companies, products, individuals, etc.). In the following, we simply call them species. Also the links may represent diverse kinds of interactions (or inputs, signal, influences, effects, etc.) among them. The directed link from species j to species i with its link weight denotes the influence of species j on species i. Each species has only one property, *fitness*, which is simply determined by the sum of its incoming interaction weights from other species in the system. Only the rule intrinsic to the system is that each species can survive as long as its fitness is greater than zero, and otherwise it goes extinct. If the minimum fitness in the system is non-positive, we delete that species (therefore totally isolated species cannot survive). Because this extinction will modify the fitness of the other species, we re-calculate the fitness and re-identify the least-fit species. We continue this deletion procedure until the minimum fitness becomes positive, meaning that the system is stable. Once the system gets to a stable state, nothing will happen in terms of this intrinsic fast process. Therefore

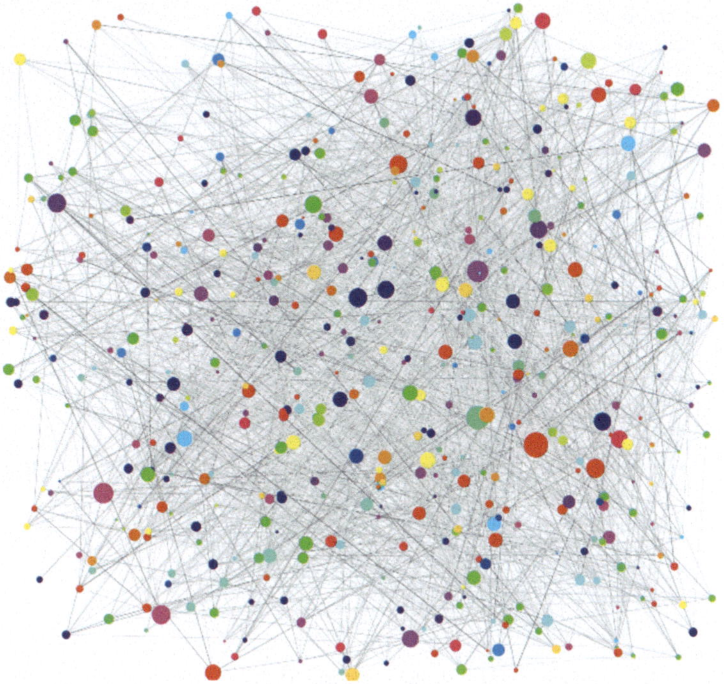

Fig. 9.1 A snapshot of ecosystem-like system obtained from the minimal model described in the Sect. 9.2.1. *Nodes* and *links* represent the general gspeciesh and ginteractionsh respectively. While the diameter of each cnode depicts the current fitness, its color is just for visibility

we proceed the time by the order of magnitude of longer unit i.e. the evolutionary time scale (in some other systems, it corresponds to the developmental time scale and so on). At each evolutionary time step t, a new species is added into the system. We establish interactions from and to the newly added species. The interacting species are chosen randomly from the resident species with equal probability, and the directions of the interactions are also determined randomly. The link weights are assigned randomly from a zero-mean distribution (for example, the standard normal distribution). Then, we re-calculate the fitness of each species to find whether the system can accommodate the new species or some species should become extinct. We repeat this addition-and-deletion steps. Note that the behavior of the system after a sufficiently large number of time steps does not depend on the initial condition. Therefore, this model has only one relevant parameter: m, the number of interactions per species. For clarity, we show below the pseudo-code of this model.

```
// Pseudo-code of the minimal model
Create an initial state with N species
Check the extinctions as described below

FOR t = 0 to t_max
   Add a new species
```

```
    FOR each of m new links
       Choose an interacting species randomly
        from resident species
       Choose the direction of the link randomly
        with equal probability 0.5 for each direction
       Assign the link weight a_ij randomly
        from a 0 mean distribution
    ENDFOR

    Flag_ext = true
    WHILE Flag_ext
       FOR each species
          f_i = 0
          FOREACH incoming links j
             f_i += a_ij
          ENDFOR
       ENDFOR
       Find the species k which has minimum fitness f_min
       IF f_min <= 0
          Delete species k
          Delete the links from/to it
       ELSE
          Flag_ext = false
       ENDIF
    ENDWHILE
    Observe the current stable community
 ENDFOR
```

9.2.2 Transition in Growth Behavior

In the present model, the essential features of the ecosystem-like systems, the introduction of a new species and the interaction-dependent survival condition for each species, are took into account. And because the both processes are introduced in neutral way, i.e. giving no apparent advantage to grow or collapse. Therefore whether the system can grow under such process will purely illuminate the relation between the system's complexity and robustness. Simulation results indeed give a fascinating answer: both of the growth and collapse can happen, depending on the only one model parameter m. The system can grow to infinitely large size if the number of interactions per species is in a moderate range (for the case of taking the standard distribution for link weights, the range is $5 \leq m \leq 18$), and, if not, it stays in a finite size (Fig. 9.2).

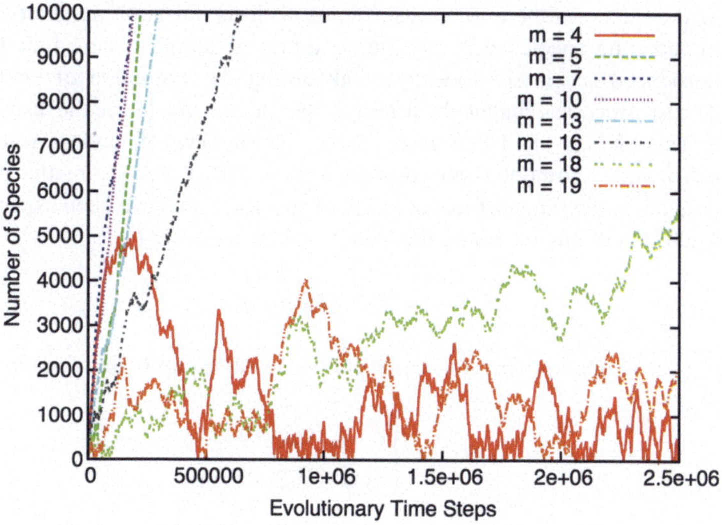

Fig. 9.2 Temporal evolutions of the number of species. The number of species diverges if $5 \leq m \leq 18$. For more precise and reliable determination of the transition point in this behavior needs systematic and longer simulations (see [11])

9.2.3 A Mean-Field Analysis and the Transition Mechanism

The first transition at between $m = 4$ and 5 turns out to be related to a kind of percolation threshold: the emergent system with too sparse interactions can have only tree-and-cycle-like network, and therefore it is too fragile to continue growing. But then why we have another transition in the denser interaction regime? This latter transition is non-trivial and novel, and therefore we mainly focus on this in this paper.

To consider the mechanism of the transition, we first investigate the topology of the emerging network. We can confirm that there is no strong structure in the emerging networks (Fig. 9.1). In other words, the structure of the emerging system remains almost random network with average degree $\sim m$. From this observation, a theoretical analysis based on a mean-field picture has been performed [11, 12]. In this theory, we only treat the distribution function of the fitness of the species in the entire community. Because the fitness distribution function (FDF) is dependent on the parameter m, we write the FDF of fitness x as $F(m, x)$. FDF of the newly introduced species, which has $m/2$ incoming links on average, is easily calculated as the positive half of the normal distribution with variance $\frac{m}{2}$:

$$F_0(m, x) = \begin{cases} 0 & (x \leq 0) \\ 2G(\sigma_m, x) & ensuremath(x \geq 0) \end{cases}, \quad \sigma_m = \sqrt{\frac{m}{2}}. \tag{9.1}$$

Where $G(\sigma, x)$ denotes the normal distribution with its deviation σ. After settling to the system, the species will experience either obtaining a new link from a newly introduced species or loosing a link during the extinction of interacting species. Those processes change the fitness of the species, and hence the distribution function. This change in FDF is found to be the one step of random walk with negative drift whose strength is proportional to $1/m$. Therefore, writing this process by an operator $\hat{\mathscr{D}}$, the (not normalized) FDF of species those have been experienced a loss or addition of one incoming link can be calculated from F_0 as,

$$F_1(m,x) = \hat{\mathscr{E}}\hat{\mathscr{D}}F_0(m,x), \tag{9.2}$$

where $\hat{\mathscr{E}}$ is the extinction operator which cut the negative part of any function:

$$\hat{\mathscr{E}}h(x) = \begin{cases} 0 & (x \leq 0) \\ h(x) & (x \geq 0) \end{cases}. \tag{9.3}$$

Note that the operators $\hat{\mathscr{D}}$ and $\hat{\mathscr{E}}$ are non-commutative. In the following we call the suffix g of F_g, the number of incoming link addition/deletion events that species has experienced, as *generation*. As we have seen, calculation of the FDF of generation g needs the FDF of younger generation, $g-1$:

$$F_g(m,x) = \hat{\mathscr{E}}\hat{\mathscr{D}}F_{g-1}(m,x), \tag{9.4}$$

Only after performing the iterative calculation, we obtain the probability distribution function of the fitness of the entire system,

$$F(m,x) = \frac{\sum_{g=0}^{\infty} F_g(m,x)}{\sum_{g=0}^{\infty} n_g(m)} \qquad \left(n_g(m) = \int_0^{\infty} F_g(m,x)dx \right), \tag{9.5}$$

which contains all the information we need under the mean-field approximation. The most important outcome from the FDF is the average probability of entire resident species going extinct during one link addition/deletion event E, which is calculated as

$$E(m) = 1 - \int_0^{\infty} \hat{\mathscr{E}}\hat{\mathscr{D}}F(m,x)\,dx = \frac{\int_0^{\infty} F_0(m,x)\,dx}{\sum_{g=0}^{\infty} n_g(m)} = \frac{1}{\sum_{g=0}^{\infty} n_g(m)}. \tag{9.6}$$

From this calculation, we find that E is a decreasing function of m. Therefore the *robustness of each species* against the disturbance increases with m.

What should be emphasized, however, is that the robustness of each species does not directly determine the *robustness of the entire system*. Let us see this using an infinitely large graph in which all the nodes have m links. The average number of species that go extinct directly because of an inclusion of the new species is simply calculated as $mE/2$. Because those extinctions may also trigger sequential extinctions, the expectation value of the total number of extinctions per inclusion of one species N_E is simply calculated from an infinite geometric series as

$$N_E = \sum_{n=1}^{\infty} \left(\frac{mE}{2} \right)^n = \frac{mE}{2 - mE}. \tag{9.7}$$

Therefore the *robustness of the entire system* is a function of mE, not the bare E. And because $N_E = 1$ means that the average number of extinctions balances with the number of inclusions in the long time average, that corresponds to the transition point of the growth behavior. In other words, the following self-consistent condition should be satisfied for the critical number of interactions per species: $m_* E(m_*) = 1$.

Let us now focus on the relevant parameter in the argument above, mE. We find that the decrease of E is slower than $1/m$ (roughly $\sim 1/\sqrt{m}$). Therefore mE is a sub-linearly increasing function of m, and it crosses the critical value 1 around $m_* = 13$. This means that the mean-filed treatment can explain the transition in the growth behavior of the system. In addition, this theory give us the simple understanding of the transition mechanism. It originates from the balance of the two effects: although having more interactions makes each species robust against the disturbances (addition and extinction of the species relating to that species), it also increases the impact of the loss of a species. In consistent with this success in explaining the transition by the mean-field analysis, we can find essentially same phase diagram in slightly modified models, such as the model with giving a randomly distributed degrees for the newly added species, the one with different distribution functions for the link weights, and so on [11].

In the classical diversity-stability relation based on the linear stability of dynamical systems, an intrinsic stability is assumed for each element to ensure the stability of each element when that has no interactions. For the system to remain stable, each element may have essentially only one interaction that is not weak comparing to the given intrinsic stability [2]. In the present mechanism, we do not assume any kind of intrinsic stability to the elements: an element with no interaction immediately goes extinct. Even so, the system with 10 or more interactions per element can grow. In this sense, the condition we have identified is very realistic. Indeed in the real systems, it is quite often to find moderately sparse networks: the average degree is in the order of 10, not order of 1, and that seems not dependent on the system size. This novel relation between the connection in the system and its robustness might be a origin of this.

9.3 The Relation with More Complex Dynamical Models

We have reviewed a novel relation between the system's robustness and the connections in it using a very simple model. In our simple model, the extinction condition $f_i \leq 0$ represents the system's intrinsic dynamics. The simplicity of the model is good in terms of universality, and hence especially good for applying to social and economic systems because it is very hard to obtain precise equation of motion or evolution rule of those. And the fact that we can find a good agreement between the model and real systems in their statistics encourages us to put more emphasis on universality. A good example is lifetime distribution function of the species [13, 14].

However, for each certain problem, we generally treat more complex models. Therefore it would be nice if we can argue more directly about the connection between our simple model and complex models. In the following, we will consider a certain class of population dynamics models and show that the necessary condition to have extinctions in it reduces to the extinction rule in the simple model.

9.3.1 The Extinction Condition in Population Dynamics Models

In many dynamical models, each element has more properties in addition to its mere existence and the interactions depend on those properties. One of the most popular class is population dynamics models, in which each element has its property, population x_i. The general form of the dynamical equation of motion of population dynamics models can be written as,

$$\dot{x}_i = f_i(x_1, x_2, \cdots, x_N), \tag{9.8}$$

where the dynamical variable $\{x_i\}$ denote the population of element i. To know which species will go extinct is generally a difficult problem, because one needs to have the trajectory. This is one of the reason why so many studies substitute the stability of the system for its linear stability, which is, strictly speaking, neither enough condition nor necessary condition to really determine the fate of the species. The necessary condition to have an extinction of certain species is relatively easier, because it is at least describable simply: the necessary condition to go extinct is to satisfy

$$\lim_{x_i \to 0} \dot{x}_i = f_i(x_1, \cdots, x_{i-1}, 0, x_{i+1}, \cdots, x_N) < 0 \tag{9.9}$$

at somewhere in the $x_i = 0$ surface. Such condition is again generally difficult to access and also different from the linear stability condition.

9.3.2 Ratio-Dependent Interactions

The interaction term in the population dynamics with the form of $f_{ij}\left(\frac{x_i}{x_j}\right)x_j$, in which the predation rate per predator j, $f_{ij}(\xi)$, is an arbitrary function of the ratio of the prey to the predator x_i/x_j, is called ratio-dependent form in theoretical ecology and regarded as a realistic model of the predation interaction [15]. A typical simple example of the form of $f(\xi)$ is

$$f(\xi) = \frac{B\xi}{A+\xi}, \tag{9.10}$$

where A and B are constants. If we neglect many-body effects such as the competition among the predators those attack the same prey, the predator's choice on multiple preys, and so on for simplicity (otherwise the dynamical equations may become implicit), the population dynamics of such systems can be written as

$$\dot{x}_i = \sum_j f_{ij}\left(\frac{x_i}{x_j}\right)x_j + \sum_k f_{ki}\left(\frac{x_k}{x_i}\right)x_i, \tag{9.11}$$

where the summations run for the predators and the preys of species i, respectively.

9.3.3 The Necessary Condition to Have an Extinction Under "natural" Ratio-Dependent Interactions

Let us next limit the case by postulating the following relatively natural features to the ratio-dependent predation rate. That is, $f(\xi)$ must go to 0 as the population of the prey goes to 0 and that must saturate at a certain value when the population of the prey is abundant, i.e.

$$\lim_{\xi\to 0} f_{ij}(\xi) = 0 \cap \lim_{\xi\to\infty} f_{ij}(\xi) = b_{ij}, \tag{9.12}$$

where b_{ij} represents the maximum predation rate on that interaction. The example we have seen in Eq. (9.10) satisfies these both features. And if we suppose it does not have any singularity around 0, we can obtain its Maclaurin series as,

$$f_{ij}\left(\frac{x_i}{x_j}\right)x_j = \left[\sum_{n=1}^{\infty} c_n^{ij}\left(\frac{x_i}{x_j}\right)^n\right]x_j. \tag{9.13}$$

Where

$$c_n^{ij} = \frac{1}{n!} \cdot \left. \frac{d^n f_{ij}(\xi)}{d\xi^n} \right|_{\xi \to 0} \tag{9.14}$$

is the coefficient of Taylor series expansion at 0. This means that the necessary condition for the extinction [Eq. (9.9)] of species i in this model is indeed not dependent on the populations of the surrounding species:

$$\begin{aligned} \lim_{x_i \to 0} \dot{x}_i &= \lim_{x_i \to 0} \left[\sum_j \left\{ \sum_{n=1}^{\infty} c_n^{ij} \left(\frac{x_i}{x_j} \right)^n x_j \right\} + \sum_k f_{ki} \left(\frac{x_k}{x_i} \right) x_i \right] \\ &= \left[\sum_j c_1^{ij} + \sum_k b_{ki} \right] x_i < 0. \end{aligned} \tag{9.15}$$

And this condition, $\sum_j c_1^{ij} + \sum_k b_{ki} < 0$, that says summation of the population-independent coefficients assigned to the interacting links should be negative, is exactly in the same class with the minimal model we introduced in Sect. 9.2.1.

9.4 Conclusion

We have reviewed the simple and universal mechanism of determining the robustness, and therefore its ability to grow, of ecosystem-like systems by introducing a simple model. It has been also shown that the necessary condition for extinctions in a certain type of dynamical models essentially result in the same condition with that of the simple model. This supports our future approach to verify the relevance of the newly found mechanism to the real phenomena.

Acknowledgements This work was partially supported by JSPS Grant-in-Aid for Scientific Research (C) Grant Number 15K05202.

References

1. MacAuthor R (1955) Fluctuations of animal populations, and a measure of community stability. Ecology 36:533–535
2. Gardner MR, Ashby WR (1970) Connectance of large dynamic (cybernetic) systems: critical values for stability. Nature 228:784
3. Bak P, Sneppen K (1993) Punctuated equilibrium and criticality in a simple model of evolution. Phys Rev Lett 71:4083–4086
4. Albert RJ, Jeong H, Barabási A-L (2000) Error and attack tolerance of complex networks. Nature 406:378–382
5. Herrmann HJ, Schneider CM, Moreira AA, Andrade JS, Havlin S (2011) Onion-like network topology enhances robustness against malicious attacks. J Stat Mech 2011:P01027
6. Taylor PJ (1988) Consistent scaling and parameter choice for linear and generalized lotka-volterra models used in community ecology. J Theor Biol 135:543–568
7. Caldarelli G, Higgs PG, McKane AJ (1988) Modelling coevolution in multispecies communities. J Theor Biol 193:345–358
8. Shimada T, Yukawa S, Ito N (2002) Self-organization in an ecosystem. Artif Life Robotics 6:78–81
9. Perotti JI, Billoni OV, Tamarit FA, Chialvo DR, Canna SA (2009) Emergent self-organized complex network topology out of stability constraints. Phys Rev Lett 103:108701
10. Murase Y, Shimada T, Ito N, Rikvold PA (2010) Random walk in genome space: a key ingredient of intermittent dynamics of community assembly J Theor Biol 264:663–672
11. Shimada T (2014) A universal transition in the robustness of evolving open systems. Sci Rep 4:4082
12. Shimada T (2015) In: Mathematical approaches to biological systems. Ohira T, Uzawa T (eds) Springer, Japan, pp 95–117
13. Shimada T, Yukawa S, Ito N (2003) Life-span of families in fossil data forms q-exponential distribution Int J Mod Phys C 14:1267–1271
14. Murase Y, Shimada T, Ito N (2010) A simple model for skewed species-lifetime distributions New J Phys 12:063021
15. Drossel B, Higgs PG, Mckane AJ (2001) The influence of predator prey population dynamics on the long-term evolution of food web structure. J Theor Biol 208:91–107

Chapter 10
Simulation of Gross Domestic Product in International Trade Networks: Linear Gravity Transportation Model

Tsuyoshi Deguchi, Hideki Takayasu, and Misako Takayasu

Abstract In this study, we introduce a model to simulate gross domestic product (GDP) for international trade network data. By applying a linear gravity transportation model, we confirm that estimated values approximately agree with the real values of GDP by tuning the model parameters. An exception is the estimated GDP of China that is about two times bigger than the real value. This discrepancy might imply that China's GDP is not saturated and it is on the way of growing.

10.1 Introduction

Today, China is becoming increasingly influential in not only the international community but also international politics and military forces. However, its presence is most notable in the international economy. In particular, with China's growing gross domestic product (GDP), the country has the potential to become the biggest economy in the near future. GDP is the most specific and popular measure in the economic statistics literature, although many have doubted China's GDP statistics [1, 2].

T. Deguchi (✉) • M. Takayasu
Department of Computational Intelligence and Systems Science, Interdisciplinary Graduate School of Science and Engineering, Tokyo Institute of Technology, 4259 Nagatsuta-cho, Midori-ku, Yokohama 226-8502, Japan
e-mail: deguchi.t.ad@m.titech.ac.jp; takayasu@dis.titech.ac.jp

H. Takayasu
Sony Computer Science Laboratories, 3-14-13 Higashigotanda, Shinagawa-ku, Tokyo 141-0022, Japan

Graduate School for Advanced Mathematical Sciences, Meiji University, 4-21-1 Nakano, Nakano-ku, Tokyo 164-8525, Japan

Department of Computational Intelligence and Systems Science, Interdisciplinary Graduate School of Science and Engineering, Tokyo Institute of Technology, 4259 Nagatsuta-cho, Midori-ku, Yokohama 226-8502, Japan
e-mail: takayasu@csl.sony.co.jp

H. Takayasu et al. (eds.), *Proceedings of the International Conference on Social Modeling and Simulation, plus Econophysics Colloquium 2014*, Springer Proceedings in Complexity, DOI 10.1007/978-3-319-20591-5_10

In this paper, we estimate and simulate countries' GDP and ranks using the linear gravity transportation model (LGTM). The gravity transportation model (GTM) is known to be effective in examining a company's transaction networks [3, 4]. The LGTM is a linearized form of the GTM and is a type of degree distribution model [5, 6].

10.2 Preceding Study

International trade networks (ITNs) are predominantly used in surveying network structures and known to follow the so-called "gravity relation" [7]. They were first examined in 2003 by Serrano and Boguñá, who presented the fundamental characteristics of ITNs for different countries [8]. Recently, physicists and network researchers have explored the structure on the basis of diverse factors, such as time series robustness, community structures, and inter-layer dependency [9–12]. Some researchers have attempted to extend ITN research to that on economic growth [13–16]. These studies contribute some interesting findings from the viewpoint of complex networks. For instance, Garlaschelli et al. found that GDP is a hidden factor that influences networks [17]. Although this fact is common knowledge in international economics, their model remains an impressive contribution to network study. Several other studies have been conducted on gravity relations [18–20].

10.3 Dataset

We adopt data from the Direction of Trades Statistics (DOTS) compiled by the International Monetary Fund (IMF) [21]. This dataset includes annual and monthly data of trades (US dollar) for countries. We use a total of 214 countries (regions) as nodes and their respective trade amounts as weighted links. The weighted links suggest bilateral trade relationships, that is, exports and imports, between countries or regions on a monthly and yearly basis, which are measured in million US dollar. DOTS also include the base data for both exports and imports. In general, the amount of export from country A to country B should be same as the amount of import to country B from country A. However, these numbers differ between the import base and export base datasets. In this case, we use the export base year dataset.

We also use GDP data from the Economic Outlook Dataset, also produced by IMF. We use this data with enough credit. However, this dataset includes only 189 countries. Thus, we arrange these datasets and aggregate ITN and GDP data.

After data processing, we used 2010 ITN and GDP data for 160 countries.

10.4 Simulation Setup

10.4.1 GDP Transaction Flow Relationship

First, we define ITN as an adjacency matrix, W, whose component, w_{ij}, represents the annual amount of transaction flow (imports) from country i to country j, measured in million US dollars. Then, we define $w_{ii} = 0$. We also introduce the binary network matrix, A, whose component $a_{ij} = 1$ when $w_{ij} > 0$ and $a_{ij} = 0$ when $w_{ij} = 0$.

The following relationship between GDP and transaction flows is often assumed in the international trade literature[7].

$$w_{ij} = G\frac{Y_i^{\alpha} Y_j^{\beta}}{R_{ij}^{\gamma}}. \tag{10.1}$$

Here, Y_i is the GDP of node i and R_{ij} is the distance between nodes i and j. The power exponent α, β, and γ are parameters. We neglect the distance R_{ij} and estimate the exponents α and β using data shown in Fig. 10.1.

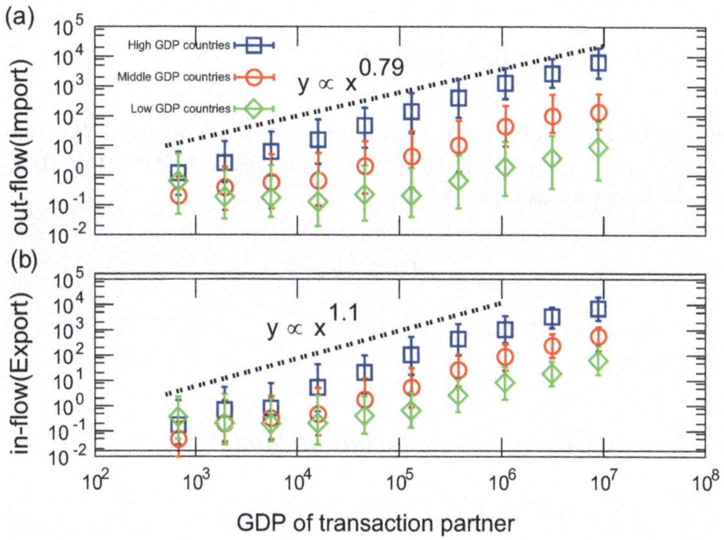

Fig. 10.1 GDP Transaction flow (import and export) relationship in a log-log plot. (**a**) Import (**b**) Export for high GDP countries (top one-third of high GDP countries), middle GDP countries (other countries not classified as high and low GDP countries), and low GDP countries (bottom one-third of high GDP countries). In this case, high, middle, and low GDP relates to the total GDP. Here, bins are defined at regular intervals in log-scale, the first and third quantiles are plotted as error bars with the median value at the center of symbols, *squares* and *circles*

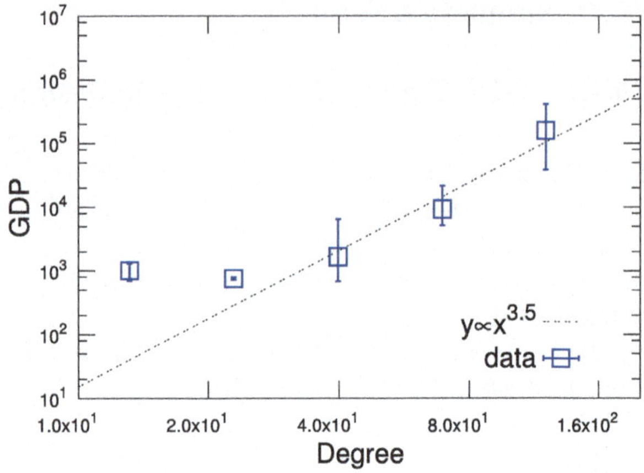

Fig. 10.2 GDP–degree relationship in a log-log plot. The first and third quantiles are plotted as error bars with the median value at the *center of squares*

$$w_{ij} \propto Y_i^{\alpha} Y_j^{\beta}. \tag{10.2}$$

Using parameter fitting, we estimate the values of α and β as $(\alpha, \beta) = (0.79, 1.1)$.

10.4.2 GDP–Degree Relationship

The number of trade parameters is higher for high GDP countries. We confirm the following power law relationship between the degrees and GDPs (Fig. 10.2). Here, we define the in-degree as $k_M = \sum_i a_{iM}$.

$$Y_M \propto k_M{}^{\zeta}. \tag{10.3}$$

Using the data, we estimate $\zeta = 3.5$ (Fig. 10.2).

10.4.3 Linear Gravity Transportation Model

From Figs. 10.1 and 10.2, we know that the amount of trades, degrees and GDPs have positive relations. Thus, we model these relations as a transportation model. In this model, we think the number of degrees directly affects the amount of trade flow, where in-degree is defined as $k_j = \sum_i a_{ij}$. And we calculate the GDPs as a result of the distributions of trades flows using gravity relation. We call this model as a LGTM, which is conceptually depicted in Fig. 10.3.

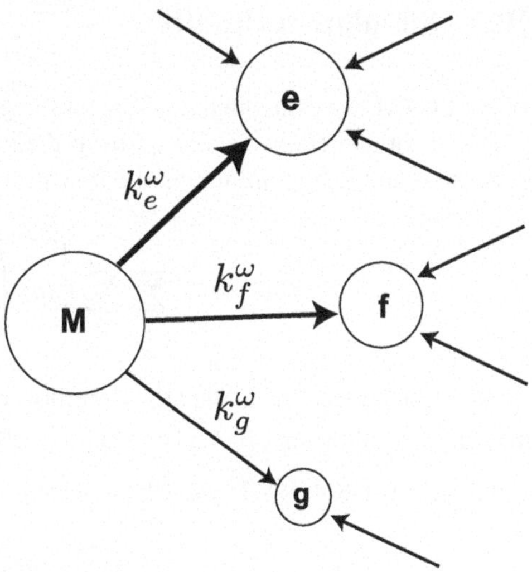

Fig. 10.3 Conceptual figure of weights of out-flow from a node in LGTM. In this case, the amounts proportional to GDPs are transported and distributed to neighbor nodes that are proportional to the ω-th power of other nodes' in-degrees

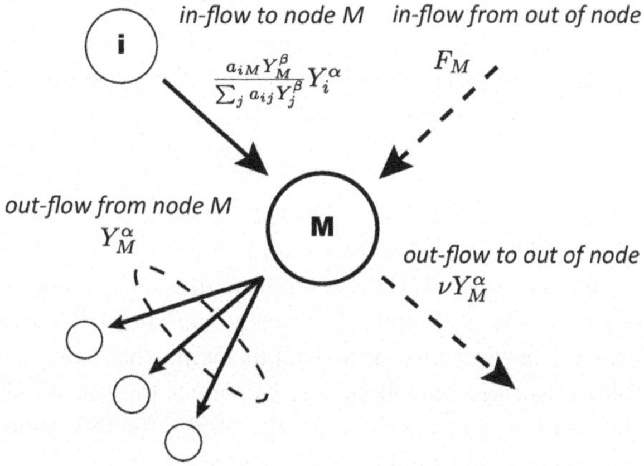

Fig. 10.4 Conceptual figure of all in-flows and out-flows for a node M in LGTM. The *black arrows* denote inflow-outflow relationships among nodes and the *dotted ones* show flow relationships outside of the network

LGTM is based on four types of flows. In the case of node M, inflow in an inter-node relationship is defined as $\frac{a_{iM} k_M^{\omega}}{\sum_j a_{ij} k_j^{\omega}} Y_i^{\alpha}$, which is affected by the degrees. Total outflow from node M in the inter-node relationship is Y_M^{α}. Inflow and outflow outside of the inter-node relationship are F_M and νY_M^{α}, respectively (Fig. 10.4). In the equilibrium, the aggregation of all flows is assumed to be zero.

$$\sum_i \frac{a_{iM} k_M^{\omega}}{\sum_j a_{ij} k_j^{\omega}} Y_i^{\alpha} - (1 + \nu) Y_M^{\alpha} + F_M = 0. \tag{10.4}$$

10.5 Simulation Results

Using LGTM, we estimate the GDPs for the given ITN data. First, we must acquire LGTM's parameters $(\alpha, \omega, \nu, F_M)$ from the real data's parameter fitting. Then, we estimate α and ω by minimizing the following function:

$$F(\alpha, \omega) = \sum_i \sum_M \left\{ log \left(\frac{w_{iM}}{\frac{a_{iM} k_M^{\omega}}{\sum_j a_{ij} k_j^{\omega}} Y_i^{\alpha}} \right) \right\}^2 . \tag{10.5}$$

Next, we derive ν from the inflow-outflow relationship in Eq. (10.6) as a transformation of equilibrium (10.4). Here, Y_M^{α} equals outflow (imports) and $\sum_i \frac{a_{iM} k_M^{\omega}}{\sum_j a_{ij} k_j^{\omega}} Y_i^{\alpha}$ equals inflow (exports). Thus, we get $\frac{1}{(1+\nu)}$ as the regression coefficient.

$$Y_M^{\alpha} = \frac{1}{(1+\nu)} \sum_i \frac{a_{iM} k_M^{\omega}}{\sum_j a_{ij} k_j^{\omega}} Y_i^{\alpha} + \frac{F_M}{(1+\nu)}. \tag{10.6}$$

Therefore, we get F_M from the equilibrium in Eq. (10.7).

$$F_M = \frac{\nu \sum_i Y_M^{\alpha}}{N} \tag{10.7}$$

We obtain the values as $(\alpha, \omega, \nu, F_M) = (0.68, 4.5, 0.032, 1.3 \times 10^2)$.

In the case of ITN, there are little differences among top countries' degrees. We introduce the preferentially selected network [6]—a simplified network produced from the original one, in which links with small contributions are removed using the following rule: For all nodes, we select the top n^{th} weight links for both inflow and outflow and cut off all other links. In this simulation, we use the case of preferentially selected network, where $n = 6$.

Under these conditions, we estimate GDP using Eq. (10.6) (Fig. 10.5). We find that the results between the real and simulated GDPs are fairly close. If the real and simulated values are proportionate, the values are assumed to be on the 45 degree line.

Next, we check the top 20 countries for both the simulated and real GDP. The results are listed in Table 10.1. In the real data, China ranks second, whereas in the simulation, it is at the top. China's estimated GDP is about two times bigger than the real GDP. This discrepancy might be caused by slow time evolution that China's GDP is on the way of approaching to the equilibrium value which is determined by the trading network structure. Simulation results are based on the equilibrium and China's GDP would be bigger in near future from the viewpoint of trade.

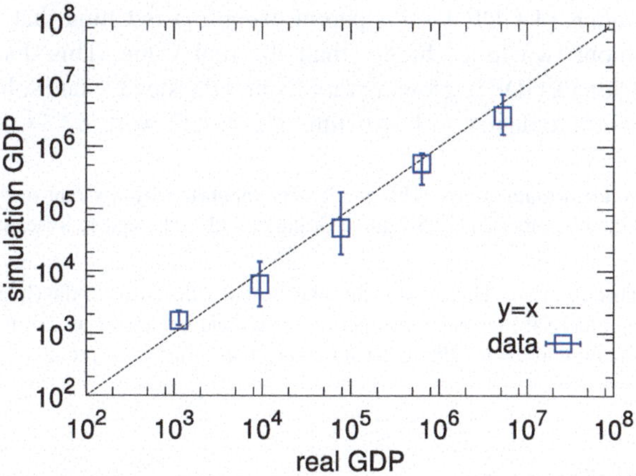

Fig. 10.5 Real and simulated GDPs estimated using Eq. (10.6) and preferentially selected network ($n = 6$). The first and third quantiles are plotted as error bars with the median value at the *center of squares*

Table 10.1 Top 20 countries (real and simulated data)

Country	Real GDP	Simulated GDP	Real rank	Simulated rank
United States	1.5×10^7	8.2×10^6	1	2
China	5.9×10^6	1.2×10^7	2	1
Japan	5.5×10^6	4.7×10^6	3	4
Germany	3.3×10^6	6.7×10^6	4	3
France	2.7×10^6	1.7×10^6	5	6
United Kingdom	2.3×10^6	1.5×10^6	6	9
Brazil	2.1×10^6	1.4×10^6	7	11
Italy	2.1×10^6	1.6×10^6	8	8
India	1.7×10^6	1.2×10^6	9	12
Canada	1.6×10^6	5.9×10^5	10	22
Russia	1.5×10^6	8.8×10^5	11	16
Spain	1.4×10^6	3.9×10^5	12	33
Australia	1.2×10^6	1.1×10^6	13	15
Korea	1.1×10^6	2.4×10^6	14	5
Mexico	1.1×10^6	1.0×10^4	15	95
Netherlands	8.4×10^5	1.7×10^6	16	7
Turkey	7.3×10^5	5.7×10^5	17	24
Indonesia	7.1×10^5	1.1×10^6	18	13
Switzerland	5.5×10^5	6.0×10^5	19	21
Saudi Arabia	5.3×10^5	4.5×10^5	20	30

In this simulation, we use parameters $(\alpha, \omega, \nu, F_M)$ in three significant digits and simulation results are represented in two significant digits

10.6 Conclusion

In this paper, we empirically introduced a linear gravity transportation model of world trade based on the network structure among countries. By tuning the model's parameters, we confirmed that estimated values approximately agree with the real values of GDP. One apparent exception is China that its estimated GDP value is about two times bigger than the real value. This discrepancy might imply that China's GDP is growing rapidly and the steady state solution of our model for given world trade network structure does not fit well.

Acknowledgements This work was supported by a Grant-in-Aid for Scientific Research (B), Grant Number 26310207 and a Grant-in-Aid for Scientific Research (C), Grant Number 24540395.

References

1. Rabinovitch S, Chinese econ stats: to doubt or not to doubt. (Financial Times, 28 June 2012), http://blogs.ft.com/beyond-brics/2012/06/28/china-stats-to-doubt-or-not-to-doubt. Accessed 30 Jan 2015
2. Rawski T (2001) China Econ Rev 12:347–354
3. Tamura K, Miura W, Takayasu H, Kitajima S, Goto H, Takayasu M (2012) Proceedings of the Asia Pacific symposium on intelligent and evolutionary systems, Kyoto, Japan. (ISBN978-4-99066920-1, 2012)
4. Tamura K, Miura W, Takayasu M, Takayasu H, Kitajima S, Goto H (2012) Int J Mod Phys Conf Ser 16:93–104
5. Watanabe H, Takayasu H, Takayasu M (2012) New J Phys 14:043034
6. Deguchi T, Takayasu H, Takayasu M. In preparation
7. Krugman P, Obstfeld M, Melitz M (2011) International economics: theory and policy, 9th edn. Pearson Education, Boston
8. Serrano M, Boguñá M (2003) Phys Rev E 68:015101
9. Garlaschelli D, Loffredo M (2005) Physica A 355(1):138–144
10. Fagiolo G, Reyes J, Schiavo S (2009) Phys Rev E 79:036115
11. Tzekina I, Danthi K, Rockmore D (2010) Eur Phys J B 63:541–545
12. Barigozzi M, Fagiolo G, Garlaschelli D (2010) Phys Rev E 81:046104
13. Hidalgo C, Klinger B, Barabási A, Hausmann R (2007) Science 317:482–487
14. Hidalgo C, Hausmann R (2009) Proc Natl Acad Sci 106:10570–10575
15. Tacchella A, Cristelli M, Caldarelli G, Gabrielli A, Pietronero L (2012) Sci Rep 2:723
16. Zaccaria A, Cristelli M, Tacchella A, Pietronero L PloS one (2014) 9:e113770
17. Garlaschelli D, Loffredo M (2004) Phys Rev Lett 93:188701
18. Fagiolo G (2010) J Econ Interact Coord 5:1–25
19. Dueñas M, Fagiolo G (2013) J Econ Interact Coord 8:155–178
20. Almog A, Squartini T, Garlaschelli D (2015) New J Phys 17:013009
21. Direction of trade statistics. In: IMF eLibrary data. http://elibrary-data.imf.org. Accessed 30 Jan 2015

Chapter 11
Analysis of Network Robustness for a Japanese Business Relation Network by Percolation Simulation

Hirokazu Kawamoto, Hideki Takayasu, and Misako Takayasu

Abstract This paper describes the application of percolation theory to a Japanese business relation network composed of approximately 3,000,000 links. In this network, we examined the process in which links are randomly removed. At the percolation transition point, we calculate the survival rate for each node as an indicator of its global network connectivity. The basic properties of each node are determined in connection with the values characterising these complex networks, such as the link number and job category. We confirm that this index has strong correlation with degree and shell number, also has significant correlation with sales and number of employee. Finally, we define the network robustness for each prefecture in Japan by using this new indicator.

11.1 Introduction

Percolation theory has been studied in the fields of physics and mathematics. Especially, many interesting properties have been revealed about the percolation transition point at which macroscopic connectivity disappears when removing its elements [1]. Because of the high versatility of this theory, it has been applied to

H. Kawamoto (✉) • M. Takayasu
Department of Computational Intelligence and Systems Science, Interdisciplinary Graduate School of Science and Engineering, Tokyo Institute of Technology, Midori-ku, Yokohama, Japan
e-mail: kawamoto@smp.dis.titech.ac.jp; takayasu@dis.titech.ac.jp

H. Takayasu
Sony Computer Science Laboratories, Shinagawa-ku, Tokyo, Japan

Meiji Institute for Advanced Study of Mathematical Sciences, Meiji University, Nakano-ku, Tokyo, Japan

Department of Computational Intelligence and Systems Science, Interdisciplinary Graduate School of Science and Engineering, Tokyo Institute of Technology, Midori-ku, Yokohama, Japan
e-mail: takayasu@csl.sony.co.jp

H. Takayasu et al. (eds.), *Proceedings of the International Conference on Social Modeling and Simulation, plus Econophysics Colloquium 2014*, Springer Proceedings in Complexity, DOI 10.1007/978-3-319-20591-5_11

a wide range of real world problems, such as electrical conduction [2] and Internet traffic congestion [3].

Since the BA model was proposed [4], percolation theory has been applied to complex networks with an inhomogeneous structure in connection with the concept of small-world [5]. Studying the percolation process in such complex networks plays an important role from the viewpoint of the fragility of a given system. It is well known that scale-free networks lose connectivity at low density if nodes are removed randomly and at high density if nodes are removed in descending order of the degree [6]. Because these studies can be viewed as a kind of stress test, percolation theory is also important for application study.

In the next section, we explain a dataset composed of about 600,000 Japanese firms and describe its basic properties as a complex network. We present the basic results of our percolation simulation in Sect. 11.3. The statistical properties of the survival rate and the theoretical analysis are provided in Sect. 11.4. In Sect. 11.5, we discuss the network robustness of the prefecture in Japan. Finally, we conclude this study and mention our plans for future work in Sect. 11.6.

11.2 Business Relation Network

The dataset we used in this study was provided by TEIKOKU DATABANK, Ltd., a Japanese credit research company. It included information about the direction of money flow, sales and employees of each firm in operation in 2011. From the point of view of a network study, the dataset provided a complex network consisting of 612,133 nodes and 3,841,496 links. As we were interested in the percolation properties of this network, we ignored the direction of the links and severed so-called dangling bonds, i.e. the bonds that could be removed from the network by the removal of a single link. We then extracted the largest strongly connected component (LSCC) from the raw network [7], and ignored the direction of each of the links for simplicity. As a result of this process, our network was composed of 327,721 nodes and 2,960,370 links. This operation enabled us to reduce the amount of numerical calculation in the following analysis.

Next, we present the basic properties of this network. The link number, namely, degree k, is distributed across a wide range, and this distribution is approximated by a power law for a large degree.

$$F(\geq k) \propto k^{-\alpha} \tag{11.1}$$

The cumulative exponent α is roughly estimated to be 1.5. Hence, the business relation network is a typical scale-free network [8]. In addition, it also has the small-world property [9].

In this network, we introduce k-shell decomposition, which is a general method intended to reveal the layer structure in a complex network [10]. Application of this method enabled this network to be decomposed into 25 layers, which are also called shells. A shell number is defined for each node, and the number of nodes with the shell number 7 is most numerous [11].

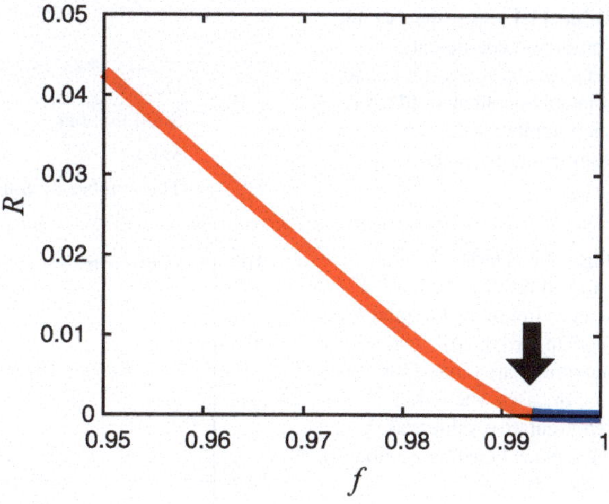

Fig. 11.1 Largest cluster size R normalised by all links in the range of f between 0.95 and 1.00. The arrow indicates the percolation transition point. The average was taken over 100 trials

11.3 Percolation Simulation

A detailed observation of the changes in the network topology became possible when links were randomly removed from the network one by one especially around the percolation transition point. In this case, we did not apply node removal as this could be viewed as a kind of correlated link removal. We calculated the largest cluster size R as an order parameter, which was defined as the ratio of the number of links in the cluster to all the initial links. Here, the control parameter f is the ratio of the number of removed links to all links. As shown in Fig. 11.1, the order parameter R is sufficiently small for f larger than f_c, which is referred to as the percolation transition point. We estimated the value of f_c as 0.994. Its value is approximately 1, but not exactly 1, and this result is consistent with the findings of previous research in which percolation simulation was applied to a complex network [12]. The properties around this point are discussed in detail from the viewpoint of statistical physics including the finite-size effect [11].

11.4 Survival Rate

11.4.1 Basic Properties of Survival Rate

In this section, we introduce the survival rate for each node and provide its basic properties. At the transition point ($f_c = 0.994$), the survival rate is defined as the ratio of the number of trials, in which the node belongs to the largest cluster, to the total number of trials. In this study, 100,000 trials were performed to estimate the value of the survival rate for each node. This parameter is widely distributed, and its large-scale behaviour approximates a power law [11]. It should be noted that this

Table 11.1 Spearman's rank correlation coefficient between the survival rate and principal parameters (degree, shell number, sales, and number of employees)

	Spearman's rank correlation coefficient
Degree	0.729
Shell number	0.765
Sales	0.378
The number of employee	0.356

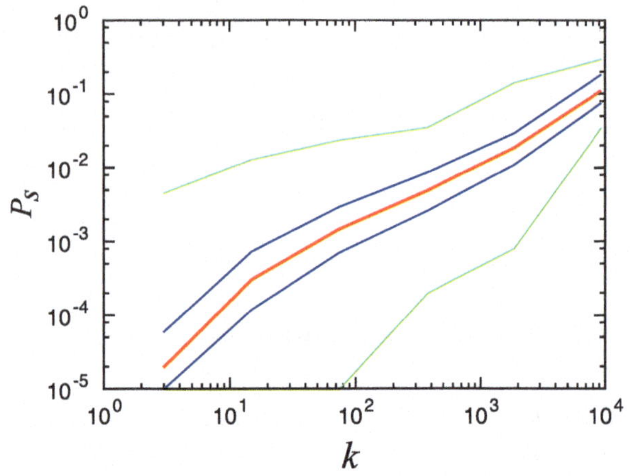

Fig. 11.2 Degree k vs the survival rate P_s in a log-log scale. Minimum, 1st quartile, median, 3rd quartile and maximum are plotted for log bin. In cases where the representative value was 0, we replaced the observation limit, 1.0×10^{-5}

index is able to characterise the global connectivity of each node in the network as we explain in Sect. 11.4.3.

Next, we discuss the correlation between the survival rate and important parameters characterising firms, such as degree, shell number, sales and the number of employees. Spearman's rank correlation coefficient was chosen for this purpose, because Pearson's correlation coefficient is susceptible to outliers. As shown in Table 11.1, there is a positive correlation between the survival rate and all the parameters, and this is especially strong for values characterising the network, such as the degree and shell number.

The correlation between degree k and the survival rate P_s was investigated in more detail. The variation of the survival rate P_s was clarified by plotting its distribution for degree k as shown in Fig. 11.2. We found that the survival rate P_s varies even in the same range of degree k. Therefore, the survival rate P_s is not completely determined by information relating to the local connectivity, such as degree k; the degree k can explain this value roughly. This fact suggests that the survival rate P_s is determined by the critical cluster, which includes the information of the whole network topology. In this sense, this robust index includes information about the global connectivity, such as the shell number, as opposed to local connectivity such as the link number.

Fig. 11.3 Nodes with small survival rate (*red circles*) and its linking nodes (*orange dots*) plotted for Hokkaido Island in Japan

11.4.2 Practical Meaning of P_s

It is important to note that the nodes with the same survival rate P_s were confirmed to have widely distributed link numbers as shown in Fig. 11.2. We subsequently investigated the features of the nodes with a high survival rate for small link numbers and those with a low survival rate for large link numbers. First, we specified a range of degree k from 1 to 10 as a set of small link numbers within a certain range of the survival rate ($5.0 \times 10^{-4} \leq P_s \leq 5.0 \times 10^{-3}$), which includes approximately 20,000 nodes. When we investigated the industry these nodes represent, it was revealed that the nodes categorised as belonging to the construction industry captured 27 % of the share, whereas the share was 21 % of the network in its initial state. This result means that nodes belonging to the construction category have a higher survival rate than nodes in other categories.

Next, we focused on nodes with a large number of links within the same range of survival rate ($5.0 \times 10^{-4} \leq P_s \leq 5.0 \times 10^{-3}$). These nodes are characterised by a low survival rate and are fragile in spite of their many links. As an example, we paid attention to the node with large k and relatively small P_s, $(k, P_s) = (448, 3.3 \times 10^{-3})$. As shown in Fig. 11.3, most of its linking nodes are located on the same island, Hokkaido, and there are only 13 links (about 3 %) connecting to firms outside this island. There are not many links connecting to nodes located outside of this island. This result suggests that this type of node bundles firms in a local region.

11.4.3 Theoretical Estimation

A theoretical estimation of the survival rate was derived by using the degree and the rates of linking nodes as follows. In the case of a node that only has one link we have the following exact relation.

$$P_{s,i} = (1 - f_c)P_{s,j} \tag{11.2}$$

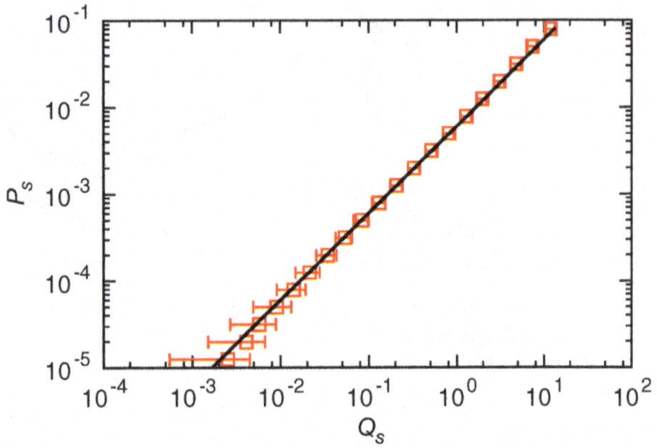

Fig. 11.4 Summation of the survival rate Q_s of nearest-neighbour nodes vs the survival rate P_s. The average is plotted in each log-scaled bin. Error bars are estimated by standard deviation in log-log scale. The line shows Eq. (11.4)

where, the subscript i represents the focusing node, and the subscript j represents the its linking node.

We next extended this formulation to the general case for nodes with multiple links. The probability of the focusing node being connected to the giant component, $P_{s,i}$ is approximated as follows.

$$P_{s,i} = 1 - \prod_{j=1}^{k} \{1 - (1 - f_c)P_{s,j}\} \tag{11.3}$$

On condition that the survival rate P_s is sufficiently small, we can approximate Eq. (11.3) by the following equation.

$$P_{s,i} \simeq (1 - f_c)Q_{s,j} \tag{11.4}$$

Here, $Q_{s,j}$ is defined as $\sum_{j=1}^{k} P_{s,j}$. This equation shows that the survival rate P_s is explained by the summation of the survival rates of linking nodes. In Fig. 11.4, we confirm that this relation is in good agreement. This examination revealed that the survival rate P_s depends on the link number k and the survival rates of the linking nodes, $P_{s,j}$. This result shows that the value $P_{s,i}$ is determined from the global network topology, and that the mean field approach used in Eq. (11.3) works well in deriving Eq. (11.4).

11.5 Network Robustness of Prefectures in Japan

By using the survival rate, we define network robustness of each prefecture as follows. We picked up the top 10,000 ranking nodes (practically consisted of 10,005 nodes counting the same ranking) by the order of survival rates, and counted the number of nodes, and normalised this by the original number of nodes in the extracted network for each prefecture. As shown in Fig. 11.5, Tokyo, Osaka, and seven other prefectures were judged to belong to the most robust class.

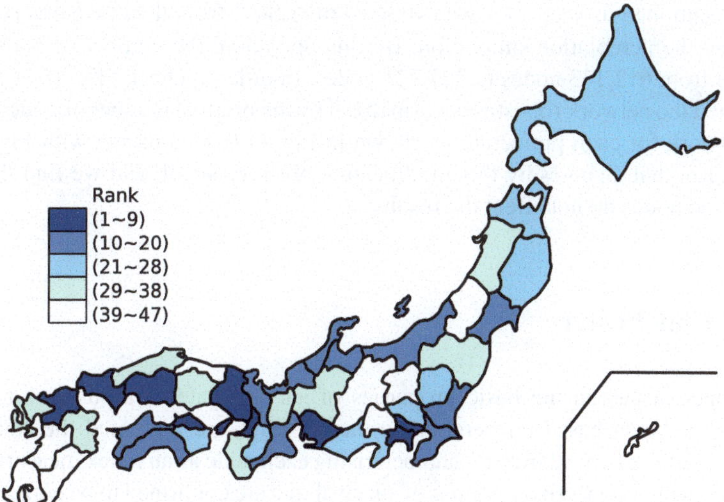

Fig. 11.5 Network robustness for each prefecture in Japan. Here, the network robustness of each prefecture is estimated by the number of top 10,005 robust nodes located in the prefecture divided by the number of nodes of the initially extracted LSCC network in the prefecture. *Colours* show the ranking of network robustness categorised into five classes, (1 ∼ 9), (10 ∼ 20), (21 ∼ 28), (29 ∼ 38), (39 ∼ 47) from the deepest to the lightest

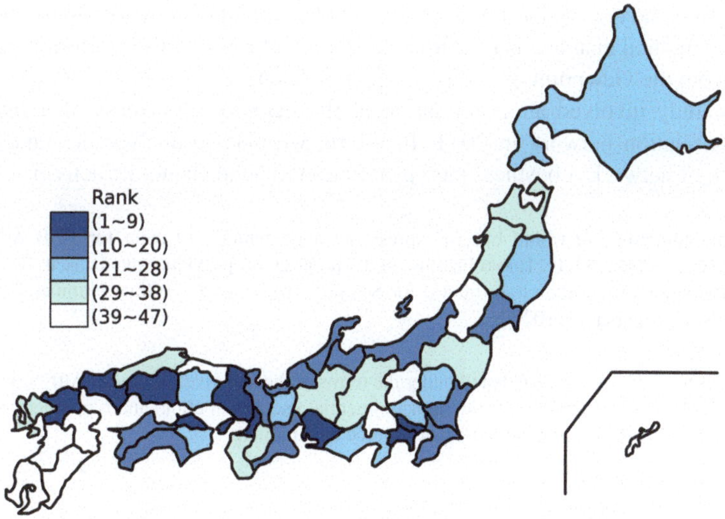

Fig. 11.6 Network robustness for each prefecture with modified normalization. Compared with Fig. 10.5 normalization by the whole number of nodes in the original raw network is applied. *Colours* show the ranking of network robustness categorised into five classes, (1 ∼ 9), (10 ∼ 20), (21 ∼ 28), (29 ∼ 38), (39 ∼ 47) from the deepest to the lightest

As mentioned in Sect. 11.2 we extracted the LSCC from the raw network when we made the percolation simulation. By this operation, the number of nodes was reduced from 612,133 nodes to 327,721 nodes. In order to check this effect, we re-calculated the network robustness normalized by the original number of nodes in the raw network for each prefecture as shown in Fig. 11.6. Comparing with Fig. 11.5, we confirm that changes by this modification are very small, and we find that the eliminated nodes do not affect the results.

11.6 Conclusion

This paper discussed the basic properties of survival rate of a business relation network in Japan based on percolation theory. First, we presented the statistical properties of the survival rate by characterising each node as an index measuring the global network connectivity. Values of survival rate are confirmed to be correlated to network connectivity indices such as degrees or shell numbers. However, as shown in Fig. 11.2, the values distribute widely for the nodes with the same degree number. It is proved in Sect. 11.4.3 that the survival rate of a node is determined by the sum of survival rates of its neighbor nodes. As the survival rates of neighbors are determined by the next neighbors, and so forth, this value reflects information about wider area's network connectivity.

We discussed regional differences from the viewpoint of network robustness. The proposed method enabled us to extract those prefectures that were determined to be robust from the viewpoint of complex network science.

This study involved an examination of the network robustness of a Japanese business relation network in 2011. In future, we plan to analyse the time series variation of network robustness paying attention to local economic activities.

Acknowledgements We would like to express our appreciation to Center for TDB Advanced Data Analysis and Modeling, Tokyo Institute of Technology for providing the datasets. This study is partially supported by the Grant-in-Aid for Scientific Research (C), Grant Number 24540395 and (B), Grant Number 26310207.

References

1. Stauffer D, Aharony A (1994) Introduction to percolation theory. Taylor and Francis, London
2. Last BJ, Thouless DJ (1971) Phys Rev Lett 27:1719
3. Takayasu M, Takayasu, Sato T (1996) Physica A 233:824
4. Barabási AL, Albert R (1999) Science 286:509

5. Watts DJ, Strogatz SH (1998) Nature 393:440
6. Albert R, Jeong H, Barabási AL (2000) Nature 406:378
7. Dorogovtsev SN, Mendes JFF, Samukhin AN (2001) Phys Rev E 64:025101
8. Miura W, Takayasu H, Takayasu M (2012) Phys Rev Lett 108:168701
9. Takayasu M, Sameshima S, Ohnishi T, Ikeda Y, Takayasu H, Watanabe K (2008) Annual Report of the Earth Simulator Center April 2007-March 2008, pp 263–268 http://www.jamstec.go.jp/esc/publication/annual/annual2007/pdf/2project/chapter3/263takayasu.pdf
10. Kitsak M, Gallos LK, Havlin S, Liljeros F, Muchnik L, Stanley HE, Makse HA (2010) Nature Phys 29:1746
11. Kawamoto H, Takayasu H, Jensen HJ, Takayasu M (2015) PLoS ONE 10(4):e0119979. doi: http://dx.doi.org/10.1371/journal.pone.0119979
12. Onnela JP, Saramäki J, Hyvonen J, Szabo G, de Menezes MA, Kaski K, Barabási AL, Kertész J (2007) New J Phys 9:179

Chapter 12
Detectability Threshold of the Spectral Method for Graph Partitioning

Tatsuro Kawamoto and Yoshiyuki Kabashima

Abstract Graph partitioning, or community detection, is an important tool for investigating the structures embedded in real data. The spectral method is a major algorithm for graph partitioning and is also analytically tractable. In order to analyze the performance of the spectral method, we consider a regular graph of two loosely connected clusters, each of which consists of a random graph, i.e., a random graph with a planted partition. Since we focus on the bisection of regular random graphs, whether the unnormalized Laplacian, the normalized Laplacian, or the modularity matrix is used does not make a difference. Using the replica method, which is often used in the field of spin-glass theory, we estimate the so-called detectability threshold; that is, the threshold above which the partition obtained by the method is completely uncorrelated with the planted partition.

12.1 Introduction

Considerable attention has been paid to the graph clustering or community detection problem and a number of formulations and algorithms have been proposed in the literature [1–5]. Although the meaning of a module in each detection method may not be equivalent, we naturally wish to know in what manner the methods perform typically and the point at which a method fails to detect a certain structure in principle [6–8]. Otherwise, we need to test all the existing methods, and this clearly requires a huge cost and is also redundant. Although most studies of the expected performance were experimental, using benchmark testing [9–11], it is expected that theoretical analysis will give us a deeper insight.

As frequently done in benchmarks, we consider random graphs having a planted block structure. The most common model is the so-called stochastic block model (or the planted partition model) [12]. Although many variants of the stochastic block model have been proposed in the literature [13–16], in the simplest case, the

T. Kawamoto (✉) • Y. Kabashima
Tokyo Institute of Technology, Yokohama, Kanagawa, Japan
e-mail: kawamoto@sp.dis.titech.ac.jp; kaba@dis.titech.ac.jp

H. Takayasu et al. (eds.), *Proceedings of the International Conference on Social Modeling and Simulation, plus Econophysics Colloquium 2014*, Springer Proceedings in Complexity, DOI 10.1007/978-3-319-20591-5_12

vertices within the same module are connected with a high probability p_{in}, while the vertices in different modules are connected with a low probability p_{out}. When the difference between the probabilities is sufficiently large, $p_{in} \gg p_{out}$, the graph has a strong block structure and the spectral method detects almost or exactly the same partition as the planted partition. As we increase the probability between the modules p_{out}, the partition obtained by the spectral method tends to very different from the planted one, and finally, they are completely uncorrelated. The point of the transition is called the *detectability threshold* [17–20]. Since we know that the graph is generated by the stochastic block model, the ultimate limit of this threshold is given by Bayesian inference and it is known that, in the case of the two-block model,

$$c_{\mathrm{in}} - c_{\mathrm{out}} = 2\sqrt{\overline{c}}, \tag{12.1}$$

where $c_{\mathrm{in}} = p_{in}N$, $c_{\mathrm{out}} = p_{out}N$ and $\overline{c}$ is the average degree. N is the total number of vertices in the graph. Equation (12.1) indicates that, even when the vertices are more densely connected within a module than between modules, unless the difference is sufficiently large, it is statistically impossible to infer the embedded structure.

It was predicted by Nadakuditi and Newman in [20] that the spectral method with modularity also has the same detectability threshold as Eq. (12.1). However, it was numerically shown in [21] that this applies only to the case where the graph is not sparse. Despite its significance, a precise estimate of the detectability threshold of the spectral method in the sparse case seems to remain missing.

In this article, we derive an estimate of the detectability threshold of the spectral method of the two-block regular random graph. It should be noted that the simplest stochastic block model, which we explained above, has Poisson degree distribution, while we impose a constraint such that the degree does not fluctuate. Therefore, our results do not directly provide an answer to the missing part of the problem. They do, however, provide a fruitful insight into the performance of the spectral method. Moreover, in the present situation, we do not face the second difficulty of the spectral method: the localization of the eigenvectors. Although the localization of eigenvectors is another important factor in the detectability problem, it is outside the scope of this article.

This article is organized as follows. In Sect. 12.2, we briefly introduce spectral partitioning of two-block regular random graphs and mention that the eigenvector corresponding to the second-smallest eigenvalue contains the information of the modules. In Sect. 12.3, we show the average behavior of the second-smallest eigenvalue and the corresponding eigenvector as a function of the parameters in the model. Finally, Sect. 12.4 is devoted to the conclusion.

12.2 Spectral Partitioning of Regular Random Graphs With Two-Block Structure

The model parameters in the two-block regular random graph are the total number of vertices N, the degree of each vertex c, and the fraction of the edges between modules $\gamma = l_{\mathrm{int}}/N$. The graph is constructed as follows. We first set module indices on the vertices, each of which has c half edges, or stubs, and randomly connect the vertices in different modules with l_{int} edges. We connect the rest of the edges at random within the same module. We repeat the process so that every edge is connected to a pair of vertices. This random graph is sparse when $c = O(1)$, because the number of edges is of the same order as the number of vertices N. We calculate the degree of correlation between the partition obtained by the spectral method and the planted partition as γ varies.

The choices of the matrix that can be used in the spectral method is wide. The popular matrices are the unnormalized Laplacian L, the normalized Laplacian $\mathcal{L}$, and the modularity matrix B. For the bisection of regular random graphs, however, all the partitions they yield have shown to be the same [22]. Thus, we analyze the unnormalized Laplacian L, since it is the simplest. The basic procedure of the spectral bisection with the unnormalized Laplacian L is quite simple. We solve for the eigenvector corresponding to the second-smallest eigenvalue of L and classify each vertex according to the sign of the corresponding component of the eigenvector; the vertices with the same sign belong to the same module. Therefore, our goal is to calculate the behavior of the sign of the eigenvector as a function of γ.

12.3 Detectability Threshold

We use the so-called replica method, which is often used in the field of spin-glass theory in statistical physics. The basic methodology here is parallel to that in [23]. Although the final goal is to solve for the eigenvector corresponding to the second-smallest eigenvalue or the statistics of its components, let us consider estimating the second-smallest eigenvalue, averaged over the realization of the random graphs. We denote by $[\dots]_L$ the random average over the unnormalized Laplacians of the possible graphs. For this purpose, we introduce the following "Hamiltonian" $H(\boldsymbol{x}|L)$, "partition function" $Z(\beta|L)$, and "free energy density" $f(\beta|L)$:

$$H(\boldsymbol{x}|L) = \frac{1}{2}\boldsymbol{x}^{\mathrm{T}} L \boldsymbol{x}, \tag{12.2}$$

$$Z(\beta|L) = \int d\boldsymbol{x}\, \mathrm{e}^{-\beta H(\boldsymbol{x}|L)} \delta(|\boldsymbol{x}|^2 - N)\delta(\mathbf{1}^{\mathrm{T}}\boldsymbol{x}), \tag{12.3}$$

$$f(\beta|L) = -\frac{1}{N\beta} \ln Z(\beta|L), \tag{12.4}$$

where $\boldsymbol{x}$ is an N-dimensional vector, $\mathbf{1}$ is a vector in which each element equals one, and T represents the transpose. The delta function $\delta(|\boldsymbol{x}|^2 - N)$ in (12.3) is to impose the norm constraint. It should be noted that the eigenvector corresponding to the smallest eigenvalue is proportional to $\mathbf{1}$ and this choice is excluded by the constraint $\delta(\mathbf{1}^{\mathrm{T}}\boldsymbol{x})$. In the limit of $\beta \to \infty$, in conjunction with the operation of $\delta(\mathbf{1}^{\mathrm{T}}\boldsymbol{x})$, the contribution in the integral of the "partition function" $Z(\beta|L)$ is dominated by the vector that minimizes the value of the "Hamiltonian" $H(\boldsymbol{x}|L)$, under the constraint of being orthogonal to the eigenvector $\mathbf{1}$ of the smallest eigenvalue. Therefore, the "partition function" is dominated by the eigenvector of the second-smallest eigenvalue and the "free energy density" $f(\beta|L)$ extracts it, i.e.,

$$\lambda_2 = 2 \lim_{\beta\to\infty} f(\beta|L). \tag{12.5}$$

The quantity we need is $[\lambda_2]_L$, the second-smallest eigenvalue averaged over the unnormalized Laplacians. However, because the average of the logarithm of the "partition function" is difficult to calculate, we recast $[\lambda_2]_L$ as

$$\begin{aligned}[\lambda_2]_L &= -2 \lim_{\beta\to\infty} \frac{1}{N\beta} [\ln Z(\beta|L)]_L \\ &= -2 \lim_{\beta\to\infty} \lim_{n\to 0} \frac{1}{N\beta} \frac{\partial}{\partial n} \ln\, [Z^n(\beta|L)]_L \,. \end{aligned} \tag{12.6}$$

The assessment of $[Z^n(\beta|L)]_L$ is also difficult for a general real number n. However, when n takes positive integer values, $[Z^n(\beta|L)]_L$ can be evaluated as follows. For a positive integer n, $[Z^n(\beta|L)]_L$ is expressed as

$$\begin{aligned}[Z^n(\beta|L)]_L &= \int \left(\prod_{a=1}^{n} d\boldsymbol{x}_a \delta(|\boldsymbol{x}_a|^2 - N)\delta(\mathbf{1}^{\mathrm{T}}\boldsymbol{x}_a) \right) \left[\exp\left(-\frac{\beta}{2} \sum_a \boldsymbol{x}_a^{\mathrm{T}} L \boldsymbol{x}_a \right) \right]_L \\ &\equiv \int \left(\prod_{a=1}^{n} d\boldsymbol{x}_a \delta(|\boldsymbol{x}_a|^2 - N)\delta(\mathbf{1}^{\mathrm{T}}\boldsymbol{x}_a) \right) \exp\left(H_{\mathrm{eff}}(\beta, \boldsymbol{x}_1, \boldsymbol{x}_2, \ldots, \boldsymbol{x}_n) \right). \end{aligned} \tag{12.7}$$

This means that $[Z^n(\beta|L)]_L$ has a meaning of a partition function for a system of n-replicated variables $\boldsymbol{x}_1, \boldsymbol{x}_2, \ldots, \boldsymbol{x}_n$ that is subject to no quenched randomness. In addition, the assumption of the graph generation guarantees that the effective Hamiltonian $\mathscr{H}_{\mathrm{eff}}(\beta, \boldsymbol{x}_1, \boldsymbol{x}_2, \ldots, \boldsymbol{x}_n)$ is of the mean field type. These indicate that $N^{-1} \ln[Z^n(\beta|L)]_L$ for $n = 1, 2, \ldots$ can be evaluated exactly by the saddle point method with respect to certain macroscopic variables (order parameters) as $N \to \infty$. After some calculations, we indeed reach an expression with the saddle point evaluation as

$$\frac{1}{N}\ln[Z^n(\beta|L)]_L = \underset{\{Q_r\},\{\hat{Q}_r\},\{\phi_a\},\{\psi_a\},\eta}{\mathrm{extr}} \left\{ NK_{\mathrm{I}}(Q_r,\hat{Q}_r) + \frac{\beta}{2}\sum_a \phi_a - \sum_{r=1,2} K_{\mathrm{II}r}(Q_r,\hat{Q}_r) + \frac{1}{N}\sum_{r=1,2} \ln K_{\mathrm{III},r}(\hat{Q}_r,\{\phi_a\},\{\psi_a\}) + \eta\gamma - \frac{1}{N}\ln \mathcal{N}_G - \ln c! \right\}, \tag{12.8}$$

where $\mathcal{N}_G$ is the total number of graph configurations and

$$K_{\mathrm{I}}(Q_r,\hat{Q}_r) = \sum_{r,s=1,2} \frac{p_r p_s}{2} \int d\boldsymbol{\mu}^{(r)} d\boldsymbol{\nu}^{(s)}\, Q_r(\boldsymbol{\mu}^{(r)}) Q_s(\boldsymbol{\nu}^{(s)}) \times \mathrm{e}^{-(1-\delta_{rs})\eta - \frac{\beta}{2}\sum_a (\mu_a^{(r)} - \nu_a^{(s)})^2},$$

$$K_{\mathrm{II}r}(Q_r,\hat{Q}_r) = p_r \int d\boldsymbol{\mu}^{(r)}\, \hat{Q}_r(\boldsymbol{\mu}^{(r)}) Q_r(\boldsymbol{\mu}^{(r)}),$$

$$K_{\mathrm{III}r}(\hat{Q}_r,\{\phi_a\},\{\psi_a\}) = \int \prod_{i\in V_r}\prod_{a=1}^{n} dx_{ia} \times \prod_{i\in V_r}\left(\hat{Q}_r^c(\boldsymbol{x}_i) \exp\left[-\frac{\beta}{2}\sum_a (\phi_a x_{ia}^2 + \psi_a x_{ia}) \right]\right). \tag{12.9}$$

In the above equations, four functions $Q_r(\mu_1^{(r)},\ldots,\mu_n^{(r)})$ and $\hat{Q}_r(\mu_1^{(r)},\ldots,\mu_n^{(r)})$ $(r = 1,2)$ play the roles of order parameters.

Unfortunately, this expression cannot be employed directly for the computation of (6) as $Q_r(\mu_1^{(r)},\ldots,\mu_n^{(r)})$ and $\hat{Q}_r(\mu_1^{(r)},\ldots,\mu_n^{(r)})$ are defined only for $n = 1,2,\ldots$. To overcome this inconvenience, we introduce the following assumption at the dominant saddle point.

> [Replica symmetric assumption] The right hand side of (12.7) is invariant under any permutation of replica indices $a = 1,2,\ldots,n$. We assume that this property, which is termed the *replica symmetry*, is also owned by the dominant saddle point of (12.8).

In the current system, this restricts the functional forms of $Q_r(\mu_1^{(r)},\ldots,\mu_n^{(r)})$ and $\hat{Q}_r(\mu_1^{(r)},\ldots,\mu_n^{(r)})$ as

$$Q_r(\mu_1,\ldots,\mu_n) = \left(\frac{cp_r - \gamma}{Np_r^2}\right)^{1/2} \int dAdH\, q_r(A,H) \left(\frac{\beta A}{2\pi}\right)^{\frac{n}{2}} \times \exp\left[-\frac{\beta A}{2}\sum_{a=1}^{n}\left(\mu_a - \frac{H}{A}\right)^2 \right],$$

$$\hat{Q}_r(\mu_1, \ldots, \mu_n) = c\left(\frac{cp_r - \gamma}{Np_r^2}\right)^{-1/2} \int d\hat{A}d\hat{H}\,\hat{q}_r(\hat{A}, \hat{H}) \times \exp\left[\frac{\beta}{2}\sum_{a=1}^{n}\left(\hat{A}\mu_a^2 + 2\hat{H}\mu_a\right)\right], \tag{12.10}$$

which yields an expression of $N^{-1}\ln[Z^n(\beta|L)]_L$ that can be extended for n of a real number. We then substitute that expression into (12.6), which finally provides

$$\begin{aligned}
[\lambda_2]_L = -\underset{\{q_r\},\{\hat{q}_r\},\phi,\psi}{\text{extr}} \Bigg\{ &\int dAdH \int dA'dH'\,\Xi(A,H,A',H') \\
&\times \frac{cp_1p_2}{2}\Bigg(\left(\frac{p_1}{p_2}+\Gamma\right) q_1(A,H)q_1(A',H') \\
&+ \left(\frac{p_2}{p_1}+\Gamma\right) q_2(A,H)q_2(A',H') \\
&+ 2\,(1-\Gamma)\,q_1(A,H)q_2(A',H')\Bigg) \\
&+ \phi \\
&- c\sum_{r=1,2} p_r \int dAdH \int d\hat{A}d\hat{H}\,q_r(A,H)\hat{q}_r(\hat{A},\hat{H}) \left(\frac{(H+\hat{H})^2}{A-\hat{A}} - \frac{H^2}{A}\right) \\
&+ \sum_{r=1,2} p_r \int \prod_{g=1}^{c}\left(d\hat{A}_g d\hat{H}_g \hat{q}_r(\hat{A}_g,\hat{H}_g)\right) \frac{\left(\psi/2 - \sum_g \hat{H}_g\right)^2}{\phi - \sum_g \hat{A}_g} \Bigg\},
\end{aligned} \tag{12.11}$$

where we set

$$\Gamma = 1 - \frac{\gamma}{cp_1p_2}, \tag{12.12}$$

$$\Xi(A,H,A',H') = \frac{(1+A')H^2 + (1+A)H'^2 + 2HH'}{(1+A)(1+A') - 1} - \frac{H^2}{A} - \frac{H'^2}{A'}. \tag{12.13}$$

The above procedure is often termed the *replica method*. Although its mathematical validity of the replica method has not yet been proved, we see that our assessment based on the simplest permutation symmetry for the replica indices offers a fairly accurate prediction for the experimental results below.

Due to the space limitation, we hereafter show only the results, omitting all the details of the calculation (see [24] for complete calculation including detailed

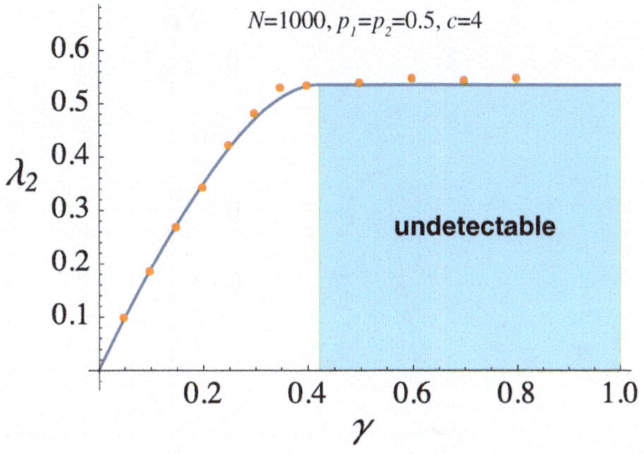

Fig. 12.1 Second-smallest eigenvalue as a function of γ. The solid line represents the estimate of the average over the realization of the graphs $[\lambda_2]_L$ and the dots represent the results of the numerical experiment of a single realization with $N = 1000$ and $c = 4$. The module sizes are set to be equal, $p_1 = p_2 = 0.5$.

derivation of (12.11)). In the limit of large size $N \to \infty$, the saddle-point analysis of (12.11) yields the solution

$$[\lambda_2]_L = \begin{cases} (1 - \Gamma)\left(c - 1 - \frac{1}{\Gamma}\right) & (1/\sqrt{c-1} \leq \Gamma), \\ c - 2\sqrt{c-1} & \text{otherwise,} \end{cases} \tag{12.14}$$

where $\Gamma = 1 - \gamma/(cp_1p_2)$; we set the size of each module as $N_1 = p_1N$ and $N_2 = p_2N$. The region of constant eigenvalue in (12.14) indicates that the second-smallest eigenvalue is in the spectral band, i.e., the information of the modules is lost there and an undetectable region exists. Therefore, the boundary of (12.14) is the critical point where the phase transition occurs. The plot of the second-smallest eigenvalue $[\lambda_2]_L$ is shown in Fig. 12.1. Although the dots represent the results of the numerical experiment of a single realization, the results agree with (12.14) quite well.

In terms of γ, the boundary of Eq. (12.14) can be recast as

$$\gamma = cf(c)p_1p_2, \tag{12.15}$$

where

$$f(c) = 1 - \frac{1}{\sqrt{c-1}}. \tag{12.16}$$

Since cp_1p_2 is the value of γ in a uniform (i.e., one-block) regular random graph, the factor $f(c)$ represents the low value of the threshold as compared to that in the uniform random case.

The distribution of the components of the corresponding eigenvector can also be obtained through this calculation. Although it cannot be written analytically, we can solve for it by iterating a set of integral equations that result from the saddle-point evaluation of the right hand side of (12.6). As shown in Fig. 12.2, the results

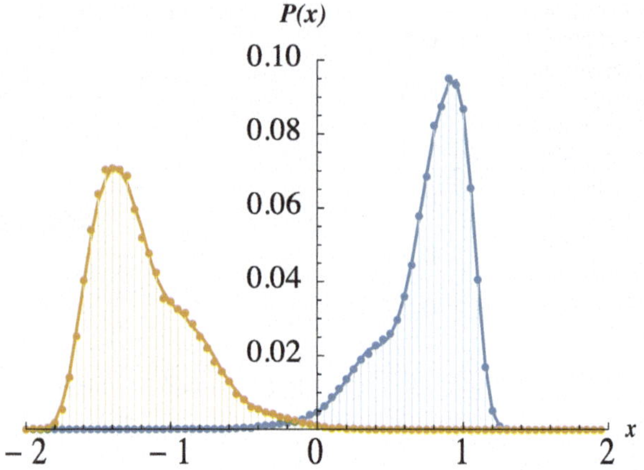

Fig. 12.2 Distributions of the elements in the eigenvector corresponding to the second-smallest eigenvector. Each plot shows the distribution of elements in each module, i.e., the distribution on the left corresponds to the module that is supposed to have negative sign elements and the distribution on the right corresponds to the module that is supposed to have positive sign elements, respectively. The *dots* represent the average results of the numerical experiments, taken over 100 samples. The ratio of the modules are set to be $p_1 = 0.6$ and $p_2 = 0.4$

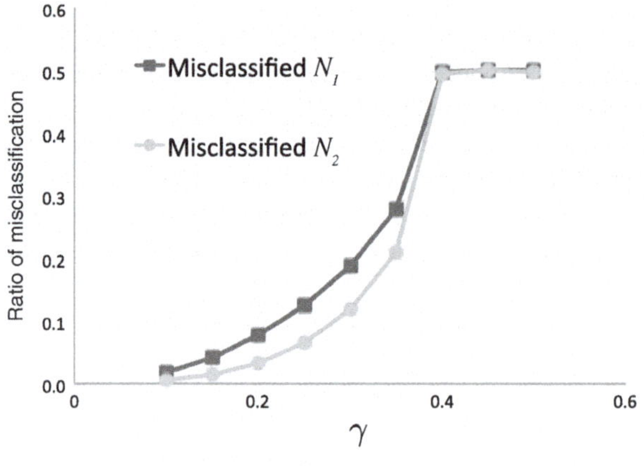

Fig. 12.3 Fraction of misclassified vertices in each module. As the parameter γ increases, the number of misclassified vertices increases polynomially

of our analysis agree with the corresponding numerical experiment excellently. In Fig. 12.2, the dots represent the average over 100 realizations of the random graphs. The ratio of misclassified vertices are shown in Fig. 12.3. It increases polynomially with respect to γ and saturates at the detectability threshold.

It should be note that, even when the number of vertices is infinity, the fraction of misclassified vertices remains finite. The misclassification of the vertices occurs because the planted partition is not the optimum in the sense of the spectral bisection. The spectral method with the unnormalized Laplacian L constitute the continuous relaxation of the discrete minimization problem of the so-called *RatioCut*. The RatioCut is lower for a partition with a sparse cut, while it penalizes for unbalanced partitions in the sense of the number of the vertices within a module; there may always exist a better cut in the sense of the RatioCut than the planted partition in the graph when γ is large.

Finally, let us compare our estimate with results of studies in the literature. In the following, we focus on the case of equal size modules, i.e., $p_1 = p_2 = 0.5$. Let the total degree within a module be K_{in} and let the total degree from one module to the others be K_{out}. Since we have $K = cN = 2(K_{\text{in}} + K_{\text{out}})$ and $K_{\text{out}} = \gamma N$, Eq. (12.15) reads

$$K_{\text{in}} - K_{\text{out}} = \frac{N}{2} \frac{c}{\sqrt{c-1}}. \tag{12.17}$$

In addition, in the limit $N \to \infty$, we have

$$K_{\text{in}} = \frac{N^2}{4} p_{\text{in}} = \frac{N}{4} c_{\text{in}}, \tag{12.18}$$

$$K_{\text{out}} = \frac{N^2}{4} p_{\text{out}} = \frac{N}{4} c_{\text{out}}. \tag{12.19}$$

Therefore, (12.17) can be recast as

$$c_{\text{in}} - c_{\text{out}} = 2\frac{c}{\sqrt{c-1}}. \tag{12.20}$$

This condition converges to the ultimate detectability threshold (12.1) in the dense limit $c \to \infty$. There exists, however, a huge gap between (12.1) and (12.20) when the degree c is small; considering the fact that the upper bound of the parameter $c_{\text{in}} - c_{\text{out}}$ is $2c$, this gap is not negligible at all. Thus, the implication of our results is that we cannot expect the spectral threshold to detect modules all the way down to the ultimate detectability threshold, even in regular random graphs, where the localization of the eigenvectors is absent.

12.4 Conclusion

In summary, we derived an estimate of the detectability threshold (12.20) of the spectral method of the two-block regular random graphs. The threshold we obtained agrees with the results of the numerical experiment excellently and is expected to be asymptotically exact in the limit $N \to \infty$. Our results indicate that the spectral method cannot detect modules all the way down to the ultimate detectability threshold (12.1), even when the degree is fixed to a constant. Since the threshold (12.20) converges to (12.1) as the degree c increases, this gap becomes negligible in the case where the degree is sufficiently large and this supports the results obtained by Nadakuditi and Newman [20].

A method for achieving the ultimate detectability threshold with the spectral method has already been proposed by Krzakala et al. [25]. They proposed using a matrix called the non-backtracking matrix, which avoids the elements of eigenvectors to be localized at a few vertices. A question about this formalism is: to what extent is the gap in the detectability in fact closed by the non-backtracking matrix as compared to the Laplacians? Our estimate gives a clue to the answer to this question. In order to gain further insight, we need to analyze the case of graphs with degree fluctuation. In that case, the methods using the unnormalized Laplacian and the normalized Laplacian will no longer be equivalent. Moreover, it is important to verify the effect of the localization of eigenvectors on the detectability. These problems remain as future work.

Acknowledgements This work was supported by JSPS KAKENHI Nos. 26011023 (TK), 25120013 (YK), and the JSPS Core-to-Core Program "Non-equilibrium dynamics of soft matter and information."

References

1. Fortunato S (2010) Phys Rep 486:75
2. Luxburg U (2007) Stat Comput 17(4):395
3. Newman MEJ (2006) Phys Rev E 74(3):036104
4. Rosvall M, Bergstrom C (2008) Proc Natl Acad Sci USA 105(4):1118
5. Palla G, Derényi I, Farkas I, Vicsek T (2005) Nature 435(7043):814
6. Fortunato S, Barthélemy M (2007) Proc Natl Acad Sci USA 104(1):36
7. Good BH, de Montjoye YA, Clauset A (2010) Phys Rev E 81(4):046106
8. Kawamoto T, Rosvall M (2015) Phys Rev E 91:012809
9. Leskovec J, Lang KJ, Anirban D, Mahoney MW (2009) Internet Math 6(1):29
10. Lancichinetti A, Fortunato S (2009) Phys Rev E 80(5), 056117
11. Aldecoa R, Marín I (2013) Sci Rep 3:2216
12. Holland PW, Laskey KB, Leinhardt S (1983) Soc Networks 5(2):109
13. Airoldi EM, Blei DM, Fienberg SE, Xing EP (2008) J Mach Learn Res 9:1981
14. Karrer B, Newman MEJ (2011) Phys Rev E 83(1):016107
15. Ball B, Karrer B, Newman MEJ (2011) Phys Rev E 84(3):036103
16. Peixoto TP (2012) Phys Rev E 85(5):056122
17. Reichardt J, Leone M (2008) Phys Rev Lett 101(7):078701
18. Decelle A, Krzakala F, Moore C, Zdeborová L (2011) Phys Rev Lett 107(6):065701
19. Decelle A, Krzakala F, Moore C, Zdeborová L (2011) Phys Rev E 84(6):066106
20. Nadakuditi RR, Newman MEJ (2012) Phys Rev Lett 108(18):188701
21. Zhang P, Krzakala F, Reichardt J, Zdeborová L (2012) J Stat Mech 2012(12):P12021
22. Newman MEJ (2013) Phys Rev E 88(4):042822

23. Kabashima Y, Takahashi H (2012) J Phys A 45(32):325001
24. Kawamoto T, Kabashima Y (2015) Limitations in the spectral method for graph partitioning: Detectability threshold and localization of eigenvectors. Phys Rev E 91(062):803 doi:10.1103/PhysRevE.91.062803
25. Krzakala F, Moore C, Mossel E, Neeman J, Sly A, Zdeborová L, Zhang P (2013) Proc Natl Acad Sci USA 110(52):20935

Chapter 13
Spread of Infectious Diseases with a Latent Period

Kanako Mizuno and Kazue Kudo

Abstract Infectious diseases spread through human networks. Susceptible-Infected-Removed (SIR) model is one of the epidemic models to describe infection dynamics on a complex network connecting individuals. In the metapopulation SIR model, each node represents a population (group) which has many individuals. In this paper, we propose a modified metapopulation SIR model in which a latent period is taken into account. We call it SIIR model. We divide the infection period into two stages: an infected stage, which is the same as the previous model, and a seriously ill stage, in which individuals are infected and cannot move to the other populations. The two infectious stages in our modified metapopulation SIR model produce a discontinuous final size distribution. Individuals in the infected stage spread the disease like individuals in the seriously ill stage and never recover directly, which makes an effective recovery rate smaller than the given recovery rate.

13.1 Introduction

Infectious diseases spread through human networks. Susceptible-Infected-Removed (SIR) model is one of the epidemic models to describe infection dynamics on a complex network connecting individuals. The ratio of the transmission rate to the recovery rate is called the basic reproduction number R_0. It is the expected number of infections caused by a typical infectious individual in a completely susceptible population [1, 2]. In the standard SIR model, the outbreak occurs when $R_0 > 1$. The likely magnitude of the outbreak, which is called the expected final size of the epidemic, depends only on R_0 [2, 3].

The spread of infectious diseases also depends on human mobility. In metapopulation SIR models, movements between different populations (groups) are taken into account [4–6]. Each node of the metapopulation network represents a group

K. Mizuno • K. Kudo (✉)
Department of Computer Science, Ochanomizu University, Tokyo, Japan
e-mail: mizuno.kanako@is.ocha.ac.jp; kudo@is.ocha.ac.jp

H. Takayasu et al. (eds.), *Proceedings of the International Conference on Social Modeling and Simulation, plus Econophysics Colloquium 2014*, Springer Proceedings in Complexity, DOI 10.1007/978-3-319-20591-5_13

of individuals. Individuals can move between two nodes connected by a link. Although the epidemic threshold is R_0 in each group, the global invasion threshold in the metapopulation system depends on the mobility rate as well as its network structure [5, 6].

In this paper, we propose a modified metapopulation SIR model in which a latent period is taken into account. Infected individuals behave like susceptible ones when they do not feel sick. They move between linked populations and spread diseases across different populations. We consider that such infected individuals are in a latent period. We assume that infected individuals become too sick to move after the latent period. Such ill individuals infect only the susceptible ones in the same population. This model is different from the SEIR model [7], which is a common epidemic model in which a latent period is incorporated as an "Exposed" state. However, it belongs to a family of generalized SIR models that include multiple infectious stages [2]. The two infectious stages in our modified metapopulation SIR model produce a discontinuous final size distribution with a jump at $R_0 = 1$.

The rest of the paper is organized as follows. The metapopulation SIR model and the modified SIR model are introduced in Sect. 13.2. We demonstrate the discontinuous final size distribution of the modified model in Sect. 13.3. The effective recovery rate, which is different from the given recovery rate, is estimated, and it is the key to find what causes the discontinuity. Discussions and conclusions are given in Sect. 13.4.

13.2 Model

First we introduce a metapopulation SIR model, which is an SIR model that is extended to metapopulation networks. In the metapopulation SIR model, each node represents a population (group) which has many individuals, and each individual is in one of three states: S (susceptible), I (infected) or R (recovered). Individuals of state S are infected by those of state I in the same population. The infection rate is given by $\alpha I_m / N_m$, where $N_m = S_m + I_m + R_m$ with S_m, I_m, and R_m being the number of susceptible, infected, and recovered individuals of population m, respectively. In other words, the rate that S becomes I depends on the transmission rate α and the proportion of I in the same population. The constant rate that I becomes R, i.e., recovery rate, is defined as β. We here assume that all individuals move between the populations connected with links in the network at a constant rate w. The travel rate

w is the same for all the individuals. The time evolution of the numbers of S, I and R in each population is described by

$$\partial_t S_n = -\alpha S_n I_n / N_n + w \sum_m (S_m - S_n), \tag{13.1a}$$

$$\partial_t I_n = \alpha S_n I_n / N_n - \beta I_n + w \sum_m (I_m - I_n), \tag{13.1b}$$

$$\partial_t R_n = \beta I_n + w \sum_m (R_m - R_n), \tag{13.1c}$$

where the summations are taken over all the populations connected with population n.

Next, we divide the infection period into two stages: an infected stage, which is the same as the previous model, and a seriously ill stage, in which individuals are infected and cannot move to the other populations. We call this model SIIR model. In this model, each individual is in one state of S (susceptible), H (infected), I (seriously ill), and R (recovered). Individuals of S in population m are infected and become H at rate $\alpha(H_m + I_m)/N_m$, where $N_m = S_m + H_m + I_m + R_m$. Individuals of H become I at a constant rate μ. Individuals of I recover and become R at a rate β. In the SIIR model, individuals of H move between the populations connected with links at a rate w, however, individuals of I do not. The time evolution of the numbers of S, H, I and R in each population is described by

$$\partial_t S_n = -\alpha S_n (H_n + I_n)/N_n + w \sum_m (S_m - S_n), \tag{13.2a}$$

$$\partial_t H_n = \alpha S_n (H_n + I_n)/N_n - \mu H_n + w \sum_m (H_m - H_n), \tag{13.2b}$$

$$\partial_t I_n = \mu H_n - \beta I_n, \tag{13.2c}$$

$$\partial_t R_n = \beta I_n + w \sum_m (R_m - R_n), \tag{13.2d}$$

where the summations are taken over all the populations connected with population n.

13.3 Final Size Distribution

The spread of a disease is expressed by attack ratio, which is the final proportion of R when I disappears in the entire metapopulation. The attack ratio plotted as the function of the basic reproduction number α/β is called a final size distribution. The final size distributions of the SIR model and SIIR models are shown in Fig. 13.1. In

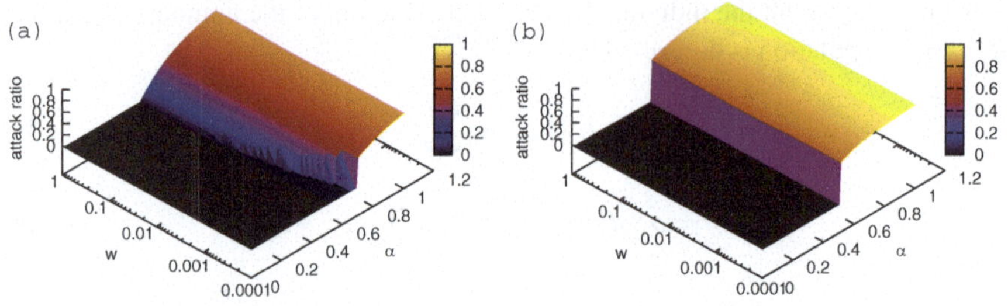

Fig. 13.1 Final size distributions of (**a**) metapopulation SIR model and (**b**) metapopulation SIIR model as the function of the transmission rate α and the travel rate w. In both cases, the recovery rate is $\beta = 0.5$

this simulation, the number of individuals in each state is taken as a real number and the time step is discrete. We use a scale-free network with 900 nodes, whose degree distribution is $P(k) \sim k^{-\gamma}$ with $\gamma = 2.5$. The essential results do not depend on γ. In the initial state, 100 susceptible individuals belong to each node except for one randomly selected node in which one infected individual is included. The global invasion does not occur when $\alpha < \beta$ in the SIIR model as well as the SIR model. The change in attack ratio is continuous at $\alpha = \beta$ in the high-w region in the SIR model, however, it is discontinuous in all region in the SIIR model. The shift of threshold in the low-w regions of the SIR model is often observed in metapopulation networks [5, 6].

In this paper, we focus on the discontinuous final size distribution of the SIIR model. The jump in the attack ratio arises from the difference between the given recovery rate and an effective recovery rate. In the SIIR model, individuals H spread the disease like individuals I and never become R directly. Therefore, the effective recovery rate β' is expected to be smaller than the given recovery rate β.

We show how to evaluate β' below. Disregarding traveling between populations, the SIIR model (13.2) is rewritten as

$$\partial_t S = -\alpha S(H + I), \tag{13.3a}$$

$$\partial_t H = \alpha S(H + I) - \mu H, \tag{13.3b}$$

$$\partial_t I = \mu H - \beta I, \tag{13.3c}$$

$$\partial_t R = \beta I, \tag{13.3d}$$

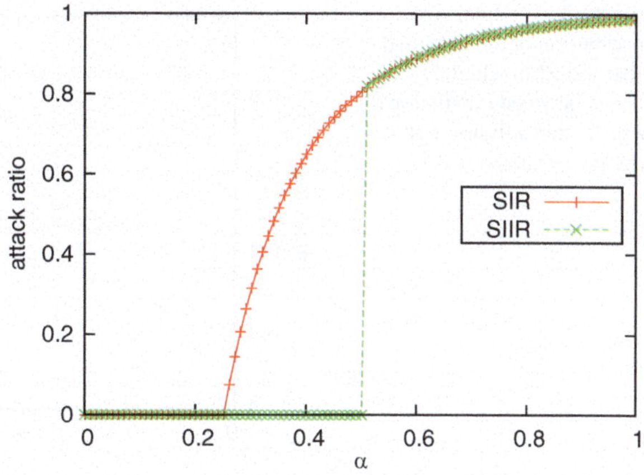

Fig. 13.2 The final size distribution as the function of the transmission rate α for the SIR model with the given recovery rate $\beta = 0.25$ is compared with that for the SIIR model with the effective recovery rate $\beta' = 0.25$, which is calculated from Eq. (13.4) with $\beta = 0.5$ and $\mu = 0.5$. Both curves agree in the region where $\alpha > 0.5$. The travel rate $w = 0.1$ for both curves

where $S = S_n/N_n$, $H = H_n/N_n$, $I = I_n/N_n$ and $R = R_n/N_n$. Combining Eqs. (13.3b) and (13.3c), we have

$$\partial_t(H + I) = \alpha S(H + I) - \beta'(H + I),$$
$$\beta' = \frac{I}{H + I}\beta.$$

We here take $\partial_t I = 0$, which leads to $H = (\beta/\mu)I$. Then, the effective recovery rate is calculated as

$$\beta' = \frac{\mu}{\beta + \mu}\beta. \tag{13.4}$$

Figure 13.2 illustrates that the evaluation of the effective recovery rate is appropriate. The simulation is performed in the same network with the same initial condition as Fig. 13.1. The travel rate is $w = 0.1$, which is in the high-w region. The attack ratio for the SIIR model is calculated for $\beta = 0.5$ and $\mu = 0.5$. In this case, the effective recovery rate is $\beta' = 0.25$. The final size distribution for the SIR model with the given recovery rate $\beta = 0.25$ agrees with that for the SIIR model in the region where $\alpha > 0.5$. This result implies the following. The effective recovery rate in the SIIR model is given by β', however, global invasion cannot occur when $\alpha < \beta$. The difference between β and β' causes the discontinuous final size distribution of the SIIR model.

Since we disregarded traveling between populations when we evaluate the effective recovery rate, the assumption that I is immobile should be irrelevant to the discontinuity in the final size distribution of the SIIR model. We now modify the SIIR model (13.2), replacing Eq. (13.2c) by

$$\partial_t I_n = \mu H_n - \beta I_n + w\sum_m (I_m - I_n). \tag{13.5}$$

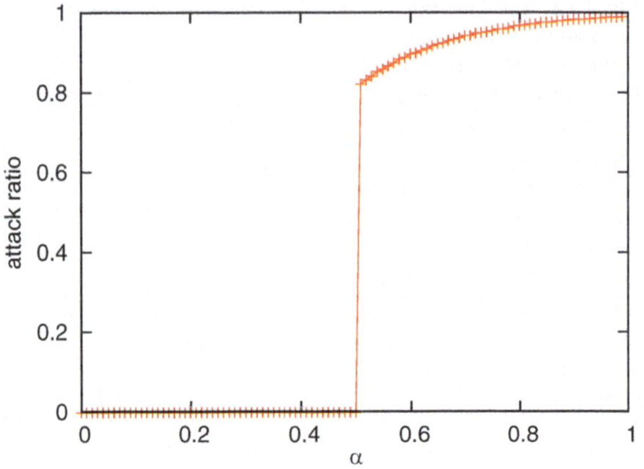

Fig. 13.3 The final size distribution of the modified SIIR model in which H moves between populations. α is the transmission rate. The travel rate $w = 0.1$

Figure 13.3 shows the final size distribution of the modified SIIR model. The simulation is performed in the same conditions as Fig. 13.2. The profile of the SIIR curve in Fig. 13.2 looks the same as the curve in Fig. 13.3. Therefore, the cause of the discontinuous final size distribution is the division of the infection period into two stages, and the mobility of I has no effect on the discontinuity.

13.4 Discussions and Conclusions

The effective recovery rate β', which is given by Eq. (13.4), can be evaluated by another way. The basic reproduction number for the generalized SIR model that includes n infectious stages is given by

$$R_0 = \sum_{i=1}^{n} \frac{\alpha_i}{\beta_i}, \tag{13.6}$$

where α_i is the transmission rate of the ith infectious stage, and $1/\beta_i$ is the mean duration of the stage [2, 8]. In our SIIR model, $\alpha_1 = \alpha_2 = \alpha$, $\beta_1 = \mu$ and $\beta_2 = \beta$, and thus, $R_0 = \alpha/\mu + \alpha/\beta = \alpha(\mu + \beta)/(\mu\beta)$. Therefore,

$$\beta' = \frac{\alpha}{R_0} = \frac{\mu\beta}{\mu + \beta}, \tag{13.7}$$

which is the same as Eq. (13.4).

In conclusion, the discontinuous final size distribution in the SIIR model is caused by the division of the infection period into two stages and the fact that the global invasion cannot occur when $\alpha < \beta$. The final size distribution depends on the effective recovery rate β', and its shape coincides with that of the SIR model with a recovery rate $\beta = \beta'$ in the region where $\alpha > \beta$.

Acknowledgements We would like to thank H. Takayasu and H. Nishiura for valuable suggestions and comments.

References

1. Anderson RM, May RM (1991) Infectious diseases of humans: dynamics and control. Oxford University Press, Oxford
2. Ma J, Earn DJD (2006) Bull Math Biol 68:679
3. Anderson D, Watson R (1980) Biometrika 67:191
4. Keeling MJ, Rohani P (2002) Ecol Lett 5:20
5. Cross PC, Lloyd-Smith JO, Johnson PLF, Getz WM (2005) Ecol Lett 8:587
6. Colizza V, Vespignani A (2007) Phys Rev Lett 99:148701
7. Schwartz I, Smith H (1983) J Math Biol 18:233
8. Hyman JM, Li J, Stanley EA (1999) Math Biosci 155:77

Part III
Interaction and Distribution

Chapter 14
Geographic Dependency of Population Distribution

Shouji Fujimoto, Takayuki Mizuno, Takaaki Ohnishi, Chihiro Shimizu, and Tsutomu Watanabe

Abstract The agglomeration effect of population, which explains why many people live near places where many other people also live, is one important interaction that influences human population. We examine the agglomeration effect by measuring the distribution of the logarithmic differences between populations living in two places separated by some distance. The shapes of the distributions of the logarithmic differences closely resemble each other without depending on the regions or the country in cases of small scale of separation distance. This result suggests a unified explanation to understand the population distributions of various regions.

S. Fujimoto (✉)
Faculty of Business Administration and Information Science, Kanazawa Gakuin University, Kanazawa, Ishikawa, Japan
e-mail: fujimoto@kanazawa-gu.ac.jp

T. Mizuno
Department of Informatics, National Institute of Informatics, Chiyoda-ku, Tokyo, Japan
Graduate University for Advanced Studies, Tokyo, Japan

The Canon Institute for Global Studies, Marunouchi, Chiyoda-ku, Tokyo, Japan
e-mail: mizuno@nii.ac.jp

T. Ohnishi
Graduate School of Information Science and Technology, The University of Tokyo, Bunkyo-ku, Tokyo, Japan

The Canon Institute for Global Studies, Chiyoda-ku, Tokyo, Japan
e-mail: ohnishi.takaaki@i.u-tokyo.ac.jp

C. Shimizu
Institute of Real Estate Studies, National University of Singapore, Singapore
e-mail: cshimizu@nus.edu.sg

T. Watanabe
Graduate School of Economics, The University of Tokyo, Bunkyo-ku, Tokyo, Japan

The Canon Institute for Global Studies, Chiyoda-ku, Tokyo, Japan
e-mail: watanabe@e.u-tokyo.ac.jp

H. Takayasu et al. (eds.), *Proceedings of the International Conference on Social Modeling and Simulation, plus Econophysics Colloquium 2014*, Springer Proceedings in Complexity, DOI 10.1007/978-3-319-20591-5_14

14.1 Introduction

Population distribution has been studied for many decades. Zipf's law [1], which argues that the size distribution of a city's population is a power-law, is known well [2–6]. However, a problem exists: how to define the area of cities when we observe population distributions. The tail of a power-law distribution is composed of megacities. By dividing megacities into several smaller cities, the distribution's tail becomes thin. Because of the different definitions of a city, population distribution is not a power-law distribution but a log-normal one [7–9]. City areas have been decided by geographical, historical, and administrative factors. Rozenfeld et al. proposed a method that decided a city's area by a city clustering algorithm [10]. In this research, we divide spatial regions by a method that ignores the shape of cities to find the properties of population distribution that do not depend on countries or local regions.

We investigated population distribution using a spatial division method by identically sized squares. This approach resembles a previous method [9]. In our case, we control the scale of the spatial division by changing the size of the squares and clarify the universal properties concerned with population agglomeration. Population's universal properties can be observed by changing the scale of the spatial division.

We introduce logarithmic differences between the nearest neighbor two square blocks in terms of population. The regional dependence of these values in terms of the shape of the distributions vanishes for small size scales. The property of the distribution of logarithmic differences is concerned with the correlation coefficient of the population in two squares. This correlation is one index to measure population agglomeration.

In this research, we investigate Japanese population data. In Sect. 14.2, we introduce eight regions to investigate local properties inside Japan. In Sect. 14.3, we compare several distributions concerned with population among these eight regions. Next we compare Japan and Europe in Sect. 14.5 and show the universal properties concerned with population in both cases.

14.2 Basic Information of Japanese Population

The Statistics Bureau of the Japanese Ministry of Internal Affairs and Communications conducts a census every five years. Much census data can be obtained in a mesh data format from its websites [11], including population data from 2000, 2005, and 2010. The mesh data are raster data that are obtained by equally dividing latitudes and longitudes. A mesh size of about $500 \times 500\,\mathrm{m}^2$ provides the highest accuracy for population. Mesh codes are assigned to each bit of data, and we can specify the data's position on the map from this code.

Table 14.1 Land use code assignment

Code : land use description	Code : land use description
1 : Paddy fields	A : Other land
2 : Other agricultural land	B : Rivers and lakes
5 : Forest	E : Beach
6 : Wasteland	F : Body of seawater
7 : Land for building	G : Golf course
9 : Trunk transportation land	0 : outside of the analysis range

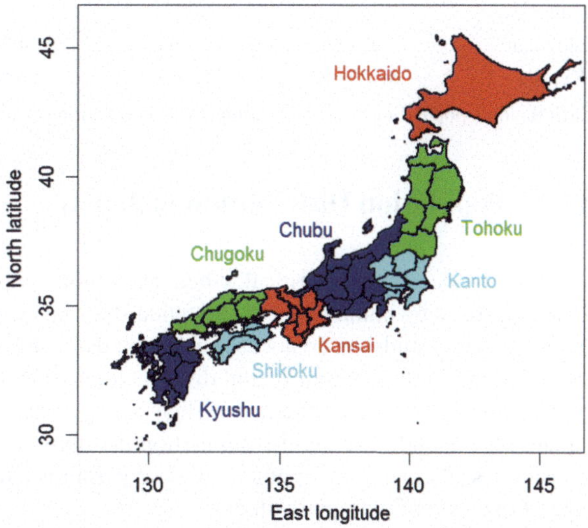

Fig. 14.1 Map of the eight regions of Japan. From north to south: Hokkaid (*red*), Tohoku (*green*), Kanto (*cyan*), Chubu (*blue*), Kansai (*red*), Chugoku (*green*), Shikoku (*cyan*) and Kyushu (*blue*). *Black lines* indicate prefecture borders

The Japanese Ministry of Land, Infrastructure, Transport and Tourism provides land use data on its website [12]. Such data are also provided in a mesh data format. A mesh of about 100 [m]×100 [m] provides the highest accuracy for land use. In these data, a land use code (see Table 14.1) is assigned to each mesh. An inhabitable place is defined as any place that is fit for humans to live in. Inhabitable areas can be estimated by subtracting such uninhabitable areas as forests and lakes from the land area. We estimated the inhabitable areas by totaling the areas whose land use codes are 1, 2, 7, 9, A, and G. Only about 33 % of Japan's land area is inhabitable because it has many mountainous areas. This percentage is smaller than European countries. For example, the inhabitable area percentages of Germany, France, and the United Kingdom are 68 %, 71 %, and 88 %, respectively. We have to use inhabitable areas instead of land areas to more precisely evaluate population density.

To investigate locality and universality, we divided Japan into the following eight regions based on traditional ways of combining several prefectures (see Fig. 14.1): Hokkaido, Tohoku, Kanto, Chubu, Kansai, Chugoku, Shikoku, and Kyushu. Table 14.2 shows the basic information of the eight regions. Population densities depend on the regions for various reasons.

Table 14.2 Basic information of eight Japanese regions

Region	Population in 2010	Land area km^2	Inhabitable area km^2	Density $/km^2$
All JP data	128,045,367	372, 907	121,941	1050.1
Hokkaido	5,506,197	83, 456	27,046	203.6
Tohoku	9,337,024	66, 890	20,306	459.8
Kanto	42,608,322	32, 424	18,256	2333.9
Chubu	21,721,795	66, 799	22,970	945.7
Kansai	22,758,142	33, 118	8,495	2679.0
Chugoku	7,563,164	31, 920	8,427	897.5
Shikoku	3,977,562	18, 805	4,860	818.4
Kyushu	4,589,693	44, 455	16,542	882.0

Each region's population density is estimated by population per inhabitable area

14.3 Population Distribution in Japan

How to divide space is critical when examining population's size distribution. Dividing space by municipal level is standard for investigating the size distributions of cities. In this study we do not use such spatial division method. We adopted square blocks of the same size as a spatial division method and divided a particular region into identical sized square lattices. Then we aggregated the population inside the square blocks and observed its population distribution. We can control the spatial division's scale using this method. We use parameter BS [km], which denotes the size of one side of the square blocks.

Figure 14.2 shows a complementary cumulative distribution function (CCDF)

$$\Pr\{X \geq x\} \tag{14.1}$$

of Japan's population in 2010. The distributions of the regions with high-density populations such as Kanto and Kansai are plotted on the right side compared to other regions. The distributions of the regions with low-density populations such as Hokkaido are plotted on the left side compared to other regions. These properties denote the distribution locality. The population distributions vary by region.

To find the distribution quantities that do not depend on the region, we focused on the population distribution's shape. For a small scale (BS $= 0.5$ [km]), the right tail of the distributions rapidly falls. As BS becomes larger, the right tail of the distributions becomes gentler. The slopes of the right tail seem close to each other for a small BS. The value of the logarithmic differences between populations whose values are close to each other seems to share similar quantities of population distribution slopes.[1]

[1]It is possible to confirm of this expectation by left figure of Fig. 14.6.

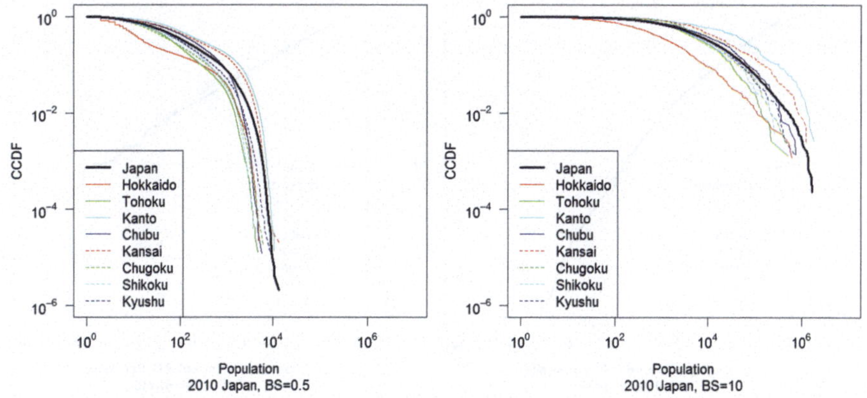

Fig. 14.2 Log-log plot of population distributions. *Left figure* is for BS = 0.5 [km]. *Right figure* is for BS = 10 [km]. Thick *black curves* show all Japanese distributions in 2010. *Thin colored curves* show all distributions of eight regions in 2010

We use $S(x, y)$ to denote the population inside a square whose vertex coordinates are (x, y), $(x + \mathrm{BS}, y)$, $(x + \mathrm{BS}, y + \mathrm{BS})$, and $(x, y + \mathrm{BS})$. The logarithmic difference between the populations of nearest neighbors in x-direction is represented by

$$\ln S(x + \mathrm{BS}, y) - \ln S(x, y), \tag{14.2}$$

and the logarithmic difference in y-direction is represented by

$$\ln S(x, y + \mathrm{BS}) - \ln S(x, y). \tag{14.3}$$

The logarithmic difference is a value that is frequently used in such time-series analyses as stock prices [13]. In this paper we apply it to spatial directions. The effects of the differences are the same regardless whether the difference direction is positive or negative in terms of the spatial direction. Next we investigate the distributions of the absolute value of the logarithmic differences.

Figure 14.3 shows the CCDF of the absolute value of the logarithmic differences between the nearest neighbor populations in Japan in 2010. For small scale (BS = 0.5 [km]), the distributions almost overlap. As BS becomes larger, the right tail of the distributions becomes gentler, and they no longer overlap.

Figure 14.4 shows the BS dependence of the moments values of the distributions of absolute value of logarithmic differences. Where n-th order moments is defined by mean of n-th powered of the stochastic variable. These values are one of the quantitative index of the overlapping of the distributions.

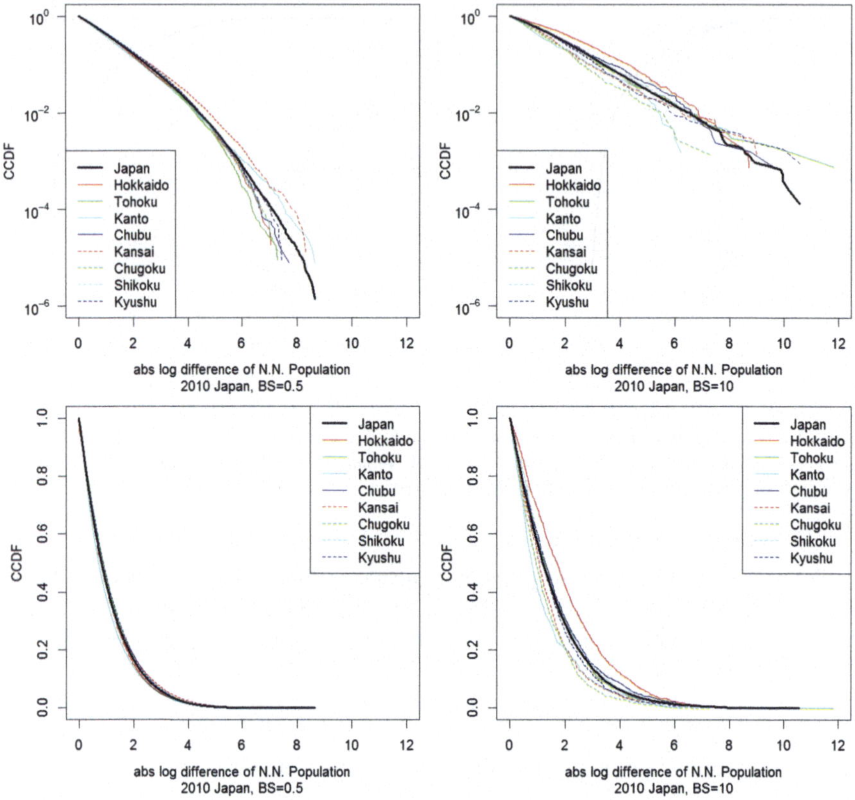

Fig. 14.3 Distributions of absolute value of logarithmic differences between nearest neighbor populations. *Top figures* are semi-log plot. *Bottom figures* are non-log plot. *Left figures* are for BS = 0.5 [km]. *Right figures* are for BS = 10 [km]. *Thick black curves* show all Japanese distribution in 2010. *Thin colored curves* show distributions of all eight regions in 2010

Figure 14.5 compares the observed distribution and the distributions represented by analytic functions. The red lines show an exponential distribution whose CCDF is defined by

$$\Pr\{X \geq x\} = \int_x^\infty \frac{1}{\mu} \exp\left(-\frac{t}{\mu}\right) dt. \tag{14.4}$$

Here parameter μ is the distribution's mean. The estimated values from the data are $\mu = 1.1022$ for BS = 0.5 and $\mu = 1.5629$ for BS = 10. The blue curves show truncated normal distribution, whose CCDF is defined by

$$\Pr\{X \geq x\} = \int_x^\infty \sqrt{\frac{2}{\pi\sigma^2}} \exp\left(-\frac{t^2}{2\sigma^2}\right) dt. \tag{14.5}$$

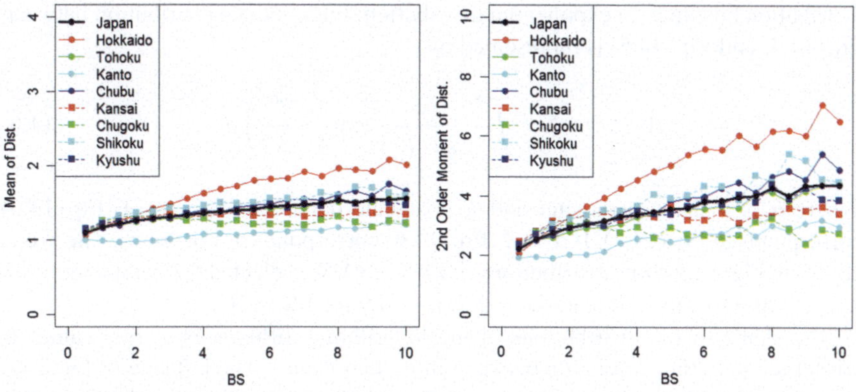

Fig. 14.4 BS dependence of the moments values of the distributions of absolute value of logarithmic differences. *Left figure* shows the 1st order moments. *Right figure* shows the 2nd order moments. *Black symbols* show the moments values of the Japanese distributions. *Colored symbols* show the moments values of the distributions of eight regions

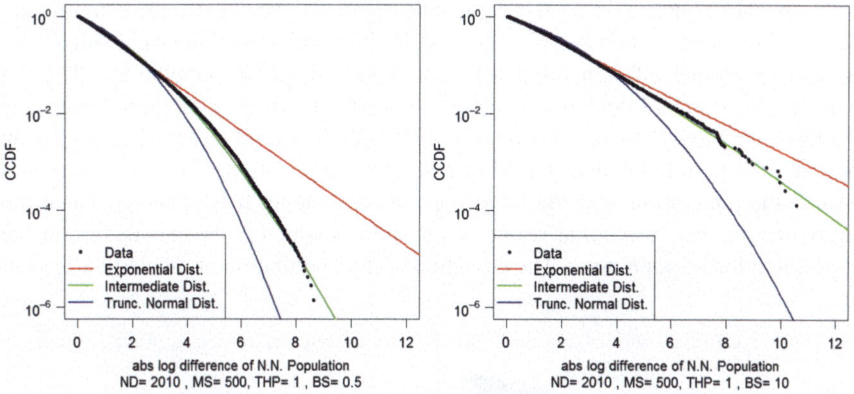

Fig. 14.5 Distributions of absolute value of logarithmic differences between nearest neighbor populations. *Left figure* is for BS = 0.5 [km]. *Right figure* is for BS = 10 [km]. *Black circles* show distributions observed from all Japanese data in 2010. *Red lines* show exponential distributions whose means match observed data. *Blue curves* show truncated normal distributions whose standard deviations match observed data. *Green curves* show intermediate distributions between *red curves* and *blue curves*

Here parameter σ is the standard deviation from the $x = 0$ of the distribution. The estimated values from the data are $\sigma = 1.4853$ for BS $= 0.5$ and $\sigma = 2.0944$ for BS $= 10$. The shape of the distributions seems to be intermediate between the exponential and the truncated normal distributions. The distributions resemble a truncated normal distribution in a small BS scale. As BS becomes larger, the distribution becomes an exponential distribution. Intermediate distribution between Eq. (14.4) and Eq. (14.5) is represented by

$$\Pr\{X \geq x\} = \int_x^{\infty} \frac{\alpha}{\lambda \Gamma\left(\frac{1}{\alpha}\right)} \exp\left(-\frac{t^{\alpha}}{\lambda^{\alpha}}\right) dt. \tag{14.6}$$

Where α is a shape parameter and λ is a scale parameter. If $\alpha = 1$, Eq. (14.6) corresponds to Eq. (14.4). If $\alpha = 2$, Eq. (14.6) corresponds to Eq. (14.5). The green curves in Fig. 14.5 show distributions of Eq. (14.6). We selected the parameters $\alpha = 1.6, \lambda = 0.9$ for BS = 0.5 and $\alpha = 1.2, \lambda = 0.9$ for BS = 10.

The shape of the distributions of the logarithmic differences of two values is concerned with the correlation between those two values. The left side of Fig. 14.6 shows a scatter plot of $\ln S(x, y)$ versus $\ln S(x + \mathrm{BS}, y)$ or $\ln S(x, y + \mathrm{BS})$. From this figure, we observe agglomeration effect that many people live near places where many other people also live. The correlation coefficient is able to interpret as an index of agglomeration effect. The right side figure's data are transformed from the left side figure's data by dilating both axis data $\sqrt{2}$ and rotating clockwise 45°. The horizontal axis of the right side figure is the logarithmic summation between the nearest neighbor populations. The vertical axis of the right side figure is the logarithmic difference between the nearest neighbor populations. The red bars are the standard deviation inside each segment, which is equally divided by the horizontal axis. The correlation of the left side figure represents the correlation between the population and the nearest neighbor population. If this correlation is strong, the population near the large population is large. It is considered that the strengthen of this correlation is one of the indices which represents degree of the agglomeration of population. The deviation of the distribution of the vertical axis of

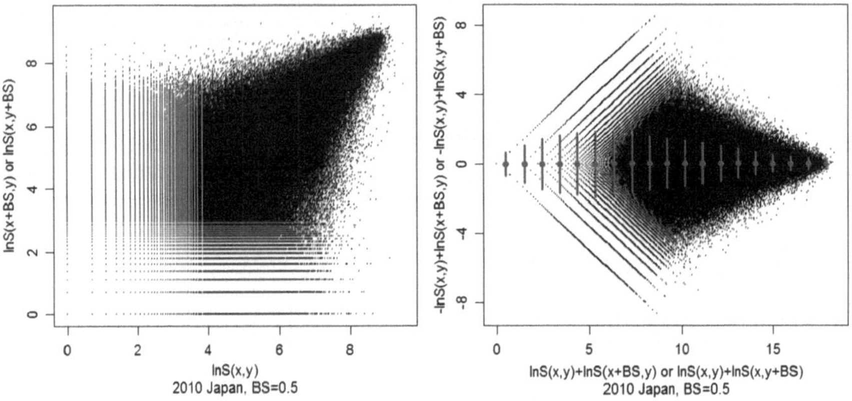

Fig. 14.6 Left side figure shows scatter plot of $\ln S(x, y)$ versus $\ln S(x + \mathrm{BS}, y)$ or $\ln S(x, y + \mathrm{BS})$. Correlation coefficient of these data is 0.69. Right side figure's data are transformed from left side figure's data by expansion and rotation. *Red circles* are means inside each segment that is equally divided by horizontal axis. *Red bars* are standard deviation inside each segment

Table 14.3 Basic information of top seven European countries by population

Region	Population in 2011	Land area km^2	Inhabitable area km^2	Density $/km^2$
All EU data	514,988,832			
Germany	80,122,036	348,560	237,800	336.9
France	62,623,425	547,557	387,537	161.6
U.K.	62,583,331	241,930	213,048	293.8
Italy	59,315,222	294,140	201,870	293.8
Spain	46,802,562	498,800	315,307	148.4
Poland	38,449,414	306,230	212,586	180.9
Romania	16,609,793	230,170	164,076	101.2

Population density in each region is estimated by population per inhabitable area

the right side figure concerns the correlation of the left side figure. The deviation of the distribution of the vertical axis of the right side figure shrinks when the correlation of the left side figure becomes strong. It is possible to estimate the degree of agglomeration of the population by observing the deviation of the distribution of the logarithmic difference.

14.4 Basic Information of European Populations

The European Union provides several kinds of statistical data from eurostat. The GEOSTAT project provides European countries' population dataset representing in a 1 km^2 grid dataset. Population data for 2006 and 2011 are available on their website [14].

The food and agriculture organization of the United Nations statistics division (FOSTAT) [15] provides land and forest area data from most countries. We can roughly estimate the inhabitable areas by subtracting forest areas from land areas.

Table 14.3 shows the basic information of the top seven European countries by population. Their population density is lower than Japan. The variation of the population density of each country is smaller than the variation of all eight Japanese regions.

14.5 Comparison between Japan and European Countries

In this section we compare Japan and European countries in terms of the distribution of log differences of population. Figure 14.7 shows the CCDF of the absolute value of the logarithmic differences between the nearest neighbor population of Japan and EU countries. The results are almost the same as those among Japan's eight

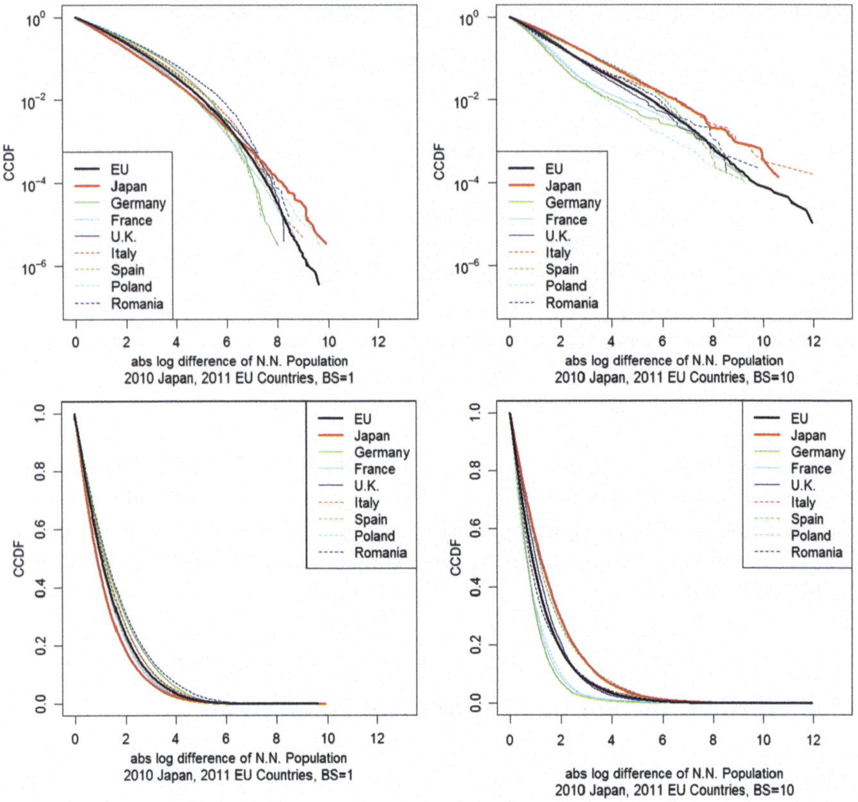

Fig. 14.7 Distributions of absolute value of logarithmic differences between nearest neighbor populations. *Top figures* are semi-log plot. *Bottom figures* are non-log plot. *Left figures* are for BS = 0.5. *Right figures* are for BS = 10. *Thick black curves* show all EU distributions in 2011. *Thick red curves* show Japanese distribution in 2010. *Thin colored curves* show all seven European countries' distributions in 2011

regions. As BS becomes larger, the right tail of the distributions becomes gentler. The overlapping of the distributions for BS = 1 is better than for BS = 10. If we observed data whose scale BS = 0.5, the overlapping would be better than for BS = 1.

The transitions of the distributions due to changes by BS are shown in Fig. 14.8. Japan's distribution shape is almost the same as that of EU at a small BS. The difference of Japan and EU becomes larger as BS increases.

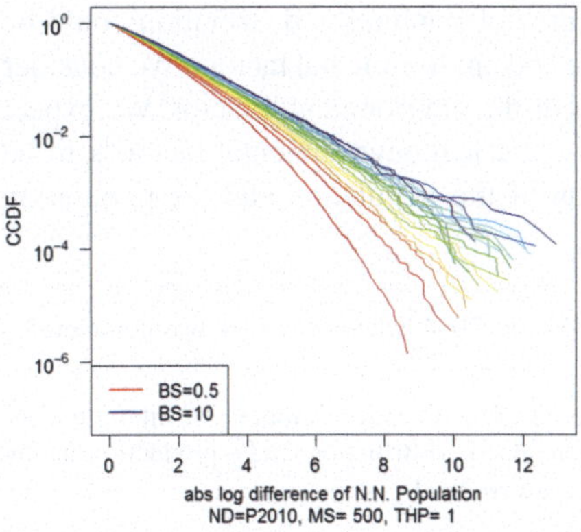

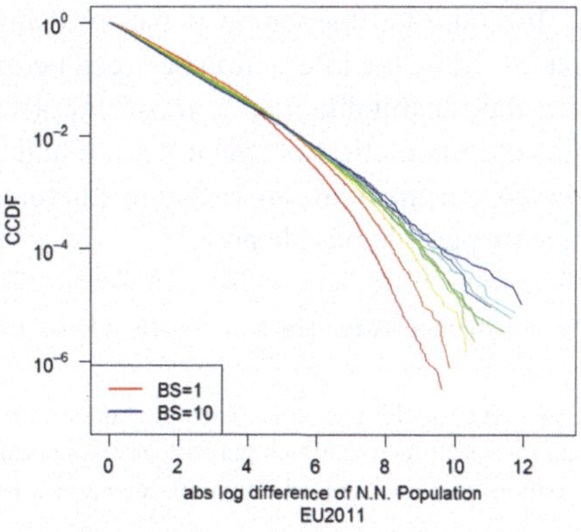

Fig. 14.8 Distributions of absolute value of logarithmic differences between nearest neighbor populations. Color gradation of curves represents size of BS. *Red*, *green*, and *blue curves*l show small, intermediate, and large sizes, respectively. *Left figure* shows distributions of Japan with BS from 0.5 to 10 by 0.5 increments. *Right figure* shows distribution of all EU with BS from 1 to 10 by 1

14.6 Conclusion

We investigated population distributions using Japan and EU data. Using a spatial division method with same size squares, we can easily control the division scale. The shape of the population distribution differs by country or region. We introduce logarithmic differences between nearest neighbor populations to identify distributions that do not depend on country or region. When the division scale is large, the distribution of logarithmic differences depends on the country or region. The local dependence of the distribution disappears as the division scale becomes smaller. The distribution's shape closely resembles a normal distribution when the division scale is small; it is close to exponential distribution when the division scale is large.

This study investigated population distributions from a universal standpoint that does not depend on country or region. In general, various interactions determine population distribution. These interactions can be divided into two types. One is internal interactions, and the other is external interactions. External interactions are such environmental elements as topography and habitability. Internal interactions are interactions between people. Our results suggest that a universal feature exists for interaction with a small-scale neighboring population.

The next stage of our study will reproduce the results of Fig. 14.8 using a simple model. If we generate population data randomly, BS dependence of the shape of the distributions of logarithmic differences are quite different from Fig. 14.8. To reproduce the BS dependence of Fig. 14.8, we have to generate population configuration which satisfy the left figures of Fig. 14.6. We will have to introduce interactions between people to generate the agglomeration effect.

It would be interesting if the local features of population distribution could be explained by the interaction between people and environmental factors. We consider that the inhabitable area is most important in the environmental factor. We expect that the interaction between people and geometrical environmental factor is to be detected from relations between fluctuation of the population and the population density per inhabitable area.

Acknowledgements The authors are grateful to SMSEC 2014, where this work was completed.

References

1. Zipf G (1949) Human behavior and the principle of least-effort. Addison-Wesley, Cambridge
2. Hill BM (1970) J Am Stast Assoc 65(331):1220
3. Kruguman P (1996) J Japanese Int Economies 10:399
4. Zanette DH, Manrubia SC (1997) Phys Rev Lett 79:523
5. Marsili M, Zhang YC (1998) Phys Rev Lett 80(12):2741
6. Gabaix X (1999) Q J Econ 114:739
7. Eeckhout BJ (2004) Am Econ Rev 94(5):1429
8. Kuninaka H, Matsushita M (2008) J Phys Soc Japan 77(11):114801
9. Holmes TJ, Lee S (2009) In: Glaeser E (ed) In the economics of agglomerations, University of Chicago Press, pp 105–131
10. Rozenfeld HD, Rybski D, Gabaix X, Makse HA (2011) Am Econ Rev 101(5):2205
11. Portal site of official statistics of japan. URL http://www.e-stat.go.jp/SG1/estat/eStatTopPortalE.do
12. National land numerical information download service. URL http://nlftp.mlit.go.jp/ksj-e/gml/datalist/KsjTmplt-L03-b.html
13. Mantegna RN, Stanley HE (2002) Nature 376:46
14. Eurostat geostat project. http://ec.europa.eu/eurostat/web/gisco/geostat-project/
15. FOSTAT. http://faostat3.fao.org/download/R/RL/E

Chapter 15
Spatiotemporal Analysis of Influenza Epidemics in Japan

Kazumi Omata and Yoshimitsu Takahashi

Abstract An influenza epidemic is a complicated phenomenon influenced by numerous social and biological factors, including geography, climate, population, transport network, and biological statuses of humans and viruses, among others. To investigate the strength of these influences, we evaluated data from influenza epidemics that occurred in Japan between April 1999 and December 2014 using wavelet analysis. We calculated wavelet transform and phase difference, which was defined as the phase difference in a prefecture other than Tokyo. The time-averaged phase differences revealed the following: (1) the epidemics were earlier and more strongly synchronized in 7 prefectures in the Kanto region, which includes Tokyo, and in 7 prefectures in the Kinki region, which includes Osaka; (2) except for these urban regions, the epidemics propagated from western to eastern prefectures, and finally to northern prefectures; (3) epidemic jumps occurred in several prefectures (e.g., Miyagi Prefecture); and (4) epidemics occasionally occurred at different times in two adjacent prefectures (e.g., Yamanashi and Tokyo). We then attempted to qualitatively deduce the causes for these observations. This study is expected to be important for integrating knowledge to derive trends in epidemics, both nationally and internationally.

15.1 Introduction

Spreading patterns of influenza epidemics, e.g., from city to city, are of great concern because they are important from the viewpoint of not only scientific interest, but also of public health [1, 2]. Public health is associated with intervention or control of epidemics, including vaccinations, masks, and the spatiotemporally

K. Omata (✉)
Center for Clinical Sciences, National Center for Global Health and Medicine, 1-21-1 Toyama, Shinjuku, Tokyo, 162-0052 Japan
e-mail: komata@hosp.ncgm.go.jp

Y. Takahashi
Department of Health Informatics, Kyoto University School of Public Health, Yoshida Konoe, Sakyo, Kyoto, 606-8501 Japan

H. Takayasu et al. (eds.), *Proceedings of the International Conference on Social Modeling and Simulation, plus Econophysics Colloquium 2014*, Springer Proceedings in Complexity, DOI 10.1007/978-3-319-20591-5_15

precise distribution of drug stockpiles. Studies of spreading patterns may also be useful for outbreaks of new types or strains of influenza. Scientific interest primarily relates to the social sciences. The manner in which epidemics propagate must reflect the static and dynamic structures of a society, i.e., forms of human social interaction such as population distribution, transport networks, availability of medical services, and personal travel.

Although epidemic spread is likely affected by stochastic factors, some regularities are evident in the epidemic data. It is important to identify such regularities through quantitative analysis, and then to investigate the causes of these regularities. To date, spatiotemporal analyses of influenza epidemics have been performed in France [3, 4], Scotland [5], 20 European countries [6], Australia [4], Brazil [7], the United States [4, 8–11], China [12], and Japan [13, 14]. These studies identified several regularities in influenza epidemics: west-to-east spread of peak influenza activity, spatiotemporal epidemic synchrony, close correlation with movement of people, relation with mortality pattern, and so on.

Our concern here is whether we can also detect such regularities in influenza epidemics in Japan. If there are regularities, are they similar to or different from those found in other countries? It is interesting to investigate relations of influenza epidemics with effects of environmental characteristics in Japan (described in Sec. 15.2). While Sakai et al. examined the influenza epidemic data in Japan from 1992 to December 1999, using the Kriging method [13], the present study has investigated data from 1999 to 2014, using wavelet analysis. The difference of the data period and analysis is also intriguing. Only a few spatiotemporal studies of influenza epidemics have been performed, and thus the present study is expected to be valuable for integrating knowledge and exploring general regularities in influenza epidemics.

15.2 Materials and Methods

15.2.1 Case Report Data

The present study analyzed weekly case report data for influenza from 46 prefectures in Japan from April 1999 to December 2014 [15] (Okinawa Prefecture was excluded because it comprises remote islands and is located in a subtropical climate with epidemic patterns vastly different from those in the other prefectures). A few typical time series are depicted in Fig. 15.1. For data collection, we used sentinel (or fixed-point) reports: reports are provided by physicians in medical institutions that are assigned and obliged by the government to report how many cases they diagnose.

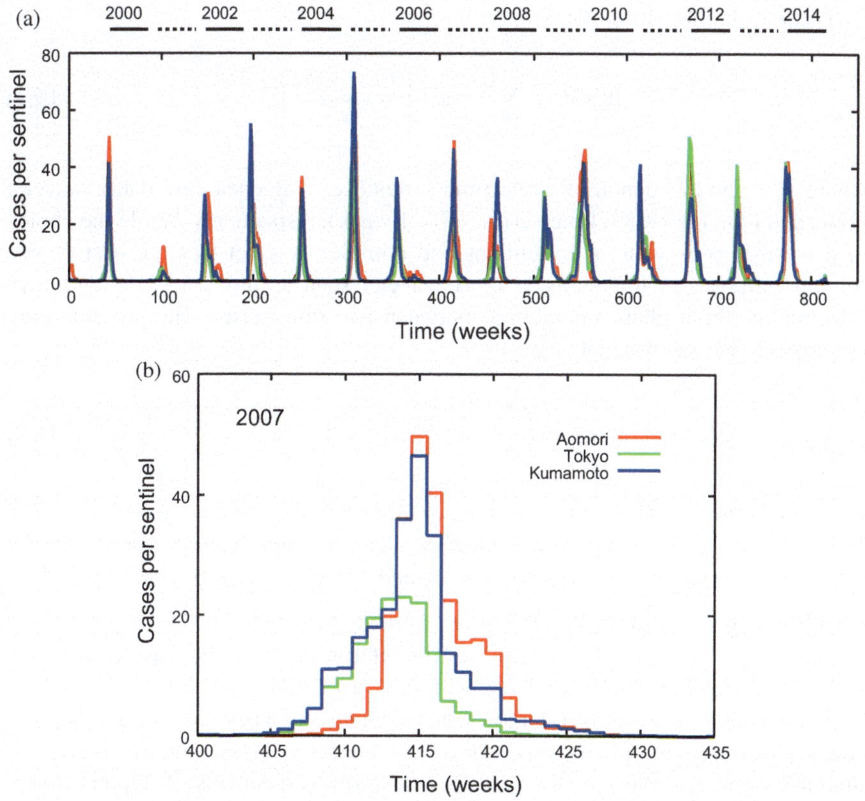

Fig. 15.1 Typical time series for Tokyo (*green*), Aomori (*red*), and Kumamoto (*blue*). (**a**) Data from April 1999 to December 2014. The numbers and bars at the top denote the year and its range, respectively. (**b**) The same data in 2007

15.2.2 Wavelet Analysis

In contrast to Fourier analysis, wavelet analysis [16–18] derives transient relationships between non-stationary time series data. Long-term variations in demography and/or climate affect the epidemic spread of infectious diseases. A few previous studies have examined various diseases using this technique [19–22]. Influenza epidemics are also greatly affected by biological factors. The strain of influenza virus varies annually, and the force of infection depends on the strain. Therefore, the magnitude of infection is different each year. Due to these non-stationary effects, wavelet analysis is well suited to the analysis of influenza epidemics.

The wavelet transform reads

$$W_n(s) = \sum_{n'=0}^{N-1} x_{n'} \psi^* \left[\frac{(n - n')\delta t}{s} \right], \tag{15.1}$$

where x_n, s, and δt denote discrete time series (i.e., influenza case data), wavelet scale, and time interval of time-series ($\delta t = 1$ week), respectively. While the choice of a wavelet ψ is wide, a continuous and complex wavelet was the best choice for our purposes in this study [19]. A wavelet such as this derives quantitative information about phase interactions between two time series. The present study employs the Morlet wavelet

$$\psi(\eta) = \left(\frac{\delta t}{s} \right)^{1/2} \pi^{-1/4} e^{i\omega_0 \eta} e^{-\eta^2/2}, \tag{15.2}$$

where ω_0 is the non-dimensional frequency taken to be 6, and $\eta \equiv (n - n')\delta t/s$. The Morlet wavelet is continuous and complex, and is frequently used. For the Morlet wavelet, the relation between frequencies and wavelet scales is given by $1/f = 4\pi s/(\omega_0 + \sqrt{2 + \omega_0^2}$ [18, 19]. When $\omega_0 \simeq 6$, the wavelet scale is inversely related to frequency, $f \simeq 1/s$, which simplifies the interpretation of the wavelet analysis, and the wavelet scale can be replaced with the wavelength.

The wavelet transform $W_n(s)$ allows the calculation of a power spectrum $|W_n(s)|^2$ and a phase as $\tan^{-1}[\Im\{W_n(s)\}/\Re\{W_n(s)\}]$, where $\Im\{W_n(s)\}$ and $\Re\{W_n(s)\}$ are imaginary and real parts of the wavelet transform, respectively. A typical power spectrum is shown in Fig. 15.4a in the Appendix. Similar to the Fourier transform, the magnitude of the spectral power indicates the strength of the periodicity of time series data; however, this periodicity is transient and does not cover the entire time range. This point is illustrated in Fig. 15.4b, which integrates the power spectra from April 1999 to December 2014. The spectra changes with time, and this represents an advantage of wavelet analysis.

While the power spectra provide us with interesting information (the intensity of the epidemics and importance of the virus type), this paper highlights the timing of the epidemics in relation to limitations in space. We calculate phases of the wavelet transform at the wavelet scale $s = 52$, i.e., during a 52-week (1 year) periodicity of influenza epidemics. The phase of a given time series can be viewed as its position in the pseudo-cycle of the series, and is parameterized in radians ranging from $-\pi$ to π [19]. Therefore, the phase is useful for characterizing phase relationships between two sets of time series data by computing the phase difference. The present study calculates the phase difference in each prefecture i compared with Tokyo, $\delta\phi = \phi_i - \phi_0$, where ϕ_i and ϕ_0 are the phases in prefecture i and Tokyo, respectively, and thereby compares and investigates the timing of the epidemics.

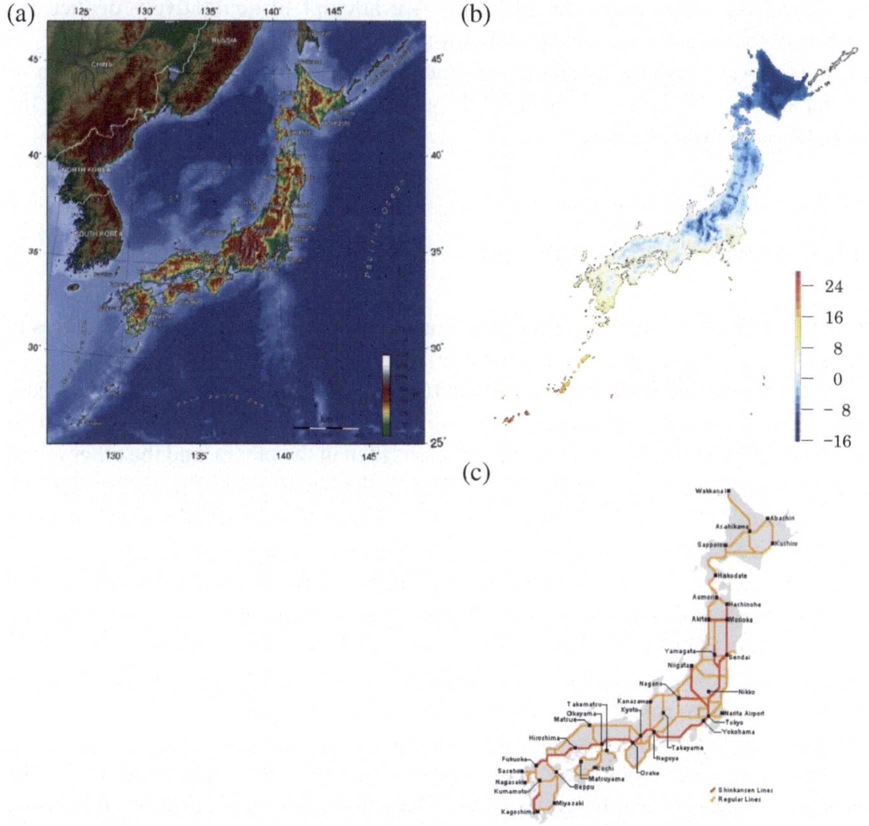

Fig. 15.2 (**a**) Schematic topographical map. (http://www.worldofmaps.net/en/asia/map-japan/topographic-map-japan.htm). (**b**) Average temperature in February. (http://www.data.jma.go.jp/obd/stats/data/mdrr/atlas/mean_temperature_02.pdf). The *color bar* indicates temperature in degrees Celsius. (**c**) Main railway network, including Shinkansen (high-speed railway) lines (*red*) and regular lines (*orange*). The line from Hakata to Kagoshima was gradually constructed starting in 2004. (http://airportguide.com/japan_rail_map.php). The URLs were accessed in December 2014

15.2.3 Supplementary Data

Other useful supplementary data are displayed in Fig. 15.2. Japan is peculiar in several respects. We consider the following environmental characteristics in this study Japan has abundant mountainous areas (70 % of the land area), but the population is concentrated in the plains (Fig. 15.2a). A large proportion of the population (more than 30 million) lives in metropolitan areas, and many cities border one another. Even though the northern and alpine regions of Japan are in a subarctic climate, most of the country is in a temperate climate (Fig. 15.2b). Japan also has a highly developed railway network that includes Shinkansen (high-speed railway) lines (Fig. 15.2c). In this study, we attempt to qualitatively deduce the effects of these factors on the spread of influenza epidemics.

This study uses the standard classification of prefectures into eight regions in Japan (see Fig. 15.3b). Table 15.1 provides a brief summary regarding the population in these regions.

15.3 Results and Discussion

Figure 15.3a shows the time-averaged phase differences between 46 prefectures in Japan. The following points were notable.

In the Kanto and Kinki regions (urban regions), the phase differences were small, and had a small range of values. This indicates two important features. One is that the epidemics occurred earlier in these regions than in the others, and the other is that the epidemics were well synchronized. The epidemics in Aichi Prefecture (ID 23), which includes Nagoya, the 4th largest city in Japan, also occurred as early as those in the above two regions; however, strong synchrony was not seen between Aichi Prefecture and its adjacent prefectures (Gifu, Shizuoka in the Chubu region, and Mie). A geographical feature of the Chubu region is the Nobi Plain, which is similar to the Kanto Plain in the Kanto region and the Osaka Plain in the Kinki region (Fig. 15.2a); however, there is no city as large as Nagoya in this region (Table 15.1). Hence, the above results suggest that although epidemics begin in large cities, a clustering of large cities is required for strong synchronization.

The phase differences show that, in the regions other than Kanto and Kinki, the epidemics spread sequentially with time, likely in the approximate order of Kyushu, Shikoku, Chugoku, Chubu, Tohoku, and Hokkaido regions. Geographically, this indicates that epidemics spread from west to east, and then northeast (Fig. 15.3b), and also see typical time-series in Fig. 15.1b). This west-to-east spread was previously reported in studies from Europe [6] and the United States [10]. They suggested that the west-to-east spread of influenza is a general phenomenon, that is supported by the results of the present study. The former [6] mentioned numerous factors as causes of this spreading pattern (population, geography, climate, etc.), and the latter [10] pointed to major traffic pathways and local contact networks. Prevailing westerlies appear to play an important role in the case of Japan, which was noted by Sakai et al. [13]. The data used in that study were cases reported from 1992 to 1999, and the Kriging method was used for analysis. The present study therefore appears to confirm a west-to-east spread. Even though Sakai et al. showed epidemic patterns spreading in concentric circles from the west-central to the east, their result is not in conflict with ours if the fluctuations of the phase differences are considered. The European study [6] also observed occasional south-to-north spread, which coincides with the spread from the Kanto to the Tohoku and Hokkaido regions observed in the present study, suggesting the possibility that a south-to-north spreading pattern is also a general phenomenon. However, although influenza epidemics occur during

Table 15.1 Approximate data based on the 2013–2014 national census

Region	Kanto	Kinki	Chubu	Kyushu	Shikoku	Chugoku	Tohoku	Hokkaido
Population (millions)	42	22	23	13	3	7	9	5
Major city	Tokyo (1)	Osaka (3)	Nagoya (4)	Fukuoka (8)	Matsuyama (27)	Hiroshima (11)	Sendai (12)	Sapporo (5)

The number in parentheses following a city name denotes the national population ranking. The other large cities are Yokohama (2), Kawasaki (9), and Saitama (10) in the Kanto region, and Kobe (6) and Kyoto (7) in the Kinki region

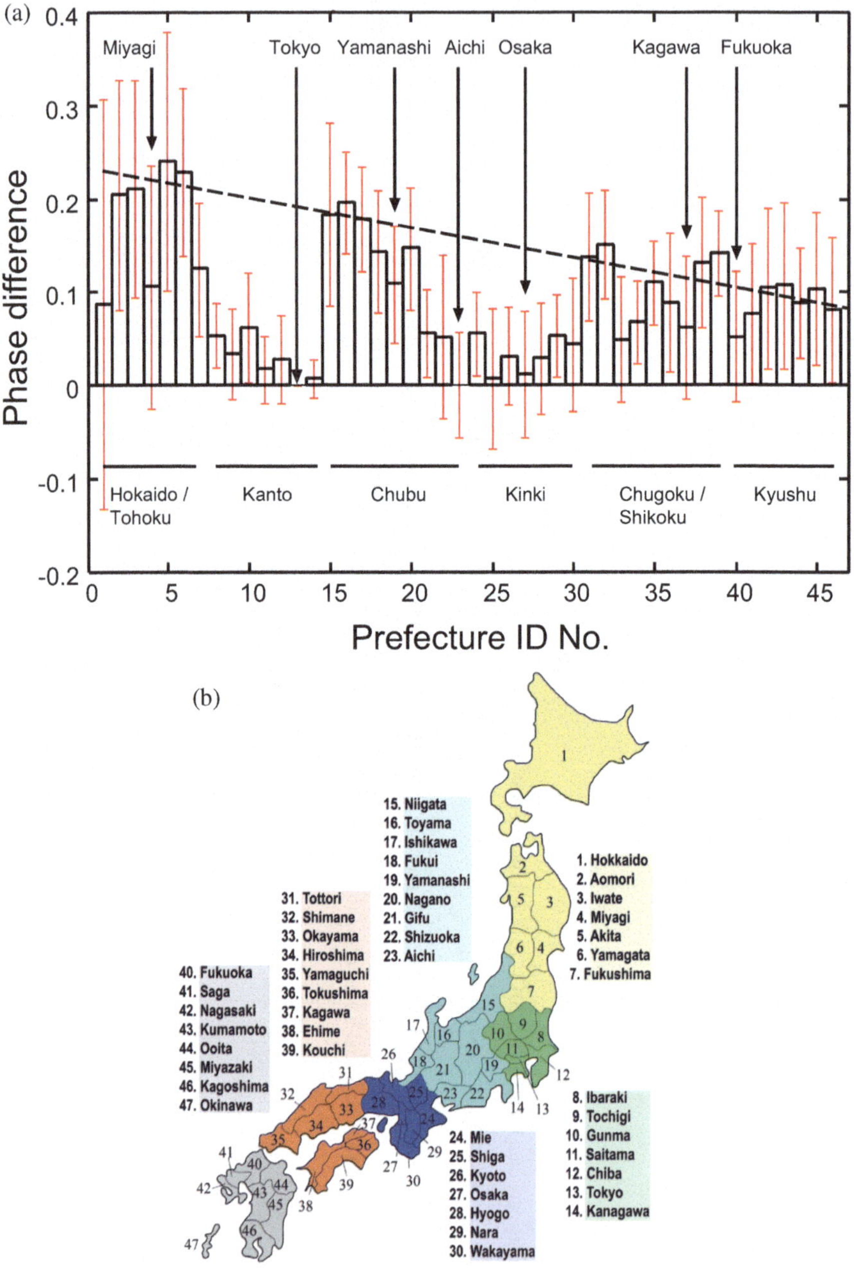

Fig. 15.3 (**a**) Time-averaged phase differences for 46 prefectures in Japan with standard deviations. The unit of the phase difference is $52/2\pi \simeq 8.3$ weeks. The broken line is a guide to eyes. (**b**) Map showing the prefecture ID numbers (http://imas.wikia.com/wiki/Amulet_Purchasing_and_Effects_Guide)

winter, it remains unclear why they spread from high-temperature regions in the south to low-temperature regions in the north (Fig. 15.2b).

An interesting point is that the phase difference in Miyagi Prefecture (ID 4), which includes Sendai, the largest city in the Tohoku region (the 12th largest in Japan), was smaller than that in the other prefectures in this region. This indicates that the epidemic timing in Miyagi Prefecture is similar to that in Tokyo Prefecture, which is likely influenced by intercity travel e.g., via the Shinkansen line (Fig. 15.2c). The geographically discontinuous epidemic spread over a large distance or against geographic constraints can be referred to as "jump spread". This result may agree with that from the study in the United States, in that the synchrony of epidemics between populous states is strong [8]. This epidemic jump can also be seen in Fukuoka (ID 40) and Kagawa (ID 37) Prefectures. The prefectural capital of Fukuoka is Fukuoka City, the 8th largest city in Japan, which is along the Shinkansen line (Fig. 15.2c). The Kyushu Shinkansen line was constructed gradually starting in 2004; hence, in the future, the value of the phase difference in the prefectures along the Kyushu Shinkansen line may converge in a small range. Although Kagawa Prefecture is separated by the Setonaikai Sea (Fig. 15.2a) and is not serviced by a Shinkansen line (Fig. 15.2c), it is directly connected to the Kinki region by highway, which illustrates the importance of personal travel in epidemic spread.

Another interesting point in the calculation of phase deference is that it may be in contrast to the cases mentioned above. Yamanashi Prefecture (ID 19) neighbors Tokyo Prefecture, and they are directly connected by a conventional railway line and highway. However, the phase difference between these two prefectures is larger than that in the other prefectures bordering Tokyo. Although Yamanashi Prefecture is not serviced by a Shinkansen line, this condition is identical to that in Chiba and Ibaragi Prefectures in the Kanto region. A probable explanation may be geographical in nature: Yamanashi Prefecture is not on the Kanto Plain, but rather in a mountainous area (Fig. 15.2a). This explanation can also be applied to Tottori (ID 31) and Shimane (ID 32) Prefectures. The southern area of these prefectures is mountainous, and forms a "wall" against Okayama (ID 33) and Hiroshima (ID 34) Prefectures (Fig. 15.2a). This wall can account for the larger phase difference in Tottori and Shimane Prefectures in the Chugoku region. These points indicate that simply attributing epidemic spread to personal travel requires caution, and a variety of heterogeneities must be considered collectively.

In the Chubu region, the fluctuation in the phase difference was small in each prefecture compared with the prefectures in the other regions. Fluctuations in phase differences indicate the interseasonal year-to-year variation of epidemic timing. In contrast, fluctuation was larger in the Tohoku and Hokkaido regions. This may be associated with the 2009 pandemic (the so-called "swine flu"), which began in Mexico and spread worldwide [23]. If these phase differences are averaged to include the values before 2009, the fluctuations in the Tohoku and Hokkaido regions are smaller (not shown). These problems regarding fluctuations in phase differences are connected with the type of virus and the emergence of antigenic variants [13]; hence, this paper only provides a brief mention of fluctuations in phase differences.

15.4 Concluding Remarks

The present study investigated the spread patterns of influenza epidemics in Japan using wavelet analysis. The phase differences were calculated for 46 prefectures. Our main findings were that epidemics occurred earlier and with greater synchronicity in the Kanto and Kinki regions, that epidemics spread west-to-east in the other regions, that epidemic jump occurred between prefectures, and that two adjacent prefectures occasionally had epidemics at different times. These results led us to deduce the following causes: personal travel, climate, and geographical constraints. The deductions presented in this paper are only qualitative, and therefore need to be confirmed by further analyses of newly available data, e.g., the number of Shinkansen passengers passing through each terminal. Although recent studies have been progressing in this direction [19], numerous factors are intercorrelated regarding the problem of epidemic spread, and careful examination is therefore necessary.

Appendix

Figure 15.4a depicts a typical power spectrum for Tokyo Prefecture in the 1st week of 2010. The large peak around $s = 52$ indicates the epidemic mode of annual synchronization. The power spectra from April 1999 to December 2014 are integrated in Fig. 15.4b. Figure 15.4a is a cross section of Fig. 15.4b. Biennial and semiannual modes can also be seen around $s = 104$ and $s = 26$, respectively. It is possible that the former is related to type B influenza and the latter to the 2009 pandemic.

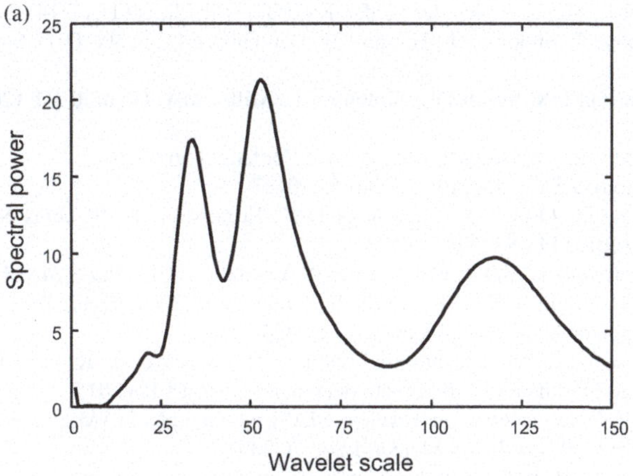

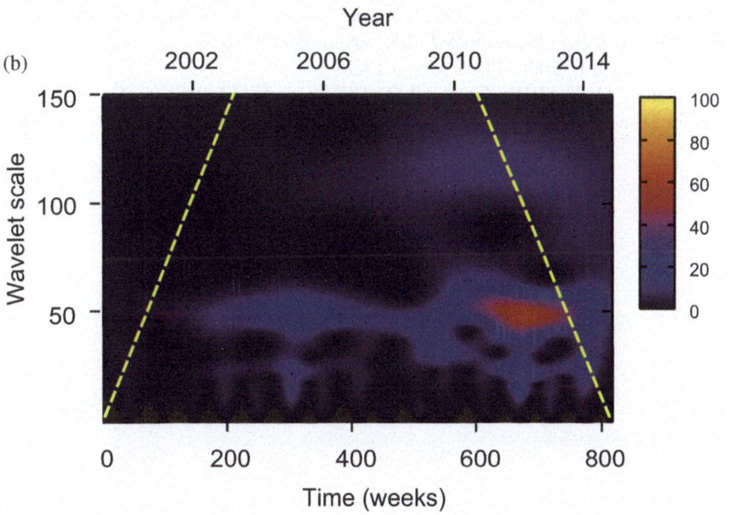

Fig. 15.4 (**a**) Power spectrum for Tokyo prefecture in the 1st week of 2010. The spectral power is normalized by the variance of the time-series. (**b**) Contour plot of power spectra for Tokyo Prefecture from April 1999 to December 2014. The color bar indicates the spectral power. The broken lines denote the edge effect of the time-series ("cone of influence" [18]). The spectral information is less accurate in the edge regions marked by these lines

References

1. Lofgren E, Fefferman NH, Naumov YN, Gorski J, Naumova EN (2007) J Virol 81:5429
2. Lipsitch M, Viboud C (2009) Proc Natl Acad Sci 106:3645
3. Bonabeau E, Toubiana L, Flahault A (1998) Proc Biol Sci 265:2421
4. Viboud C, Bjornstad ON, Smith DL, Simonsen L, Miller MA, Grenfell BT (2004) Emerg Infect Dis 10:32
5. Mugglin AS, Cressie N, Gemmell I (2002) Stat Med 21:2703

6. Paget J, Marquet R, Meijer A, van der Velden K (2007) BMC Infect Dis 7:141
7. Alonso WJ, Viboud C, Simonsen L, Hirano EW, Daufenbach LZ et al (2007) Am J Epidemiol 165:1434
8. Viboud C, Bjornstad ON, Smith DL, Simonsen L, Miller MA, Grenfell BT (2006) Science 312:447
9. Greene SK, Ionides EL, Wilson ML (2006) Am J Epidemiol 163:316
10. Wenger JB, Naumova EN (2010) PLoS ONE 5:e10187
11. Stark JH, Sharma R, Ostroff S, Cummings DAT, Ermentrout B, Stebbins S, Burke DS, Wisniewski SR (2012) PLoS ONE 7:e34245
12. Yu H, Alonso WJ, Feng L, Tan Y, Shu Y, Yang W, Viboud C (2013) PLoS Med 10:e1001552
13. Sakai T, Suzuki H, Sasaki A, Saito R, Tanabe N, Taniguchi K (2004) Emerg Infect Dis 10:1822
14. Onozuka D, Hagihara A (2008) Epidemiology 19:824
15. Infectious Disease Surveillance Center, in Infectious Diseases Weekly Report (IDWR) Japan, http://www.nih.go.jp/niid/en/of subordinate document. Cited 15 Jan 2015
16. Daubechies I (1992) Ten lectures on wavelets. SIAM monographs SIAM, Philadelphia
17. Lau KM, Weng H (1995) Bull Am Meteorol Soc 76:2391
18. Torrence C, Compo GP (1998) Bull Am Meteorol Soc 79:61
19. Cazelles B, Chavez M, de Magny GC, Guegan J, Hales S (2007) J Roy Soc Interface 4:625
20. Grenfell BT, Bjornstad ON, Kappey J (2001) Nature 414:716
21. Chaves LF, Pascual P (2006) PLoS Med 3:1320
22. Yang L, Wong CM, Yau EHY, Chan KP, Ou CQ, Peiris JSM (2008) PLoS ONE 3(1):e1399
23. Centers for Disease Control and Prevention (CDC) (2009) MMWR Morb Mortal Wkly Report 8:467

Chapter 16
A Universal Lifetime Distribution for Multi-Species Systems

Yohsuke Murase, Takashi Shimada, Nobuyasu Ito, and Per Arne Rikvold

Abstract Lifetime distributions of social entities, such as enterprises, products, and media contents, are one of the fundamental statistics characterizing the social dynamics. To investigate the lifetime distribution of mutually interacting systems, simple models having a rule for additions and deletions of entities are investigated. We found a quite universal lifetime distribution for various kinds of inter-entity interactions, and it is well fitted by a stretched-exponential function with an exponent close to 1/2. We propose a "modified Red-Queen" hypothesis to explain this distribution. We also review empirical studies on the lifetime distribution of social entities, and discuss the applicability of the model.

Y. Murase (✉)
RIKEN Advanced Institute for Computational Science, 7-1-26, Minatojima-minami-machi, Chub-ku, Kobe, Hyogo, 650-0047, Japan

CREST, Japan Science and Technology Agency 4-1-8 Honcho, Kawaguchi, Saitama, 332-0012, Japan

T. Shimada
Department of Applied Physics, School of Engineering, The University of Tokyo, 7-3-1 Hongo, Bunkyo-ku, Tokyo 113-8656, Japan

CREST, Japan Science and Technology Agency 4-1-8 Honcho, Kawaguchi, Saitama, 332-0012, Japan

N. Ito
RIKEN Advanced Institute for Computational Science, 7-1-26, Minatojima-minami-machi, Chub-ku, Kobe, Hyogo, 650-0047, Japan

Department of Applied Physics, School of Engineering, The University of Tokyo, 7-3-1 Hongo, Bunkyo-ku, Tokyo 113-8656, Japan

CREST, Japan Science and Technology Agency 4-1-8 Honcho, Kawaguchi, Saitama, 332-0012, Japan

P.A. Rikvold
Department of Physics, Florida State University, Tallahassee, FL 32306-4350, USA

H. Takayasu et al. (eds.), *Proceedings of the International Conference on Social Modeling and Simulation, plus Econophysics Colloquium 2014*, Springer Proceedings in Complexity, DOI 10.1007/978-3-319-20591-5_16

16.1 Introduction

Society is a system of diverse coexisting entities showing a high turnover of its membership [1]. Examples of such entities include enterprises, products, and media contents. Lifetime distributions of these entities are one of the most fundamental properties of such systems. Thus, understanding of these distributions will reveal the underlying social dynamics. For example, the lifetime of products would be strongly related to the market trend, and the lifetime of enterprises can be a crucial condition for the stability of economic activity and employment. Although several models have been proposed to fit lifetime distributions, most of these models do not explicitly take into account the interactions between entities. Lifetime distributions of mutually interacting systems are not fully understood even though these interactions often play a significant role in actual society as we see, for example, in cascading bankruptcies of enterprises. In this short article, we investigate the lifetime distributions for "ecosystem" like systems, where diverse entities undergo competition for survival.

Several theoretical models have been proposed for lifetime distributions. The simplest assumption is that an "extinction" (the elimination of an entity) occurs randomly with a constant rate, i.e., characterized by a Poisson process [2]. The lifetime distribution of "species" (a social entity) is then a simple exponential function. Although this assumption is mathematically simple, it is radical from a sociological point of view because no evolutionary advantage or aging effect is taken into account. This hypothesis is known as the Red-Queen hypothesis or Van-Valen's law in evolutionary ecology. On the other hand, if mortality rate is dependent on age, the lifetime distribution deviates from a simple exponential function [3, 4]. If a long-lived species has an evolutionary advantage, the probability that a species goes extinct will be a decreasing function of its age. This assumption seems reasonable because a species which has been successful in surviving is expected to have some superior properties. The decreasing mortality function yields a lifetime distribution with a heavier tail than the corresponding exponential function. If a long-lived species has higher mortality than a younger species, perhaps due to its aging or degradation, the lifetime distribution will decay faster than the exponential one. Another simple model is the return-time distribution of random walks [5]. Assuming that "fitness" of each species, which may be population or any other measure of the distance from extinction, follows a neutral random walk, the lifetime distribution will be modeled by a return time distribution, which is known to be approximately $t^{-3/2}$. If we assume a critical branching process instead of a random walk, we find a t^{-2} power law. A more general theory which combines a random walk and a branching process was also proposed [6].

All the above models consider a stochastic process of one species. Lifetime distributions of mutually interacting species have been investigated mainly in the context of biological coevolution [7, 8]. Among the simplest models of a mutually interacting system are the so-called self-organized criticality (SOC) models, which predict power laws [9]. In addition to these simplistic models, population dynamics models or individual based models are also proposed with the aim of bridging the ecological and evolutionary timescales. These include the tangled-nature models [10–17], the web-world model [18, 19], and others [20, 21]. All those models have population dynamics of each species (or birth-death process at an individual level) and rules for the emergence and extinctions of species. Some of these show a power law lifetime distribution t^{-2} [12–17] while others show a curved line that lies somewhere between a power law and an exponential distribution: concave on a log-log plot and convex on a semi-log plot [17, 20]. Interestingly, these seem to be classified into a few universality classes regardless of the apparent diversity of the models [17]. In the models which add new species with randomly determined interaction coefficients, a skewed lifetime distribution is universally observed under various population dynamics equations. This type of addition of new species is called "migration" because a new species is not correlated with the current species at all. On the other hand, with the "mutation" model, where a new species appears as a result of a modification of a current species, t^{-2} power law is robustly observed. Even though the models have quite different numbers of species, types of interactions, and network topologies, they share similar species-lifetime distributions implying the existence of universality.

In this article, we mainly focus on the skewed profile found for migration rules because it is the simplest model to add a new species. We will show the origin of the skewed profile by introducing a simple graph model. In the next section, the model definition and its typical results are given. In Sect. 16.3, the origin of the skewed profile is explained using what we call the modified Red-Queen hypothesis. Then, in Sect. 16.4 we review the empirical data observed in society and discuss their relation with the model. The last section is devoted to conclusions.

16.2 Dynamical Graph Model

In order to investigate the lifetime distributions of mutually interacting systems, we propose a simple dynamically evolving model which was originally introduced for biological community assembly [22]. A system is represented by a weighted and directed network, which self-organizes by successive migrations and eliminations (extinctions) of nodes. Each node i has a state variable called "fitness" f_i, which is defined as the sum of the weights of incoming links, i.e., $f_i = \sum_j a_{ij}$, where a_{ij} is the weight of a link from node j to i. Node i can survive if its fitness is larger than or equal to zero, otherwise it is eliminated from the system.

At each time step, a new node is added to the system. New links between existing nodes and the new node are randomly assigned with probability c, whose weights are randomly drawn from the Gaussian distribution with mean 0 and variance 1. After a migration, the species with minimum fitness is identified and is eliminated from the system if the minimum fitness is negative. Since the extinction of a node affects the fitness of other species, successive extinctions can happen. This process

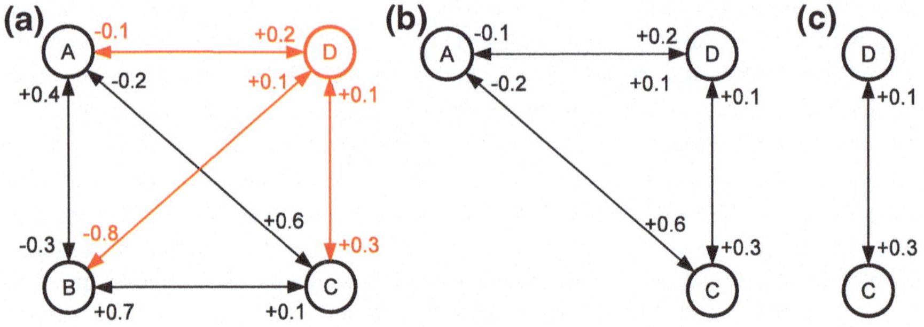

Fig. 16.1 An example of the model dynamics. *Nodes and arrows* denote species and interactions, respectively. (**a**) Before the migration of species D, species A, B, and C coexist (**b**) When species D immigrates into the community, species B goes extinct (**c**) Then, another resident species A goes extinct due to the extinction of species B. After extinctions of species A and B, all remaining species (C and D) have positive fitness values. The figure is taken from [22]

is repeated until all the fitness values in the system become non-negative for each time step. (See Fig. 16.1.)

This simple model reproduces a characteristic skewed lifetime distribution found for population dynamics models with migration rules. As shown in Fig. 16.2, the distribution is neither a simple exponential nor a simple power law distribution. It is well fitted by a stretched exponential function with an exponent close to 1/2. Note that the number of species N fluctuates in a finite range and the statistics are taken from a statistically stationary state. Since this model shares a similar profile to ones for more complicated population dynamics models, this model is expected to capture the essential aspects of the skewed lifetime distribution.

16.3 Modified Red-Queen Hypothesis

The lifetime distribution corresponding to a stretched exponential function with exponent $1/2$ is explained by what we call the modified Red-Queen hypothesis. This hypothesis assumes that the mortality of each species is not dependent on its age but on the number of species in the system. Let us assume that N fluctuates in a finite range, and that the probabilities that N increases or decreases are independent of N. In other words, we assume that N follows a random walk with a negative drift. (Without a negative drift, we would get a divergence of N.) These assumptions can be obtained by a mean-field analysis [23, 24]. Based on these assumptions, the extinction probability of a species is proportional to $1/N$, meaning that the typical species lifetime τ is proportional to N. Because a stable distribution of N is an exponential function under these assumptions, a species-lifetime distribution is then expressed by a superposition of exponential functions of time scale τ with an exponential weight:

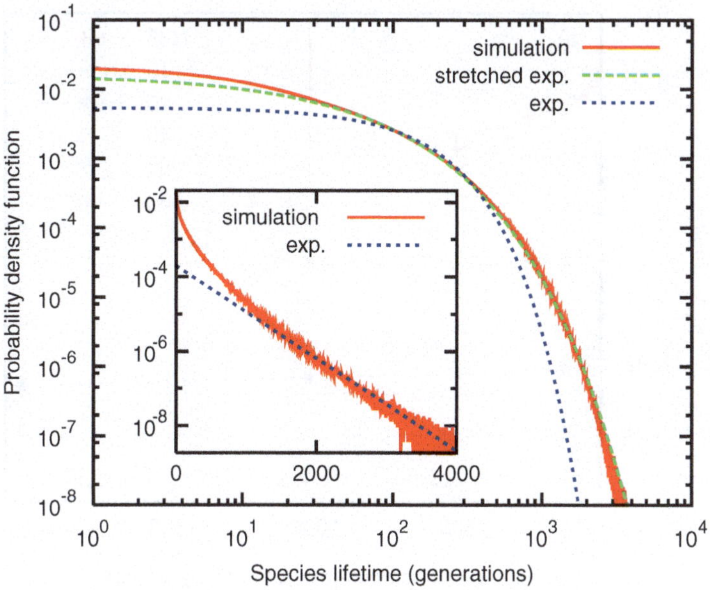

Fig. 16.2 Species-lifetime distribution on log-log scales and linear-log scales (inset) for the dynamical graph model. Fitting curves are also shown with a stretched exponential function and a simple exponential function. In the inset, a fitting with a simple exponential function is also shown as a guide to the eye. This figure is modified based on a figure in [22]

$$p(t) = \int_0^\infty \frac{\exp(-t/\tau)}{\tau} b \exp(-b\tau) d\tau \tag{16.1}$$

$$\approx \sqrt{\pi}(bt)^{-1/4} \exp(-2\sqrt{bt}) \qquad (t \gg 1), \tag{16.2}$$

where b is a coefficient which depends on the probability that N decreases. This set of assumptions is called the modified Red-Queen hypothesis because the mortality is independent of the age of species.

These assumptions are validated by calculating mortality against communities obtained by simulations. We calculated the probability that each resident species goes extinct in the next time step by adding a test immigrant to the snapshot of the simulations. The relationship between mortality and the age of a species is shown in Fig. 16.3 for several numbers of species N. For all N, mortalities show a sharp decrease at $t \approx 0$ and then converge to a constant value which is approximately proportional to $1/N$. These results are in good agreement with the modified Red-Queen hypothesis.

The reason that the mortality does not depend on the age t but on N is explained by the time evolution of the fitness distribution. Let us calculate the probability distribution of the fitness of a species at age t, $P_t(f)$, under the assumption that N is fixed, i.e., one immigrant comes to the system and one resident species goes extinct

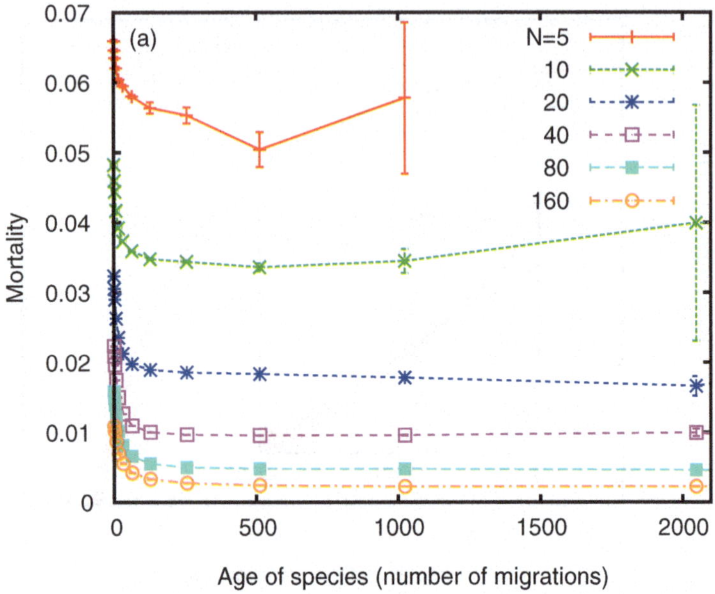

Fig. 16.3 Average probability that a species goes extinct (mortality) as a function of the age for communities of $N = 5$, 10, 20, 40, 80, and 160. This figure is taken from [22]

at each time step. For simplicity, we discuss the case that $c = 1$. The distribution of the fitness of a newcomer which succeeded in migrating, $P_0(f)$, is given by

$$P_0(f) = \begin{cases} 2\mathcal{N}(0, N) & (f \geq 0) \\ 0 & (f < 0) \end{cases}, \tag{16.3}$$

where $\mathcal{N}(0, N)$ denotes the Gaussian distribution with mean 0 and variance N, because the fitness is the sum of N Gaussian random numbers. At the next time step, the fitness changes due to the migration and the extinction. While the interaction coefficient with an immigrant species is given by $\mathcal{N}(0, 1)$, the interaction coefficient with the resident species should have a positive mean because the sum of the incoming links is positive. Therefore, we assume that the distribution of the interaction with the resident species is $\mathcal{N}(\mu_t/N, 1)$ where μ_t is the mean of $P_t(f)$. In total, the distribution of total change in fitness for each time step is the convolution of $\mathcal{N}(0, 1)$ and $\mathcal{N}(-\mu_t/N, 1)$, i.e., $\mathcal{N}(-\mu_t/N, 2)$. Under this assumption, the time evolution of $P_t(f)$ is calculated as follows:

$$P_{t+1}(f) = \begin{cases} C_{t+1} \times [P_t * \mathcal{N}(-\mu_t/N, 2)](f) & (f \geq 0) \\ 0 & (f < 0) \end{cases}, \tag{16.4}$$

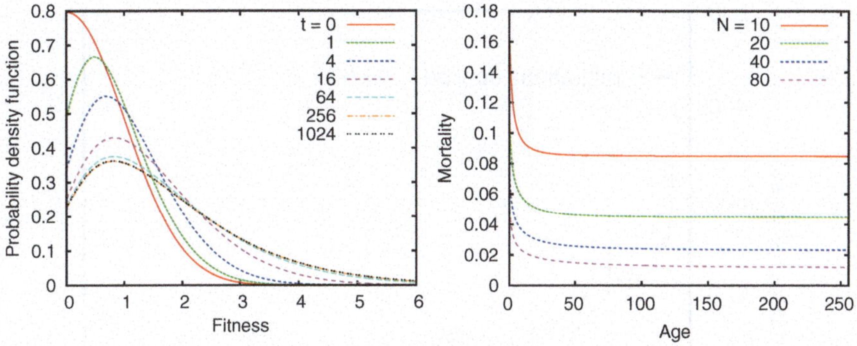

Fig. 16.4 (*left*) Time evolution of the probability density function of the fitness, $P_t(f)$, under the fixed-N approximation for $N = 10$. (*right*) The probability of the species that goes extinct as a function of age. The probability is calculated under the fixed-N approximation for several N

where C_t is the normalization coefficients and the operator $*$ denotes the convolution. The coefficient C_t is determined so that the positive part of the convolution function is normalized. From this equation, mortality at each time step, $m(t)$ is also calculated as the ratio of the negative part of the convolution function. Numerically calculated time evolution of $P_t(f)$ and $m(t)$ are shown in Fig. 16.4. As t increases, $P_t(f)$ and $m(t)$ approach a constant profile and a constant value, respectively. The value to which $m(t)$ converges is inversely proportional to N. All these are consistent with the simulation results and the modified Red-Queen hypothesis.

While the mortality converges to a constant value for large t, it shows a sudden drop at $t \sim 0$. It means species that have just entered the system are more susceptible to extinctions than long-lived species. This fact implies that the species compositions depend on the way they are constructed. For example, let us prepare an initial state with one thousand species whose coupling constants are randomly assigned from a Gaussian distribution with mean zero. After starting the dynamics of extinction, the community immediately loses approximately half of the species, while the rest of the species can coexist. The community constructed in this way (here, we call it "prepared" community) is qualitatively different from the communities constructed by repetitive migration-extinction processes (here, we call it "trial-and-error" community). This difference is observed in the robustness against the migrations of new species. Figure 16.5 shows the time evolution of N for a prepared community which starts from $N = 1000$. Although 423 species survived at $t = 0$, most of these species went extinct as soon as migrations of new species were started. Thus, prepared communities are fragile against migration of new species because the fitness distribution in prepared communities has a peak at around zero. On the other hand, the trial-and-error communities are robust against migrations. The number of species N for the trial-and-error communities keeps fluctuating around a constant value under successive migrations. This is an important implication that the model for the society should include a kind of trial-and-error construction process, otherwise we might miss an important point.

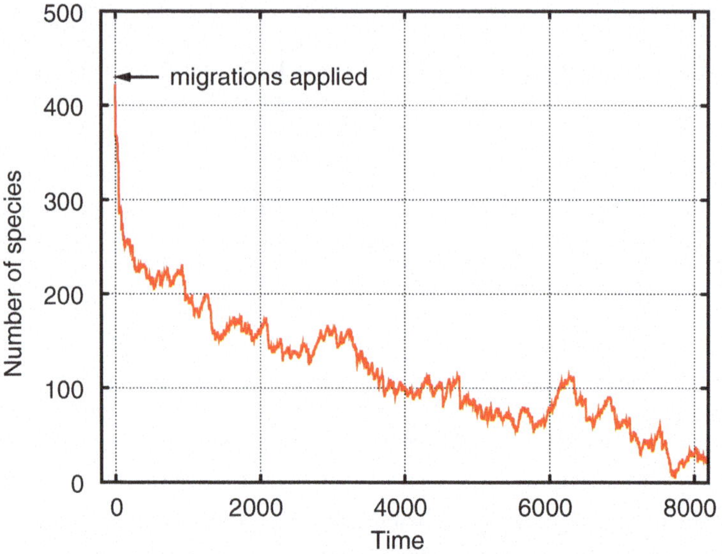

Fig. 16.5 Time development of the number of species. A community with one thousand species whose interactions were randomly assigned was used as an initial state. The number of coexisting species was 423 after eliminating the unfit species from the initial state. Then, migrations are applied. Due to the migrations, the number of species drops immediately. Connectance $c = 0.2$ was used

Finally, it should be emphasized that the modified Red-Queen hypothesis is quite different from the assumption of age-dependent mortality. Actually, assuming the mortality function proportional to $t^{-1/2}$ can yield a Weibull distribution with exponent $1/2$, thus a similar lifetime distribution is obtained [4]. However, these two assumptions have a crucial difference when predicting the extinction risk of a species. The extinction risk can be estimated based on species age if an age-dependent mortality is valid, although this is not the case under the modified Red-Queen hypothesis. Furthermore, the modified Red-Queen hypothesis answers why the exponent is $1/2$ while the age-dependent mortality can assume an arbitrary exponent.

16.4 Relation with Empirical Data

In this section, we review several empirical data sets and discuss the applicability of our models.

The first example is product life cycles in a Japanese convenience store chain [25]. It was found that the lifetime distributions of noodles, juice, and sandwiches show similar curves. Interestingly, these curves are fitted by the Weibull distribution with an exponent close to $1/2$, which is quite similar to the distribution found in our model. Moreover, this skewed profile is found also in biological coevolution. The data estimated from the fossil record show a skewed profile [3]. Although a q-exponential fit is proposed in [3], the stretched exponential function fits the data as well. It indicates the universality shared by sociological and biological systems.

The second example is the lifetime distribution of Japanese firms. The lifetime distribution of Japanese firms which went bankrupt in 1997 is well approximated by a simple exponential function [26], hence it seems consistent with the (non-modified) Red-Queen picture. Thus, we can conclude that the age-dependent mortality picture is not valid at least for Japanese firms. On the other hand, these data are not sufficient to reject the modified Red-Queen hypothesis even though the distribution is not skewed. This is because the period of the measurement is only one year, which is much shorter than the typical lifetime. A measurement on longer time scales will deepen our understanding of the bankruptcy dynamics of enterprises. Another possible reason for the inconsistency with our model is that the model assumes that the migration frequency of new species is independent of N. If we assume an N-dependent migration rate, we will get a different lifetime distribution. Data on the occurrence frequency of new enterprises would clarify this point.

Several lifetime distributions of media contents have also been investigated. Media contents can be interacting with each other as most of these are competing for the attention of potential consumers.

The first data set of media contents is movie popularity [27]. The cumulative distribution of the persistence time of a movie fits a stretched exponential form with an exponent about 1.6, indicating that it decays faster than exponentially. This quick decay is explained by a few observed stylized facts and an assumptions that a movie is withdrawn when the gross income per theater gets below a threshold value. The observed statistics shows a $1/t$ decay in gross income of a movie per theater, indicating that the mortality of a movie increases with age due to aging. That is why we see a faster decay than a simple exponential function.

Another example of media contents is comic series. Figure 16.6 shows the cumulative lifetime distributions of comic series that have run in three major Japanese weekly comic magazines. We defined the lifetime of each comic series as the duration between the first and the final issues, and collected data from a Wikipedia article which contains the list of the start and end dates for each series. Since the termination of a series should strongly depend on its popularity, the series are competing with each other for a limited number of concurrent series. As we see in the figure, all these distributions show approximate exponential decays. Thus, the Red-Queen picture looks the most reasonable hypothesis for comic series. This is clearly different from the movie duration. We conjecture the reason of the difference is that the comic series has a new story every week while a movie is not renewed. One of the possible reasons that the dynamical graph model is not applicable is that the number of concurrent series does not show fluctuations, which is a key for the modified Red-Queen hypothesis.

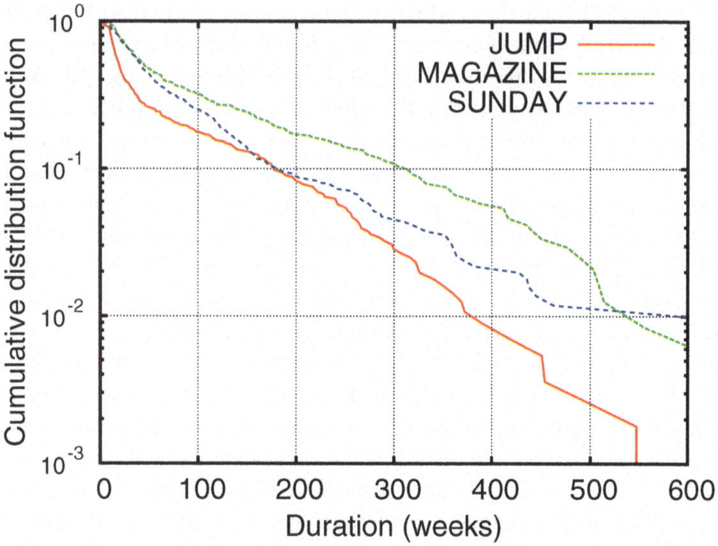

Fig. 16.6 The cumulative distribution functions of lifetimes of the series that have run in Weekly Shōnen Jump, Weekly Shōnen Magazine, and Weekly Shōnen Sunday. The lists were obtained from the Japanese pages of [28–30] in December 2009. Currently running series are not included in the statistics

Another data set of persistence in social media can be found for Twitter [31]. A recent study showed that the length of a topic sequence follows an approximate t^{-2} power law. This is the most heavy-tailed distribution among the above. One of the characteristic points of Twitter trends is that a trend can be recurrent, i.e., the trends appear more than once. The dynamical graph model clearly misses this point. Therefore, the model is not appropriate for Twitter trends.

16.5 Conclusion

In this article, we investigated the lifetime distribution for an interacting systems. A skewed lifetime distribution is robustly observed for a wide range of models, and it fits a stretched exponential function with exponent $1/2$. We proposed the dynamical graph model and revealed that this profile is explained by the modified Red-Queen hypothesis. This hypothesis is a novel idea and its meaning is completely different from the age-dependent mortality assumption even though both yield similar profiles.

We also reviewed empirical data sets and discussed the applicability of the model. While some data sets are similar to our model, dissimilar data sets are also common. It is clear that the dynamical graph model is one of the simplest models for mutually interacting systems, hence it is not applicable to all the data. Based on this study, we expect further exploration both of theoretical models and empirical data.

A theoretically important question is to find other universality classes and identify key factors which change these classes. For example, it is not clear what is the fundamental difference between the dynamical graph and the SOC models. Effects of the network topology is another big open question. The dynamical graph model is essentially described by an Erdös-Rényi random network, however real-world networks have internal structures. Studies of scale-free networks or modular networks are expected. We will get a much deeper insight into the dynamics of diverse and open systems, when these questions are answered.

Acknowledgements The systematic simulations in this study were assisted by OACIS [32]. P.A.R. is supported in part by U.S. NSF Grant No. DMR-1104829.

References

1. Sen P, Chakrabarti BK (2013) Sociophysics: an introduction. Oxford University Press, Oxford
2. Van Valen L (1973) Evol Theor 1(1):1
3. Shimada T, Yukawa S, Ito N (2003) Int J Mod Phys C 14(9):1267
4. Lawless JF (2011) Statistical models and methods for lifetime data, vol. 362. Wiley, New York
5. Van Kampen NG (2007) Stochastic processes in physics and chemistry. North-Holland, Netherlands
6. Pigolotti S, Flammini A, Marsili M, Maritan A (2005) Proc Natl Acad Sci USA 102(44):15747
7. Drossel B (2001) Adv Phys 50(2):209
8. Newman M, Palmer R (2003) Modeling extinction. Oxford University Press, New York
9. Bak P, Sneppen K (1993) Phys Rev Lett 71(24):4083
10. Hall M, Christensen K, di Collobiano SA, Jensen HJ (2002) Phys Rev E 66(1):011904
11. Christensen K, di Collobiano SA, Hall M, Jensen HJ (2002) J Theor Biol 216(1):73
12. Rikvold PA, Zia RKP (2003) Phys Rev E 68(3):031913
13. Sevim V, Rikvold PA (2005) J Phys A Math Gen 38(43):9475
14. Rikvold PA (2007) J Math Biol 55:653
15. Rikvold PA, Sevim V (2007) Phys Rev E 75(5):051920
16. Murase Y, Shimada T, Ito N, Rikvold PA (2010) Phys Rev E 81(4):041908
17. Murase Y, Shimada T, Ito N, Rikvold PA (2010) J Theor Biol 264(3):663
18. Caldarelli G, Higgs PG, McKane AJ (1998) J Theor Biol 193(2):345
19. Drossel B, Higgs PG, McKane AJ (2001) J Theor Biol 208:91
20. Shimada T, Yukawa S, Ito N (2002) Artif Life Robotics 6:78
21. Tokita K, Yasutomi A (2003) Theor Pop Biol 63:131
22. Murase T, Shimada T, Ito N (2010) New J Phys 12(6):063021
23. Shimada T (2014) Sci Rep 4:4082
24. Shimada T, Ohira T, Uzawa T (eds) (2015) Mathematical approaches to biological systems. Springer, Tokyo Heidelberg New York Dordrecht London, pp 95
25. Mizuno T, Takayasu M (2014) Progr Theor Phys Suppl 179:71
26. Fujiwara Y, Physica A 337:219 (2004)

27. Pan RK, Sinha RK (2010) New J Phys 12(11):115004
28. URL https://ja.wikipedia.org/wiki/%E9%80%B1%E5%88%8A%E5%B0%91%E5%B9%B4%E3%82%B8%E3%83%A3%E3%83%B3%E3%83%97%E9%80%A3%E8%BC%89%E4%BD%9C%E5%93%81%E3%81%AE%E4%B8%80%E8%A6%A7
29. URL https://ja.wikipedia.org/wiki/%E9%80%B1%E5%88%8A%E5%B0%91%E5%B9%B4%E3%82%B5%E3%83%B3%E3%83%87%E3%83%BC%E9%80%A3%E8%BC%89%E4%BD%9C%E5%93%81%E3%81%AE%E4%B8%80%E8%A6%A7
30. URL https://ja.wikipedia.org/wiki/%E9%80%B1%E5%88%8A%E5%B0%91%E5%B9%B4%E3%83%9E%E3%82%AC%E3%82%B8%E3%83%B3%E9%80%A3%E8%BC%89%E4%BD%9C%E5%93%81%E3%81%AE%E4%B8%80%E8%A6%A7
31. Asur S, Huberman BA, Szabo G, Wang C (2011) In: Proceedings of the 5th international AAAI conference on weblogs and social media, pp 434–437
32. Murase Y, Uchitane T, Ito N (2014) Phys Proc 57:73

Chapter 17
Firm Age Distributions and the Decay Rate of Firm Activities

Atushi Ishikawa, Shouji Fujimoto, Takayuki Mizuno, and Tsutomu Watanabe

Abstract In this study, we investigated around one million pieces of Japanese firm-size data, which are included in the database ORBIS, and confirmed that the age distribution of firms approximately obeys an exponential function. We estimated the decay rate of firms by comparing their activities in 2008 and 2013 and found that it does not depend on firm age and can be regarded to be constant. Here, decay rate of firms denotes the state transition probability of firm activities. These two observations are qualitatively consistent when the number of newly founded firms is nearly constant. This phenomenon is analogous to nuclear decay. We quantitatively confirmed this consistency by comparing the parameters of exponential age distribution with the decay rate of firm activities. At the same time, using this result, we estimated the number of firms founded annually and the decay rate of firm activities in Japan before World War II.

17.1 Introduction

There are various subjects for econophysics research (see, e.g., [1–3]). One main argument, which has a deep relationship with macroeconomics, is power-law distribution that is confirmed in the large-scale range of various firm-size data [4–6]. Log-normal distribution is also frequently noted in mid-sized variables. These two

A. Ishikawa (✉) • S. Fujimoto
Kanazawa Gakuin University, Kanazawa, Japan
e-mail: at.ishikawa@gmail.com; sjfjmt@gmail.com

T. Mizuno
National Institute of Informatics, Tokyo, Japan

The Graduate University for Advanced Studies [SOKENDAI], Tokyo, Japan

JST PRESTO, Tokyo, Japan
e-mail: mizuno@nii.ac.jp

T. Watanabe
University of Tokyo, Tokyo, Japan
e-mail: watanabe1284@gmail.com

H. Takayasu et al. (eds.), *Proceedings of the International Conference on Social Modeling and Simulation, plus Econophysics Colloquium 2014*, Springer Proceedings in Complexity, DOI 10.1007/978-3-319-20591-5_17

laws, which are observed in firm-size variables at a point in time, are related to the short-term statistical laws observed in firm-size variables at two successive years [7–14]. The short-term laws denote (quasi-) inverse symmetry and (Non-) Gibrat's law [15, 16]. To the best of our knowledge, however, long-term statistical laws by decade have not been sufficiently investigated.

Several studies reported that the age distribution of firms obeys an exponential function [17, 18]. Since some firms still exist that were founded over 100 years ago, the age distribution of firms, which is observed at a point in time, must be related to long-term statistical laws. In this study, we investigate the age distribution of firms by employing an exhaustive database of Japanese firms.

The rest of this paper is organized as follows. Section 17.2 describes our database. In Sect. 17.3, first we confirm that the age distribution of firms, observed in 2008, approximately follows an exponential function. Second, by classifying firms into age-rank bins, we compare the activities of firms in 2008 and 2013 and find that their decay rates do not depend on firm age. This observation is new to the best of our knowledge. In Sect. 17.4, we point out that the two laws observed in Sect. 17.3 are closely related to each other when the number of newly founded firms is nearly constant. This is analogous to nuclear decay [19]. We quantitatively verify this relation by comparing the parameters of exponential age distribution with the decay rate of firm activities. At the same time, using this result, we estimate the number of firms founded annually and the decay rate of firm activities in Japan before World War II. The last section concludes this paper.

17.2 Data

We employ the database ORBIS provided by Bureau van Dijk [20] that contains the activity data of 2,739,268 firms and the financial data of 984,502 firms in 2008 Japan. It also includes the activity data of 3,419,105 firms in 2013 Japan. Since the number of active firms in Japan is considered to be around several million [21], this database is exhaustive.

The activities of firms in the database ORBIS are largely categorized into the following three states:

- **Active**
  Active, Active (payment default), Active (insolvency proceedings), Active (dormant), Active (branch)
- **Inactive**
  In liquidation, Bankruptcy, Dissolved (merger or take-over), Dissolved (demerger), Dissolved (liquidation), Dissolved (bankruptcy), Dissolved, Inactive (branch), Inactive (no precision)
- **Unknown**

We define the decay of firms as a transition from active to inactive or unknown (including Not Available). Note that in this definition, the decay of firms necessarily means bankruptcy. In the database ORBIS, the same vendor researched firm data both in 2008 and 2013 Japan and the response rate was sufficiently high; therefore, this definition is appropriate in Japan. In the database, 2,099,834 active firms existed in 2008, among which 1,888,294 firms remained active in 2013. 211,540 firms decayed in this 5-year period. Of the total, 1 firm was In liquidation, 14 firms were Bankruptcy, 58 Dissolved (merger or take-over), 1 Dissolved (liquidation), 6 Dissolved, 49 Inactive (branch), 207,553 Inactive (no precision), 3,645 Unknown, and 213 NA.

17.3 Data Analysis

17.3.1 Age Distribution of Firms

To calculate the age of firms T, we must identify a starting point in time for every firm. In this study, we regard a firm's incorporation as its starting point. Since firms have financial statement from their year of foundation, we denote their age of foundation as $T = 1$. In our analyses, since we are interested in the number of firms as a function of firm age T to consider decay rate of firms, we call $N(T)$ the age distribution of firms. The cumulative age distribution of firms is defined by

$$N_{>}(T) = \int_{T}^{\infty} dTN(T). \tag{17.1}$$

Figure 17.1 depicts cumulative age distribution $N_{>}(T)$ of firms whose year of foundation can be identified. The data in Fig. 17.1 are cumulated to smooth the dispersion. As shown in Fig. 17.1, the cumulative age distribution of firms approximately obeys an exponential function:

$$N_{>}(T) \propto e^{-\lambda T}. \tag{17.2}$$

Here, λ is a constant parameter. Maximum age T in Fig. 17.1 is 150 years. This corresponds to the starting point of the modernization of Japan around the Meiji Restoration (1868) that marked the return of imperial power in Japan. In Fig. 17.1, the difference in the data around 60 years corresponds to World War II. If we apply Eq. (17.2) to the entire range in Fig. 17.1, the parameter is estimated as $\lambda \sim 0.078$. This value is close to 0.053 reported in [18], and is also close to around 0.05 reported in [17].

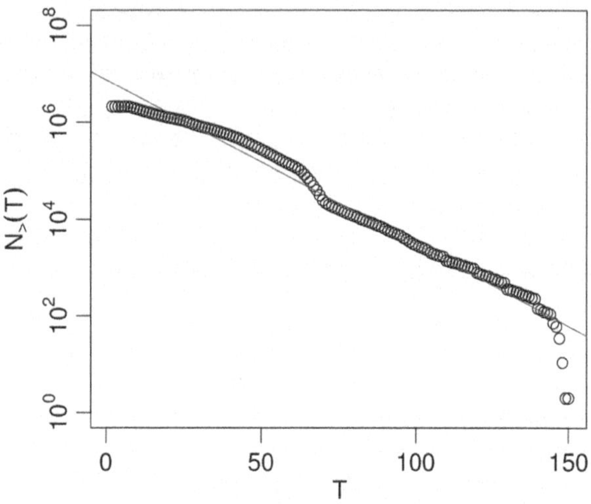

Fig. 17.1 Cumulative age distribution $N_>(T)$ of firms in Japan included in database ORBIS. Horizontal axis is firm age T, and vertical axis is number of firms whose age exceeds T

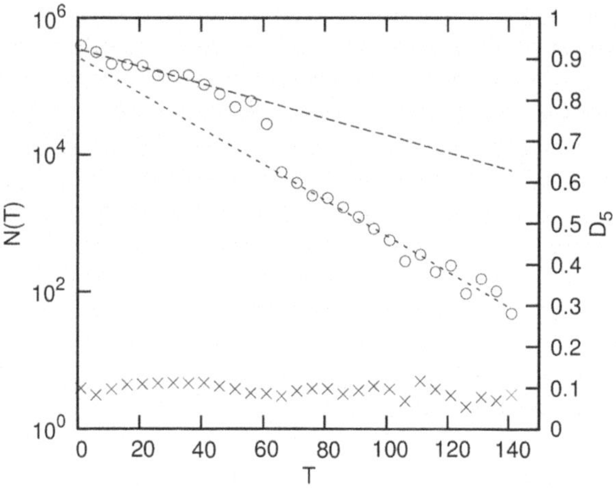

Fig. 17.2 Number of firms $N(T)$ classified into age-rank bins of 5-year widths is depicted as *circles* (∘). Decay rate of firm activities in 5-year D_5 is depicted as *crosses* (×) by right axis

17.3.2 Decay Rate of Firm Activities

We classify the firms into age-rank bins and compare their activities in 2008 and 2013. By classifying the active firms in 2008 into age-rank bins with 5-year width T, the number of firms $N(T)$ and the decay rate in five years D_5 are depicted in Fig. 17.2. Interestingly, the decay rate in the five years of active firms in 2008 does not depend on firm age T and is constant. The average value is estimated by

$$D_5 = 0.093 \pm 0.015. \tag{17.3}$$

In Fig. 17.2, there is a difference of $N(T)$ around $T = 60$. This corresponds to the decay of many firms during World War II. The number of firms established before and after World War II decreased following an exponential function. However, the parameters of the two exponential functions are different.

The exponential decrease of $N_>(T)$ and $N(T)$ is closely related to the constant decay rate of firm activities that does not depend on the age of firms. In the next section, we discuss this point.

17.4 Consistency of Laws

We clarify the relation between the two laws (17.2) and (17.3) observed in the previous section. One is the exponential decrease of the (cumulative) age distribution of firms. The other is their constant decay rate that does not depend on firm age. First, by analogy to nuclear decay, we analytically relate the two laws. After that, we quantitatively verify the relation by comparing parameter λ, which is estimated by the exponential decrease of the number of firms, with the decay rate of firm activities in 5-year period D_5.

Using constant parameter N_0, the age distribution of firms is written as

$$N(T) = N_0 e^{-\lambda T}. \tag{17.4}$$

Parameter N_0 is the number of firms newly founded at $T = 0$. Recently in Japan, since the number of newly founded firms has not changed drastically [22], the expression of Eq. (17.4) is acceptable as the number of firms at a point in time T. Equation (17.4) leads to a survival rate where firms are active in time T and in time $T + 1$ as follows:

$$\frac{N(T+1)}{N(T)} = e^{-\lambda}. \tag{17.5}$$

From this survival rate, the decay rate of firm activities in year D_1 is expressed as

$$D_1 = 1 - e^{-\lambda}. \tag{17.6}$$

Since λ is a constant parameter, Eq. (17.6) denotes that the decay rate of firm activities does not depend on age of firms T and is constant. This is an analytical consequence from the exponential decrease of the number of firms. Since this equivalence is well known in nuclear decay [19], we call λ a decay constant, as in physics. When λ is sufficiently smaller than 1, Eq. (17.6) leads to a λ that nearly equals the decay rate of firm activities in year D_1: $\lambda \sim D_1$.

Next, we quantitatively verify that D_1, which is estimated by the decay rate of firm activities, agrees with λ, obtained as the parameter of the age distribution of firms. From Eqs. (17.3) and (17.9), D_1 is evaluated as

$$D_1 = 0.020 \pm 0.003. \tag{17.7}$$

This means that 2 % firms annually decay at any age of the firms.

By applying the exponential function to the whole range of the cumulative age distribution of firms (Fig. 17.1), the decay constant is estimated as $\lambda \sim 0.078$, which is different from D_1. This disagreement reflects that in Japan, λ and N_0 changed before and after World War II when many firms decayed. Because D_1 is measured between 2008 and 2013, we must estimate λ by measuring the recent age distribution of firms, which is not cumulated, and compare λ with D_1.

As shown in Fig. 17.2, by applying the exponential function to the range from $T = 1$ to 45 in which N_0 is relatively stable, the decay constant is estimated as

$$\lambda = 0.029 \pm 0.03. \tag{17.8}$$

This value is close to D_1. From this agreement, we quantitatively confirmed the relation between the two laws concerning the decay of firms. This value is also close to around 0.03 which is reported in [17] as a value with respect to a recent few decades.

Furthermore, Fig. 17.2 shows that parameter λ^B of the exponential function, which is followed by firms founded before World War II, is larger than λ after World War II. In Fig. 17.2, λ^B is evaluated by

$$\lambda^B = 0.060 \pm 0.03. \tag{17.9}$$

The equivalence between λ and D_1 claims that decay rate λ^B of the firms before World War II is approximately twice decay rate λ after World War II. Figure 17.2 also shows that in Japan, number of firms ${N_0}^B$ annually founded before World War II approximately equals N_0 after World War II:

$$ {N_0}^{\mathrm{B}} \sim N_0. \tag{17.10}$$

17.5 Conclusion

In this study, by employing an exhaustive database of firms in Japan, we showed that the number of firms exponentially decreases with respect to their age from their foundation. When the number of firms established annually is nearly constant, this exponential decrease means that their decay rate does not depend on the their age. We confirmed this feature and classified firms into age-rank bins and investigated the decay rate and showed that the parameter of the exponential age distribution of firms agrees with the decay rate, as we analytically expected. The decay rate of firm activities before World War II is estimated to be approximately twice that after World War II in Japan. This confirms that the social structure of Japan dynamically changed after World War II.

The study of bankruptcy prediction of firms has a long history and there are various articles and practical applications [23–29]. Most of the models do not include

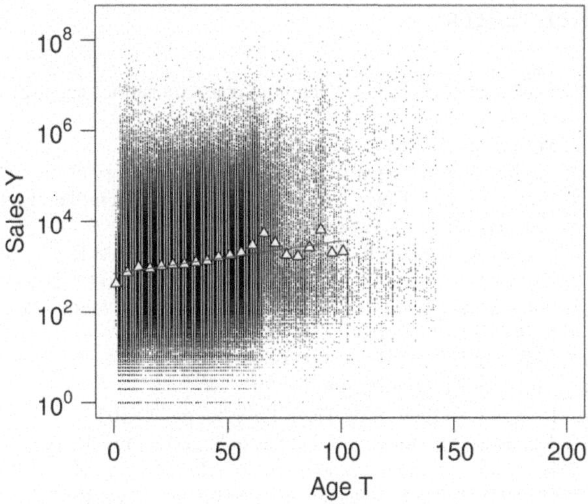

Fig. 17.3 Scatterplot between firm age T and sales Y. By classifying sales of firms in 2008 Japan included in database ORBIS into age-rank bins with 5-year widths, logarithmic average of sales in each bin is depicted by *triangles* (△)

the firm age as a parameter explicitly. For instance in [23], the model includes five parameters; (*Working Capital*)/(*Total Assets*), (*Retained Earnings*)/(*Total Assets*), (*Earnings before Interest and Taxes*)/(*Total Assets*), (*Market Value of Equity*)/(*Book Value of Total Liabilities*), and (*Sales*)/(*Total Assets*). Therefore, the main claim of this paper is consistent with them. However, parameters included in the models presumably depend on firm age and most of the models showed that the bankruptcy probability is smaller as the firm age is higher. This discrepancy probably comes from the data processing method that we consider not only bankruptcy but also dissolved (merger or take-over) as decay of firm activities.

We investigated the age distribution of firms to understand the long-term statistical laws observed by firm-size data. We can also observe that firm age exponentially correlates with the logarithmic average of sales at that age, as shown in Fig. 17.3. This relation significantly connects the age distribution of firms with the power-law distribution [4–6] of firm-size data. Furthermore, this exponential correlation can be interpreted as the average growth of sales. On the other hand, we found that the listed firms grow under power law by employing the database of listed firms both in the United States and Japan [30]. In the near future, we want to clarify these long-term statistical laws of firms, such as firm age distribution and firm growth, and confirm the relations among long- and short-term statistical laws, such as (quasi-) inverse symmetry and (Non-) Gibrat's law.

Acknowledgements This study was supported by JSPS KAKENHI Grant Number 24510212, 24710156.

References

1. Stanley HE (1971) Introduction to phase transitions and critical phenomena. Clarendon Press, Oxford
2. Mantegna RN, Stanley HE (1995) Nature 376:46
3. Saichev A, Malevergne Y, Sornette D (2009) Theory of Zipf's law and beyond. Springer, Berlin
4. Pareto V (1897) Cours d'économie politique. Macmillan, London
5. Newman MEJ (2005) Contemp Phys 46:323
6. Clauset A, Shalizi CR, Newman MEJ (2009) SIAM Rev 51:661
7. Fujiwara Y, Souma W, Aoyama H, Kaizoji T, Aoki M (2003) Phys A 321:598
8. Fujiwara Y, Guilmi CD, Aoyama H, Gallegati M, Souma W (2004) Phys A 335:197
9. Ishikawa A (2006) Phys A 367:425
10. Ishikawa A (2006) Phys A 371:525
11. Ishikawa A (2007) Phys A 383:79
12. Ishikawa A (2009) Prog Theor Phys Suppl 179:103
13. Tomoyose M, Fujimoto S, Ishikawa A (2009) Prog Theor Phys Suppl 179:114
14. Ishikawa A, Fujimoto S, Mizuno T (2011) Phys A 390:4273
15. Gibrat R (1932) Les inégalités économiques. Sirey, Paris
16. Sutton J (1997) J Econ Lit 35:40
17. Fujiwara Y (2004) Phys A 337:219
18. Miura W, Takayasu H, Takayasu M (2012) Phys Rev Lett 108:168701
19. Rutherford E, Soddy F (1903) Philos Mag 6:576
20. Bureau van Dijk, http://www.bvdinfo.com/Home.aspx
21. Statistics Bureau, Ministry of Internal Affairs and Communications, http://www.stat.go.jp/english/index.htm
22. Small and Medium Enterprise Agency, http://www.chusho.meti.go.jp/sme_english/index.html
23. Altman EI (1968) J Financ 23:589
24. Evans DS (1987) J Polit Econ 95:657
25. Evans DS (1987) J Ind Econ 35:567
26. Dunne T, Roberts M, Samuelson L (1989) J Law Econ 32:233
27. Variam J, Kraybil D (1992) Econ Lett 38:31
28. Shirata CY (1995) J Risk Manag 23:117
29. Kouki M (2009) J Middle East Financ Econ 14:26
30. Ishikawa A, Fujimoto S, Mizuno T, Watanabe T, in preparation

Chapter 18
Empirical Analysis of Firm-Dynamics on Japanese Interfirm Trade Network

Hayato Goto, Hideki Takayasu, and Misako Takayasu

Abstract We analyze Japanese interfirm trade network data for 20 years from the viewpoint of the metabolism of scale-free network evolution. We find that the preferential attachment effect of established firms is stronger than that of merged firms. This shows that merging firms should choose counterparties using delicate business strategies that may not be related to the degree. We also find that the distribution of lifespan of links is approximated well by an exponential function with the characteristic time of 6 years. The results imply the link creation and deletion is well characterized by a Poisson process.

18.1 Introduction

In 2009, the Federal Reserve Board of Governors implemented bank stress testing to check banks' asset health.[1] The results could indicate what kind of reaction big banks would have under certain bad situations. Because banks are among the most

[1] http://www.federalreserve.gov/newsevents/press/bcreg/20090507a.htm.

H. Goto (✉) • M. Takayasu
Department of Computational Intelligence and Systems Science, Interdisciplinary Graduate School of Science and Engineering, Tokyo Institute of Technology, 4259-G3-52, Nagatsuta-cho, Midori-ku, Yokohama 226-8503, Japan
e-mail: goto@smp.dis.titech.ac.jp; takayasu@dis.titech.ac.jp

H. Takayasu
Sony Computer Science Laboratories, 3-14-13, Higashigotanda, Shinagawa-ku, Tokyo 141-0022, Japan

Graduate School of Advanced Mathematical Sciences, Meiji University, 4-21-1 Nakano, Nakano-ku, Tokyo 164-8525, Japan

Department of Computational Intelligence and Systems Science, Interdisciplinary Graduate School of Science and Engineering, Tokyo Institute of Technology, 4259-G3-52, Nagatsuta-cho, Midori-ku, Yokohama 226-8503, Japan
e-mail: takayasu@csl.sony.co.jp

H. Takayasu et al. (eds.), *Proceedings of the International Conference on Social Modeling and Simulation, plus Econophysics Colloquium 2014*, Springer Proceedings in Complexity, DOI 10.1007/978-3-319-20591-5_18

important economic agents, it would be useful to assess the robustness of each bank and to consider this for precautionary actions in each situation.

Likewise, it is important for Japanese society to grasp the extent of each firm's capacity to deal with certain kinds of stresses because firms are also among the most important economic agents. Furthermore, the assessments would be particularly applicable to the real world if we could evaluate each firm as a part of the system that considered relationships among economic agents. As Fujiwara [1] commented, outer stresses could have very serious effects on firms' performances. This is in addition to serious effects from inner stresses, based on research about the reasons for bankruptcies in Japan.

Based on this discussion, some studies evaluate robustness using the system of economic agents. Iyetomi and his collaborators [2] developed an agent-based model to simulate firms' dynamics, which was constructed based on the relationship between firms and banks. In addition, Fujiwara and his collaborators [3] also analyzed lending networks between large firms and banks to evaluate their robustness in Japan. Those analyses would be beneficial to understand the strength of Japanese society from a macro viewpoint.

Here, as a first step toward developing a model that quantitatively estimates robustness of relational economic systems, we empirically analyze statistical properties of time evolution of real Japanese interfirm trade networks using a large time-series firm database.

18.2 Large Time-Series Firm Database

18.2.1 Japanese Interfirm Trade Networks

Business practices in Japan are unique. When building trustworthy relationships or managing credit risk, Japanese people first tend to gather their business partners' detailed corporate information. Then, professional third-party organizations are used to search their partners' credit status. TDB is one of the largest corporate research providers in Japan; it has assessed the credit status of firms for 115 years. Their credit research reports include detailed information of the financial statements of firms, their history, business partners, management and banking transactions.

In Sect. 18.3 of this study, we use three kinds of time-series data provided by TDB because the data have been stored using digitalization for several decades. As summarized in Table 18.1, the following data types are used: interfirm trade network data (which link the direction from consumers to suppliers, and have been stored from 1994 to 2014); interfirm bank trade network data (which link the direction from banks to firms, and have been stored from 1981 to 2014); and bankrupt firms data (which have been stored from 1980 to 2014).

Table 18.1 Dataset detail

Type	Time series	Total nodes	Total links
Trade network	From 1994 to 2014	19,527,573	68,608,558
Bank–trade network	From 1981 to 2014	Bank: 49,461, firm: 2,519,473	77,878,253
Bankrupt firms data	From 1980 to 2014	494,890	–

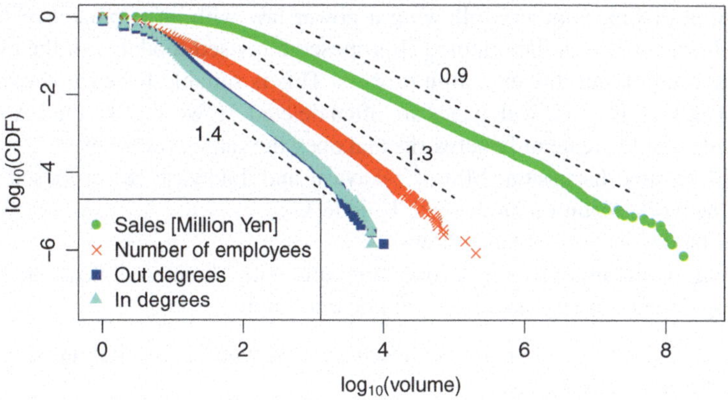

Fig. 18.1 Cumulative distribution of number of links, sales and number of employees in log-log scale (*green circles* for sales, *red crosses* for number of employees, *blue squares* for out degrees and *light blue triangles* for in degrees; the *dashed line* indicates the relationship with 0.9, 1.3, 1.4 and 1.4, respectively, for 2013)

18.2.2 Basic Properties

In 2007, M. Takayasu and her collaborators [4] published the first paper about the basic properties of the Japanese interfirm trade network. Using a different database, they found that those distributions follow power laws.

First, we confirm consistent power law distributions of link numbers, sales, and employees with exponents of about 1.4, 0.9, and 1.3, respectively, for 2013, as shown in Fig. 18.1. As for links on the Japanese interfirm trade network, there are directed money flows from buyer firms to supplier firms, that is, each node has in and out degrees. Therefore, we confirm consistent power law distributions of in and out link numbers and both distributions follow power-laws with exponents of about 1.4. Moreover, all of those distributions are almost identical in other years. Recently, Mizuno, Watanabe and Souma reported asymmetric distributions of in and out degrees based on the TDB data; however, their results cannot be reproduced [5].

18.3 Empirical Data Analysis

In 1999, Barabasi and Albert [6] proposed a simple model (BA model) of network growth realizing a power law with the concept of preferential attachment in which a new link is more likely to be attached to a node that has a larger number of links. Their simple algorithm creates an ever-growing network with cumulative distribution of link numbers following a power law with an exponent of two. In 2006, Moore and his collaborators [7] proposed a revised model with the event of node annihilation into the evolution process. This is suitable for such networks as *WorldWideWeb* because web pages are often deleted. However, like the BA model, their model could create only networks that obey the same power law.

Based on this discussion, Miura, Takayasu and Takayasu [8] proposed a new general network evolution model (MTT model).

Their model is described as follows:

Starting with any given network structure with directed links, one of the following three processes are chosen at one time step:

1. With probability a, one node is chosen randomly and it is annihilated together with the connecting links.
2. With probability b, one node is created. The new node has one in-link and one out-link. The partner nodes connecting to these new links are chosen randomly following the preferential attachment rule which is explained in the third section.
3. With probability c, a pair of nodes are chosen randomly and they coagulate making one node with conserving links. Merging nodes are chosen randomly following the preferential attachment rule which is also explained in the third section.

The parameters satisfy, $a + b + c = 1$, and the processes are repeated.

Their model seems to be suited to apply business network evolution for the following two reasons. First, their model realizes a statistically steady state in which cumulative distribution of the number of links of networks obeys a power law with an exponent close to the empirical value of 1.4 by tuning the parameters. Second, their stochastic model directly treats firm events of establishment, bankruptcy and merger that can be compared with the real data.

For these reasons, we analyze the data from the viewpoint of the MTT model.

1. Exponent of preferential attachment
   As for the analysis of the "Exponent of preferential attachment," the MTT model realizes cumulative distribution of number of links of business networks when both preferential exponents of creation nodes and coagulation nodes are the same. Here, we consider not only establishments but also mergers as the event of node creation; that is, we analyze both preferential exponents of creation nodes (establishments) and coagulation nodes (mergers).

2. Properties of bankrupt firms
   As for the analysis of the "Properties of bankrupt firms," the MTT model assumes an independent annihilation node in its process. We check this in the data.
3. Lifespan of trades
   Finally, we analyze the "Lifespan of trades" using trade network data and bank–trade network data. Through business activities, firms try to forge new relationships with others to earn money and to exchange old trades with ones that have better conditions. That is, links in the Japanese interfirm trade network also follow a constant metabolism, as do nodes. The MTT model indirectly assumes that the metabolism of links is replaced randomly by using the process of annihilation nodes. With regard to research analysis, there are some economic studies on these topics in the form of firm–firm trades and firm–bank trades in economic field [5, 9].

18.3.1 Exponent of Preferential Attachment

As mentioned above, preferential attachment is a key concept that means that a new link is more likely to be attached to a node with a larger number of links. Although the MTT model considered the exponents of preferential attachment of creation nodes and those of coagulation nodes as different parameters, the model assumed that both parameters had the same values when they were simulated to reproduce the properties of the Japanese interfirm trade network. Therefore, we check this assumption empirically.

Here, we follow the manner of adopting the MTT model in order to observe the preferential exponent.

$$\kappa(k) = \int_0^k \frac{Q(k)}{N(k)} dk \sim k^{\alpha+1} \tag{18.1}$$

where $Q(k)$ is the probability of connecting to an old firm with degree k and $N(k)$ is the number of nodes with degree k. We observe the following integrated attachment rate function to reduce fluctuation, as introduced by Jeong et al. [10]. As shown in Fig. 18.2, all of the obtained $\kappa(k)$ of established firms and that of merged firms are approximated by power laws with exponents of 2.4 and 1.5, respectively. In addition, both distributions of in degrees and out degrees are roughly the same.

Miura et al. [8] used the data that cannot be distinguished between established firms and merged firms. Therefore, they decided to set each preferential exponent $\lambda_b = \lambda_c = 1$ for their simulation. In this study, we use the data that can be distinguished between established firms and merged firms. We find that the preferential attachment's effect of established firms is stronger than that of merged firms with exponents of about 1.4 and 0.5, respectively. This makes sense because

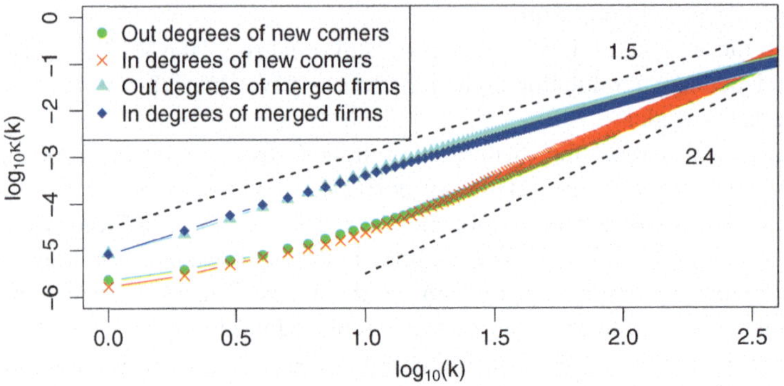

Fig. 18.2 Cumulative distribution of number of trades of firms that were attached by newcomers or merged firms in log-log scale (*green circles* for out degrees of firms that were attached by newcomers and *red crosses* for in degrees, *light blue triangles* for out degrees of firms that were attached by merged firms and *blue squares* for in degrees). The *bottom-dashed* guideline shows a line segment with *slope* $\simeq$ 2.4 that fits well with the newcomers' distribution from 1.0 to 2.5 of the horizontal axis (*Adjusted R squared* > 0.99) and the *top-dashed* guideline shows a line segment with *slope* $\simeq$ 1.5 that fits well with merged firms' distribution from 0.0 to 2.5 of the horizontal axis (*Adjusted R squared* > 0.99)

merger firms should choose counterparties with delicate business strategies that may not be related to the degree.

18.3.2 Properties of Bankrupt Firms

In the annihilation process of the MTT model, a node is chosen randomly. To verify the validity, Miura and his collaborators observed and empirically checked the lifespan of firms. As a result, the distribution was confirmed to be approximated well by an exponential function and they confirmed the validity of this process. Similarly, we observe the lifespan of about 500,000 bankrupt firms with data stored from 1990 to 2014. By comparison, we observe the years of existence of about 1,400,000 firms in 2013.

Figure 18.3 shows both distributions are roughly the same and well characterized by an exponential function, $P \geq (t) \propto \exp(\frac{-t}{\tau})$, where $\tau \simeq 23$ years ($\tau \simeq 19$ in Miura et al. [8]). It seems to be reasonable to annihilate a node randomly in the network evolution process from the viewpoints of lifespan of firms.

We know that bankruptcy is caused by various kinds of factors. For instance, we compare the distribution of sales of nonbankrupt firms with that of sales just before bankruptcy. As shown in Fig. 18.4, both distributions follow power laws, but the exponents are 1.3 and 0.9, respectively, and so, firms with obviously low sales tend to go bankrupt more than firms with high sales. Although random choice of the annihilation node is reasonable from the viewpoint of the lifespan of firms,

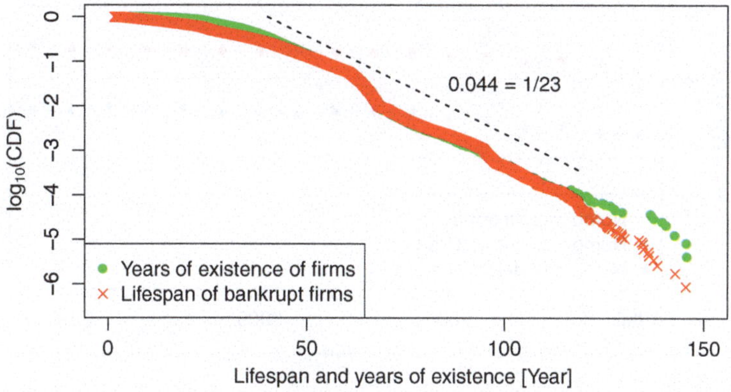

Fig. 18.3 Cumulative distribution of lifespan of firms in semi-log scale (*green circles* for the years of existence of firms for 2013 and *red crosses* for the lifespan of bankrupt firms for 1990–2014). The *dashed* guideline shows an exponential distribution with $\tau \simeq 23$

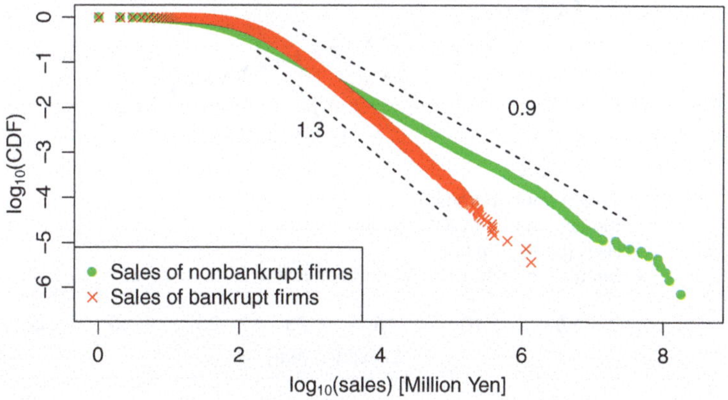

Fig. 18.4 Cumulative distribution of sales in log-log scale (*green circles* for sales of nonbankrupt firms for 2013 and *red crosses* for sales just before bankruptcy from 1990–2014). The *dashed* guidelines show power laws with 0.9 and 1.3

it is not suitable from a practical point of view. As for the model for Japanese interfirm network evolution, we should take account of various states of firms by using financial statements.

18.3.3 Lifespan of Trades

To clarify the behavior of links from the viewpoint of metabolism, here, we estimate the lifespan of links on the Japanese interfirm trade network.

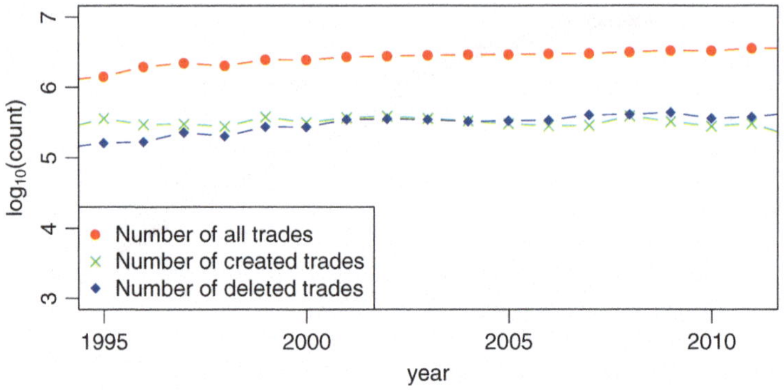

Fig. 18.5 Time evolution of the numbers of all trading links, newly created links and deleted links in semi-log scale (*red circles* for the number of all trades, *green crosses* for the number of created trades and *blue squares* for the number of deleted trades)

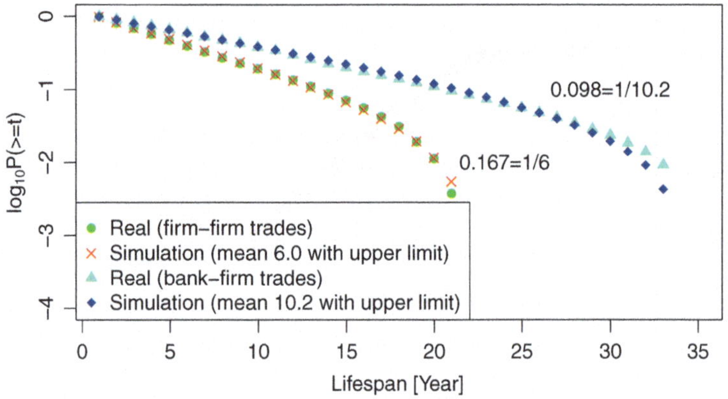

Fig. 18.6 Distribution of lifespan of firm–firm trades and bank–firm trades in semi-log scale (*green circles* for real trades for 1994–2014, *red crosses* for simulation of exponential distribution with the mean 6.0 and upper limit, *light blue triangles* for real bank–trades for 1981–2014 and *purple squares* for simulation of exponential distribution with the mean 10.3 and upper limit)

First, we stack all time-series Japanese interfirm trade network data from 1994 to 2014 and then, we estimate the start date and end date for each trade link. As shown in Fig. 18.5, the numbers of all trade links, created links and deleted links per year are almost steady. We find that about 15 % of trades are replaced each year.

Moreover, we observe the lifespan of trades by using stacked trade network data by the green circles in Fig. 18.6. These are expected to be characterized by the following exponential function just like the case of the lifespan of firms.

$$P_{firm-firm}(\geq t) = \exp(\frac{-t}{\tau}) \tag{18.2}$$

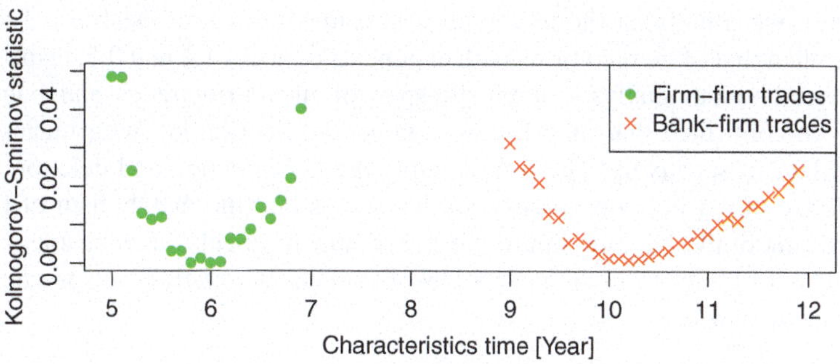

Fig. 18.7 Two-sample Kolmogorov-Smirnov statistical distribution between real data and simulation (*green circles* for firm–firm trades for 1994–2014 and *red crosses* for bank–firm trades for 1981–2014). As for firm–firm trades, Kolmogorov-Smirnov statistics take the minimum values 0.0006 when the characteristic time takes 6.0. As for bank–firm trades, Kolmogorov–Smirnov statistics take the minimum values 0.0010 when the characteristic time takes 10.2

The value of τ can be estimated by the survival rate of links per year, $1 - 0.15$, which should be given by $\exp(-\frac{1}{\tau})$. Then, we have $\tau \simeq 6$ years. We check the validity of this result by simulating the lifespan of trades following Eq. (18.2) and observing lifespan with the time window of 20 years, that is, the length from 1994 to 2014. As for the evaluation function to estimate the characteristics time τ, we use the two-sample Kolmogorov–Smirnov test [11], which can be used to test whether two distributions differ. The definition of the test statistic is $4D^2\frac{n_1 * n_2}{n_1+n_2}$, where D is a maximum vertical deviation between two distributions and n_1 *and* n_2 are the number of samples of each distribution. That is, if there is a small difference between two distributions, the test statistics take a small value. Figure 18.7 shows the statistical distributions of firm–firm trades by the green circles. It seems to be reasonable that the characteristic time is $\tau = 6.0$ for the firm–firm trade distribution. The red crosses in Fig. 18.6 show the simulation results with the characteristic time $\tau = 6.0$, which fits well with the real data.

In addition, we analyze the lifespan of bank–firm trades. The light blue triangles in Fig. 18.6 show the lifespan of that by using stacked data from 1981 to 2014. As shown by the statistical distributions of the red crosses in Fig. 18.7, it seems to be reasonable that the characteristic time is $\tau = 10.2$. The purple squares in Fig. 18.6 show the simulation results with the characteristic time, which fits well with the real data. We derive the following exponential distribution, meaning that about 10 % of bank–firm trades are replaced in each year.

18.4 Conclusion

In this study, we empirically analyzed time-series Japanese interfirm trade networks from the viewpoint of scale-free network evolution and we found some new properties of the Japanese interfirm trade network. As for our analysis of preferential exponents, we found that the preferential attachment effect of established firms is stronger than that of merged firms with exponents of about 1.4 and 0.5, respectively. With regard to our analysis of the lifespan of firm–firm trades and bank–firm trades, we confirmed that they follow exponential distribution with means about 6.0 and 10.2, respectively. The results imply the link creation and deletion is well characterized by a Poisson process, so this shows that the metabolism of links is replaced randomly. For our future work, we aim to check the robustness of the metabolism of trades from the viewpoints of firm characteristics such as sales scale and business categories.

Acknowledgements We would like to express our appreciation to Center for TDB Advanced Data Analysis and Modeling, Tokyo Institute of Technology for providing the datasets. This study is partially supported by the Grant-in-Aid for Scientific Research (C), Grant Number 24540395 and (B), Grant Number 26310207.

References

1. Fujiwara Y (2008) Chain of firms' bankruptcy: a macroscopic study of link effect in a production network. Adv Complex Syst 11(05):703–717
2. Iyetomi H et al (2009) Agent-based model approach to complex phenomena in real economy. Prog Theor Phys Suppl 179:123–133
3. Fujiwara Y et al (2009) Structure and temporal change of the credit network between banks and large firms in Japan. Economics E-Journal 3:7
4. Takayasu M et al (2007) Massive economics data analysis by econophysics methods-the case of companies' network structure. Ann Rep Earth Simul Cent, p 263. http://www.jamstec.go.jp/ceist/j/outreach/publication/annual/annual2008/pdf/2project/chapter3/P237Takayasu.pdf
5. Mizuno T, Wataru S, Tsutomu W (2014) The structure and evolution of buyer-supplier networks. PLoS One 9(7):e100712
6. Barabási A-L, Albert R (1999) Emergence of scaling in random networks. Science 286(5439):509–512
7. Moore C et al (2006) Exact solutions for models of evolving networks with addition and deletion of nodes. Phys Rev E 74(3):036121
8. Miura W, Takayasu H, Takayasu M (2012) Effect of coagulation of nodes in an evolving complex network. Phys Rev Lett 108(16):168701
9. Ono A et al (2010) Current status of firm-bank relationships and the use of collateral in Japan: an overview of the Teikoku databank data. In: Hitotsubashi University Design of Interfirm Network WP Series, vol 4
10. Jeong H, Néda Z, Barabási A-L (2003) Measuring preferential attachment in evolving networks. Europhys Lett 61(4):567
11. Massey FJ Jr (1951) The Kolmogorov-Smirnov test for goodness of fit. J Am Stat Assoc 46(253):68–78

Chapter 19
Direct Participants' Behavior Through the Lens of Transactional Analysis: The Case of SPEI®

Biliana Alexandrova-Kabadjova, Antoaneta Serguieva, Ronald Heijmans, and Liliana Garcia-Ochoa

Abstract This paper presents a methodology to study the flow of funds in large value payment systems (LVPSs). The presented algorithm separates the flow of payments in two categories: (1) external funds, i.e. funds transferred from other financial market infrastructures (FMIs) or provided by the central bank and (2) the reuse of incoming payments. Our method further studies the flow of intraday liquidity under the framework of its provision within the Mexican FMIs. The aim is to evaluate the impact of the intraday liquidity provision, and understand how liquidity is transmitted to participants in the Mexican LVPS SPEI®.

19.1 Introduction

The worldwide economic crisis has revealed that liquidity problems of (large) banks can occur suddenly, and with serious consequences for the (global) financial stability. The Lehman Brothers' collapse in 2008 is the most recent and widely referred to example. The interest, by both academics and financial authorities (such as central banks), in intraday liquidity management has gained momentum since then. Studying intraday liquidity flows, gives valuable insight into: (1) the provision of liquidity and the level of efficient use, (2) potential liquidity risks in settling payment obligations, and (3) the degree of interdependencies between financial

B. Alexandrova-Kabadjova (✉) • L. Garcia-Ochoa
Banco de México, Mexico City, Mexico
e-mail: balexandrova@banxico.org.mx; lilianago@banxico.org.mx

A. Serguieva
University College London, London, UK

King's College London, London, UK
e-mail: a.serguieva@ucl.ac.uk; k1346646@kcl.ac.uk

R. Heijmans
De Nederlandsche Bank, Amsterdam, The Netherlands
e-mail: ronald.heijmans@dnb.nl

H. Takayasu et al. (eds.), *Proceedings of the International Conference on Social Modeling and Simulation, plus Econophysics Colloquium 2014*, Springer Proceedings in Complexity, DOI 10.1007/978-3-319-20591-5_19

market infrastructures (FMIs) in terms of liquidity, in particular between the large value payment and securities settlement systems (LVPSs and SSSs). Central banks can use this insight to improve the intraday liquidity provision, to enhance the legal documentation of FMIs and to improve the implementation of the Principles of Financial market Infrastructures (PFMIs, see [1]).

To smoothen settlement of transactions in the FMIs, central banks provide (intraday) liquidity to participants in their LVPS. This liquidity together with liquidity from their payment obligations flow through different FMIs. Then liquidity is redistributed among participants either as transfers, payment obligations or secured/unsecured lending/borrowing among them. By studying liquidity flows central banks gain insight in the emerging network among participants revealing the structure of interdependency among financial institutions. Given that some of these institutions are direct participants in more than one FMI, the overall network of funds transfers among FMIs must be taken into account. Furthermore, central banks obtain information on the behavior of the participants related to the intraday liquidity management, i.e. the decisions during the day with respect to the number/value of payment obligations. We have identified three factors affecting the decision on how many payment orders will be sent for settlement by a participant throughout the day: (1) the amount of central bank money this participant has access to, (2) the amount of funds in terms of borrowing the institution can obtain from other participants, if required, and (3) the volume of payments received due to existing obligations either towards the participant or to its clients in a particular moment of the day.

Most countries in the industrialized world have implemented real time gross settlement (RTGS) systems (see [2]). In comparison to deferred net settlement (DNS) systems, RTGS eliminate settlement and credit risk that could arise between participants in the netting system. Nevertheless RTGS increase pressure on the level of intraday liquidity used to fulfill payment obligations, for that reason the most analysis on large value payment systems focuses on RTGS systems rather than netting. This paper defines a methodology to study the flow of funds observed in LVPSs related to funds transferred from other FMIs or provided by a central bank (first factor above), and the reuse of incoming payments (the second and third factors above). Under the specific operational rules used by the Mexican LVPS SPEI®, an algorithm is developed, using individual transaction data, to distinguish to what extent incoming payments are used to cover obligations. The time-scale of the algorithm is determined by the settlement rules of the system, which is every 3 s. In this study we present different time profiles, which are created under the same frequency of seconds. For more details of the operation framework, please refer to [3]. In comparison to other RTGS, SPEI® processes a high volume of transactions in real time settlement—on average 853,000 daily during 2013. From those operation around 91 % correspond to payments with value lower than 10,000 EUR.

The outline of this paper is as follows. Section 19.2 provides a brief literature overview. Section 19.3 describes the three different liquidity related issues: (1) the mechanism of liquidity provision, (2) distinguishing between the use of external funds and incoming payments, and (3) the profiles of daily, weekly and hourly activities in the Mexican LVPS SPEI®. Section 19.4 concludes and gives final remarks.

19.2 Literature Review

The focus of this paper is on liquidity flows and intraday patterns in (the Mexican) LVPS. This section reviews recent literature on LVPS flows and pattern, along with further research topics providing insight into FMI (data) analysis.

Armantier et al. [4] seek to quantify how the changing environment in which Fedwire (the largest LVPS of the United States) operates has affected the timing of payment value transferred within the system. They observe several trends in payment timing from 1998 to 2006. After 2000, the peak in payment activity shifts to later in the day. Indeed, post-2000, a greater concentration of payments occurs after 17:00. At the same time, however, several factors have been associated with increased payment activity early in the day, such as (1) the creation of the Continuous Linked Settlement (CLS) Bank, an institution that settles U.S. dollar payments early in the morning, (2) changes to the Clearing House Interbank Payments System's (CHIPS) settlement practices and (3) expanded Fedwire operating hours. Despite these developments, they find that the distribution of payment activity across the day still peaks more in the late afternoon.

Becher et al. [5] investigate the factors influencing the timing and funding of payments in the CHAPS Sterling system (the British LVPS), drawing where appropriate on comparisons with payment activity in Fedwire. Their results show that the settlement of time-critical payments in CHAPS supplies liquidity early in the day. Liquidity can be recycled to fund less urgent payments. CHAPS throughput guidelines also provide a centralized coordination mechanism that essentially limits any tendency toward payment delay. The relatively small direct membership of CHAPS further facilitates coordination, enabling members to maintain a constant flux of payments during the day.

In RTGS systems the liquidity demand is relatively high, due to the fact that each transaction is settled individually, and it is preferable that banks do not (unnecessarily) delay payments, as this will directly have an effect on the liquidity position of their counterparties. Participants choosing to pay after receiving its incoming payments are known as freeriders. Diehl [6] addresses the question of how free riding in LVPSs should be measured properly. He developed several measures for identifying free riding, which can be measured at individual bank level. His analysis shows that a combination of at least two measures is recommended for capturing the effects of free riding. Diehl's results are based on nine important banks in the German part of TARGET (the European LVPS system). The evaluated measures show stable payment behavior of most participants over time. However, his results also show some remarkable regime shifts, which indicate interesting insights about a single participant.

Massarenti et al. [7] studied the intraday patterns and timing of all TARGET2 interbank payments, and the evolution of settlement delay. One of their results shows that the first hour and a half and the last hour are the most crucial times during the system's operating hours, in terms of the numbers of payments processed (opening hours) and value of payments (closing hour). Their analysis provides only a system-wide view and can be used by operators and overseers of central banks.

Heijmans and Heuver [8] looked at the different liquidity elements which can be identified from LVPS transaction data. They developed a method using LVPS transaction data to identify liquidity stress in the market as a whole and at individual bank level. The stress indicators look at liquidity obtained from the central bank (monetary loans), unsecured interbank loans (value and interest rate), the use of collateral and the payments on behalf of their own business and of their clients.

In the case of Mexico, two simulation studies have been conducted by Alexandrova-Kabadjova and Solís-Robleda [3, 9] to analyze the settlement rules of SPEI® and the liquidity management of commercial banks that are direct participants. The authors examined the behavior of the selected financial institutions related to the reuse of incoming payments. They found that despite the growing volume of low value payments processed in real time through SPEI®, settlement is performed efficiently. Furthermore, the authors argue that observed patterns in the commercial banks' intraday behavior, particularly in the hours low value payments are sent, could be a sign of coordination among participants. However, a study accounting for all direct participants in SPEI® has not been done.

All of these papers look at (transaction level) data of one FMI, i.e. in these papers an RTGS system. However, they do not consider operational level per type of direct participants. This paper aims to identify the behavior that characterizes different types of participants in SPEI®.

19.3 Intraday Liquidity Flows

SPEI® is an LVPS and is operated by Banco de México (BdM). It operates under an RTGS scheme [10], as a bilateral and multilateral net settlement process is executed at the latest every 3 s after receiving a new instruction. Low and large value payments between banks and third parties are processed simultaneously in SPEI®. The SIAC is system that grant collateralized overdrafts. CLS is the Continuous Linked Settlement system that processes foreign exchange settlements under a Payment versus Delivery (PvD) scheme. DALÍ is a Security Settlement System (SSS) (see [11]).

Figure 19.1a schematically presents the different sectors of direct participants and infrastructures involved in the liquidity provision. Commercial banks (CB) and development banks (DB), which are credit institutions (CI), have access to central bank money in SIAC, as well as accounts in SPEI® and DALÍ. Brokerages (B) have accounts in SIAC, SPEI® and DALÍ, but do not have direct access to central bank money, whereas other NBFI have accounts only in SPEI®. In addition,

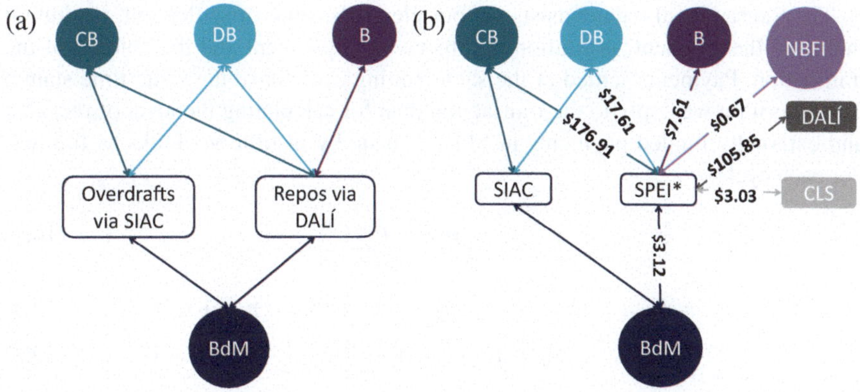

Fig. 19.1 Intraday liquidity flows. (**a**) Liquidity provision. (**b**) SPEI® volume in 2013

credit institutions and brokerages could also obtain intraday liquidity from the BdM through government debt repos, which are carried within DALÍ. Figure 19.1b presents the weighted links that connect SPEI® with the following FMIs: (1) SIAC, (2) CLS and (3) DALÍ.

Martinez-Jaramillo et al. [12], Bravo-Benitez et al. [13], and Alexandrova-Kabadjova et al. [14] show how the central bank gives shape to the complex interaction structure to these liquidity channels. The institutions with direct access to liquidity provided by the BdM through SIAC, evaluate in advance the level they need to settle obligations throughout the day in SPEI® and DALÍ. Based on these pre-evaluated amounts, they then transfer from their accounts in SIAC to their accounts in SPEI®, and use the accounts in SPEI® and DALÍ to settle their obligations throughout the day.

19.3.1 An Algorithm for Use of External Funds vs. Incoming Payments

Given the time structure of the transactional data in SPEI® (semi RTGS), it requires more liquidity than a (Deferred) Net Settlement (DNS) scheme. Nevertheless, payments are settled under bilateral and multilateral netting. For that reason it is important to measure how much this liquidity saving mechanism reduces pressure on the amount of external funds used. Given the high number of retail payments settled during a day, we further assume participants use incoming payments to cover other obligations. This process also reduces the demand for external funds. RTGS share common features around the globe, nevertheless operational and settlement rules and institutional frameworks exhibit significant differences among country. The algorithm we apply for this study is specifically designed for the rules and institutional framework, under which operates SPEI®.

The transactional data consists of four elements: time of settlement, institution that send the payment, institution that receive the payment and the amount of the transaction. Payments settled at the same netting cycle have the same time stamp. The algorithm we apply to the transaction data for calculating the level of recycled and externally funded payments in SPEI® made by institutions looks as follows: For each cycle:

$$A_{it} = P_{it}^{rec} - P_{it}^{sent} \tag{19.1}$$

$$F_{it} = F_{it-1} - (S_{it-1} + A_{it}), S_{it} = 0, \;\; if(S_{it-1} + A_{it}) < 0 \tag{19.2}$$

$$S_{it} = (S_{it-1} + A_{it}), \; if(S_{it-1} + A_{it}) >= 0 \tag{19.3}$$

where I is the set of participants in SPEI® and T the set of cycles in 1 day. Further, P_{it}^{rec} and P_{it}^{sent} are the sum of the incoming and outgoing payment amounts by each $i \in I$ in each cycle $t \in T$, respectively. S_{it} is the positive balance for each $i \in I$ in each cycle $t \in T$, given that $S_{i0} = 0$ for all i. F_{it} is the amount of funds each $i \in I$ in each cycle $t \in T$ has according to the transaction data, given that $F_{i0} = 0$ for all i.

F_{i0} and S_{i0} are set to 0, as the liquidity provision is on a daily basis and does not transfer to the next day. S_{it} is an auxiliary variable, i.e. it keeps track of the incoming funds that are used to cover other obligations in a way that if there is no availability of such funds, the value of S_{it} is zero at any time during the day. The value of F_{it}, which is a cumulative variable, on the other hand represents the overall need of external funds, e.g. F_{iD} gives the sum of external funding that institution i really needed during that day, as D denotes the last settlement period t in the set T of settlement cycles for the day. From historical data, we have been able to identify for each real cycle the transactions that have been settled together, such that each cycle t corresponds to a settlement cycle executed by SPEI® with the exact transactions corresponding to that cycle. In Fig. 19.2, $\sum_{i \in CI}(F_{iD})$ is the daily levels of external funds throughout the year 2013 for the dark grey area, whereas $\sum_{i \in CI} \sum_{t \in T} P_{it}^{sent} - \sum_{i \in CI}(F_{iD})$ will produce the light grey area, where CI stands for credit institutions, corresponding to the transactions covered with incoming payments by credit institutions. This area is labeled in the figure as "Recycling CI".

19.3.2 *Different Time-Scale Profiles of SPEI® Participants*

Figure 19.2 illustrates the use of external funds vs. incoming payments in SPEI®, on a daily basis for 2013. The black and lightest grey areas on the bottom show the need for external funding for CI and NBFI, respectively, whereas the dark grey and medium grey areas on the top show the proportion of incoming payments used to cover outstanding obligations for CI and NBFI. The proportion of the use of incoming payments is calculated as the total payments made by direct participants in SPEI® minus the evaluated by the algorithm needed external funding on that

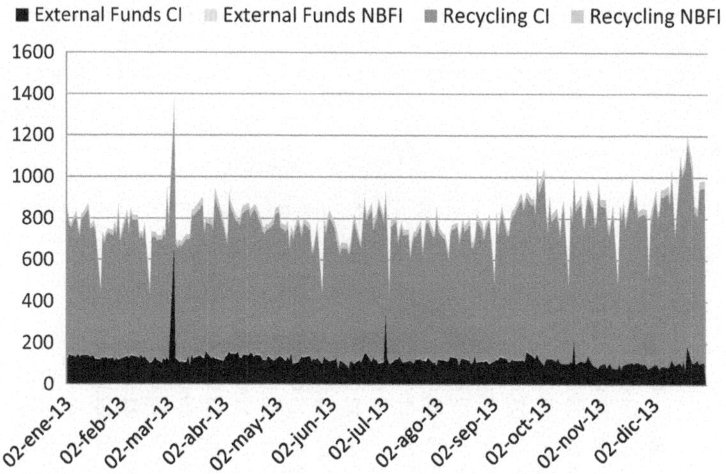

Fig. 19.2 Daily pattern of funding vs. recycling per type of institutions

day. Overall the percentage corresponding to external funds vs. incoming payments represents 15–85 %, respectively. We further observe that there are two spikes, corresponding to the daily patterns of March 4th and July 2nd for the study period. According to the results presented in [15], the increase in the value of external funds in both cases was used to cover payments initiate by participants and not for third party payments.

Figure 19.3 presents the accumulated hourly profile by type of SPEI® participants specifically for CI and NBFI, for the year of 2013. The meaning of colours is the same as in Fig. 19.2. The majority of payments are executed between 7.00 and 20.00 h. However, some payments are made by between 24:00 and 2:00. These payments are related to settlements in CLS and their coordination with the European market.[1]

Further, we observe that the higher volume of externally funded transaction by CI is settled between 9:00 and 12:00 (shown in the figure as the black area). Only Brokerages can get external funds to SPEI® for covering their obligations in this system from DALÍ. Nevertheless, this represents a very low volume of the overall transactions in SPEI®. For this reason the detailed intraday information, analysed and presented in Fig. 19.3, captures the externally financed obligations in SPEI® of NBFIs, as the very thin brightest grey area above the area of external funds used by CI. On the other hand, the internally financed or recycled payments of these institutions are captured by the dark and medium grey areas on the top of the figure. Obligations covered by CI from incoming payments represent the highest volume

[1]The CLS coordinates the transfer of currencies, and related payments appear as transactions between SPEI® participants and CLS. The system operates in Mexico between 24:00 and 5:00, which corresponds to 7:00–12:00 Central European Time.

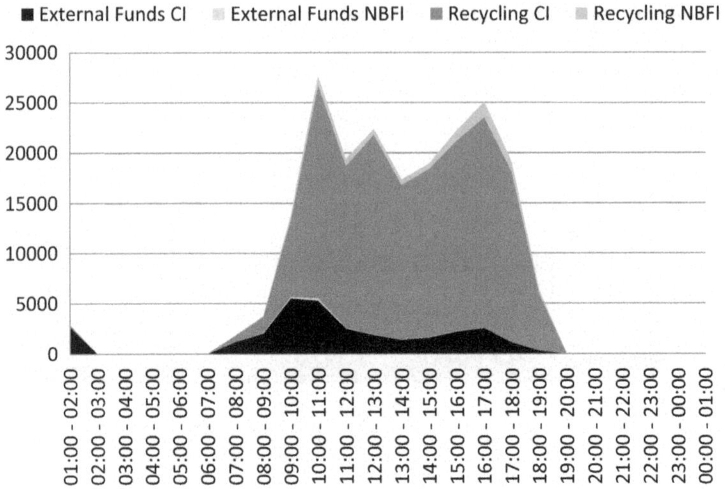

Fig. 19.3 Hourly pattern of funding vs. recycling per type of institutions

of settled transaction between 8:00 and 19:00, accounting for about 85 % of all transactions. Three peaks of internally financed obligations are observed—at 10:00, at 13:00 and at 18:00. The volume of payments covered by NBFI is stable during the daily hours, with an increase between 15:00 and 18:00 h. The analysis further shows that unsecured lending is taking place in SPEI® from CI to NBFI, which implies that the area shown as "Incoming payments NBFI" also includes those transactions and they are costly for the NBFI. In future work, we will focus on extending the methodology towards detection of transactions redistributing liquidity.

Figure 19.4 shows the average weekly pattern observed during the year 2013 of interactions among direct participants in SPEI®. We notice overall the behavior of direct participants is stable. We also observe that the actions taken by CI related to external funds usage are repetitive during the week. The black area at the bottom of Fig. 19.4 corresponds to externally financed obligations of CI—CBs and DBs. Those funds come from SIAC. Contrary, the external funds of NBFI flow from DALÍ, those funds are shown within a very thin brightest grey area. Next, direct participants' internally financed obligations are presented in the upper part of the figure.

Figure 19.5 introduces the average monthly pattern of the external and internal funding dynamics of the participants. Here, with the exception of the second day of month, we observe externally financed payments of CI are relatively stable on average monthly basis. The volume of internally covered payments by CI exhibits a more variable pattern during the month, with the highest peak presented at the end. Finally, the volume of transactions that NBFI covered with incoming payments is stable on average at monthly basis.

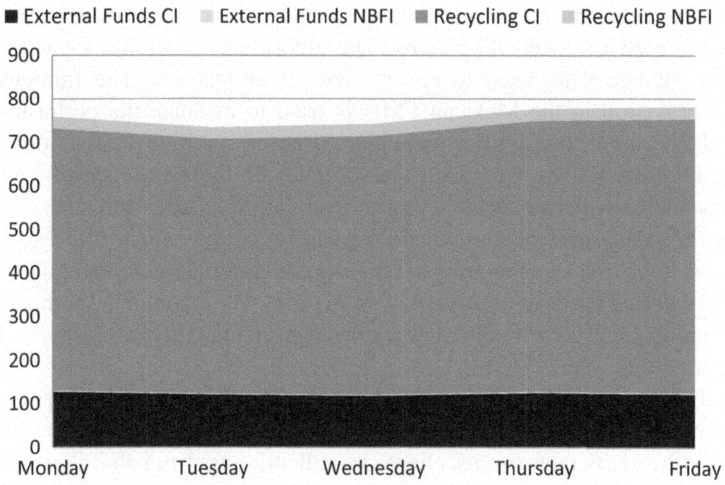

Fig. 19.4 Weekly pattern of funding vs. recycling per type of institutions

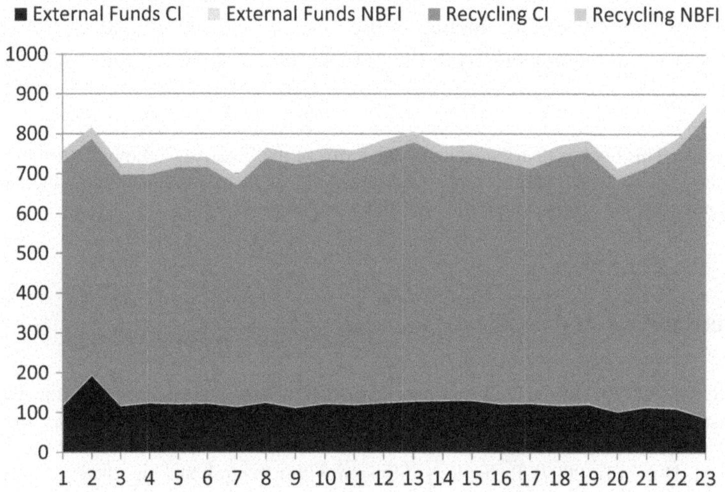

Fig. 19.5 Monthly pattern of funding vs. recycling per type of institutions

19.4 Summary and Conclusion

Liquidity issues have been one of the strongest lines of research for studies related to payment systems for over a decade now. This focus reveals the significant role of liquidity for the sound functioning of financial market infrastructures. The analysis in this paper contributes to this body of research. We presented an algorithm to study intraday payment flows in a system, in which around 91 % of the transactions correspond to retail payments. The algorithm separates funds transferred from other

FMIs (or provided by a central bank) from the incoming payments from other participants used to cover obligations. The algorithm determines to what extent incoming payments are used to cover payment obligations. The framework of liquidity provision in the Mexican FMIs is used to evaluate the performance of the developed algorithm, with a particular focus on the high transaction volume processed by the LVPS. We found that despite the high presence of low value payments and considering all direct participants in SPEI, the settlement rules are efficient (see [16]). Nevertheless there is room for improvement if the volume of retail payments is expected to grow in the future and participants would like to keep the use of external funds in the lowest possible level. We recommend comparing the current settlement rules with alternative rules that include the possibility for higher volume of netting.

The algorithm provides information about the participants' behavior related to the use of incoming payments. However, the information about who initiates an operation (the participant or its client) is still missing from the algorithm. We need to evaluate to what extent payments initiated by a third party increase the demand on liquidity or help to reduce the pressure on it through recycling. In addition, we need to gain more insights into the mechanism for redistribution of funds among participants, by including the unsecured/secured lending transactions into the analytical framework from the perspective of systemic relevance analyzed in [14].

References

1. CPSS, IOSCO (2012) Principles for financial market infrastructures: disclosure framework and assessment methodology. Bank for International Settlements and International Organization of Securities Commissions, Basel
2. Bech M, Hobijn B (2007) Technology diffusion within central banking: the case of real-time gross settlement. Int J Cent Bank 3:147
3. Alexandrova-Kabadjova B, Solís-Robleda F (2013) In: Alexandrova-Kabadjova B, Martinez-Jaramillo S, Garcia-Almanza A, Tsang E (eds) Simulation in computational finance and economics: tools and emerging applications. IGI Global, Hershey, pp 46–68
4. Armantier O, Arnold J, McAndrews A (2008) Changes in the timing distribution of fedwire funds transfers. FRBNY Econ Policy Rev September-2008:83
5. Becher C, Galbiati M, Tudela M (2008) The timing and funding of CHAPS sterling payments. FRBNY Econ Policy Rev September-2008:113
6. Diehl M (2013) Measuring free riding in large value payment systems: the case of TARGET2. J Financ Mark Infrastructures 1(3):31
7. Massarenti M, Petriconi S, Lindner J (2012) Intraday patterns and timing of TARGET2 interbank payments. J Financ Mark Infrastructures 1(2):3
8. Heijmans R, Heuver R (2014) Is this bank ill? The diagnosis of doctor TARGET2. J Financ Mark Infrastructures 2(3):3

9. Alexandrova-Kabadjova B, Solís-Robleda F (2012) In: Hellqvist M, Laine T (eds) Diagnostics for the financial markets - computational studies of payment system. Bank of Finland, Helsinki, pp 431–460
10. CPSS (1997) Real-time gross settlement systems. Bank for International Settlements, Basel
11. CPSS (2011) Payment, clearing and settlement systems in the CPSS countries. Bank for International Settlements, Basel
12. Martinez-Jaramillo S, Alexandrova-Kabadjova B, Bravo-Benitez B, Solórzano-Margain JP (2014) An empirical study of the Mexican banking system's network and its implications for systemic risk. J Econ Dyn Control 40:242
13. Bravo-Benitez B, Alexandrova-Kabadjova B, Martinez-Jaramillo S (2014) Centrality measurements of the Mexican large value payment system from the perspective of multiplex networks. Comput Econ 1–29. doi:10.1007/s10614-014-9477-0
14. Alexandrova-Kabadjova B, Serguieva A, García-Ochoa L (2014) In: Subrahmanian VS, Chen SH (eds) Proceedings of the 2014 international conference on behavior, economic and social computing (BESC). IEEE Press, pp 1–7
15. Alexandrova-Kabadjova B, Garcia Ochoa L (2015) The tale of two networks in SPEI: insights from structural indicators. In: Proceedings from payment and settlement system simulation seminar, Bank of Finland (forthcoming)
16. Allsopp P, Summers B, Veale J (2009) The evolution of real-time gross settlement: access, liquidity and credit, and pricing. The world bank financial infrastructure series, pp 1–53, February 2009

Part IV
Traffic and Pedestrian

Chapter 20
Pedestrian Dynamics in Jamology

Daichi Yanagisawa

Abstract In this paper, some achievements of research on pedestrian dynamics in Jamology are reviewed. The author focuses on three situations, i.e., one-dimensional unidirectional flow, egress process and queuing process. Experimental, theoretical and simulation results, which give us some prescriptions of easing jam in the situations above, are presented.

20.1 Introduction

"Jamology" is an interdisciplinary study on self-driven particles such as vehicles, pedestrians, ants, molecular motors, and many others [1]. It has characteristics of both physics and engineering. Its goal is not only elucidation of collective phenomena of self-driven particles, but also development of solutions for jam, which disrupts smooth flow.

Dynamics of pedestrians, which has been vigorously studied in traffic engineering, architecture and psychology, is also one of the main research topic in Jamology. Pedestrians (self-driven particles) do not obey the law of action and reaction; therefore, Newtonian mechanics does not work effectively. Moreover, it is almost impossible to predict the movement of individual pedestrian in detail since he/she has own will. In spite of these difficulties, researchers have developed new theories and models [1, 2], and studied macroscopic collective behaviors of pedestrians when the destination of pedestrians is clear. Some solutions for ease congestion (jam) are also considered.

In this paper, the author reviews the achievements of research on one-dimensional unidirectional flow, egress (evacuation) process and queuing process. In congested unidirectional flow, slow rhythm improves pedestrian flow [3]. A simple egress model succeeds in explaining the effect of competitive and cooperative behavior at an exit, which has been previously studied by experiment and simulation

D. Yanagisawa (✉)
Department of Aeronautics and Astronautics, School of Engineering, The University of Tokyo, 7-3-1, Hongo, Bunkyo-ku, Tokyo, 113-8656, Japan
e-mail: tDaichi@mail.ecc.u-tokyo.ac.jp

H. Takayasu et al. (eds.), *Proceedings of the International Conference on Social Modeling and Simulation, plus Econophysics Colloquium 2014*, Springer Proceedings in Complexity, DOI 10.1007/978-3-319-20591-5_20

[4, 5]. In queuing process, the effect of walking distance, which is necessary for modeling pedestrians, are introduced to the original queueing model in the queueing theory[6]. Our extended model succeeds to suggest a suitable type of queuing system as a function of the parameters in queuing systems [7].

20.2 Effect of Rhythm on Unidirectional Flow

Unidirectional flow is one of the fundamental situations in pedestrian dynamics; however, it still includes complex phenomena such as overtaking and movement in lateral directions. Therefore, researchers often consider an ideal condition, i.e., one-dimensional circuit where overtaking is prohibited as in Fig. 20.1a. Then they investigate fundamental diagram (FD) as in Fig. 20.1b, which is a relation between flow and density of pedestrians [8, 9]. We mainly see two phases in FD. One is free-flow phase, where flow becomes large against the increase of density. The other is congested phase, where flow becomes small against the increase of density.

In [10] the effect of music on an individual pedestrian has been studied experimentally. Inspired by this research, we analyze the effect of rhythm on crowded pedestrians experimentally and reveal that *slow* rhythm increases pedestrian flow in congested situations without any danger.

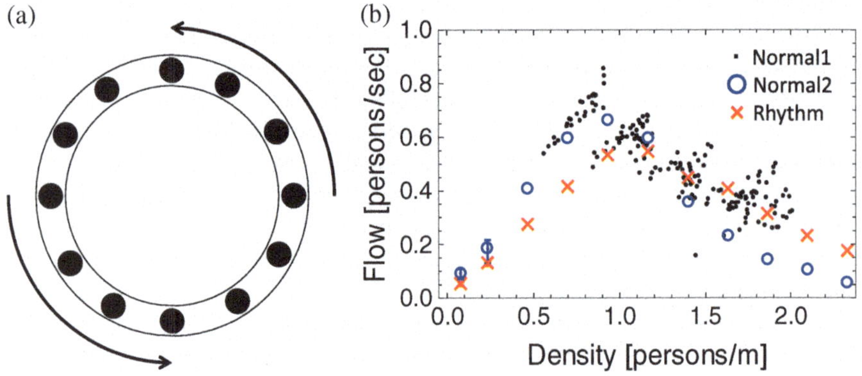

Fig. 20.1 (**a**) Schematic view of one-dimensional unidirectional pedestrian flow in a circuit. (**b**) Fundamental diagram. Normal1 is data in [8]. Normal2 and Rhythm are data in [3]

20.2.1 Experimental Setup

We constructed a circuit whose inner and outer radii were $r_i = 1.8\,[\mathrm{m}]$ and $r_o = 2.3\,[\mathrm{m}]$, respectively. The participants of the experiment, who were male university students between 18 and 39 years old, walked the circuit in the counter-clockwise direction. In the beginning of each trial, we briefly instructed participants to distribute homogeneously in the circuit without signs on the floor or measuring the distance between each participant.

We executed 11 kinds of density conditions. The number of the participants in the circuit in each condition was $N = \{1, 3, 6, 9, \cdots, 30\}$. The conditions $N = 1$ and $N = 3$ were tried three times with different participants, and the other conditions were tried once. Each trial was more than 1 min in the cases $N \geq 3$. The global density is calculated as

$$\frac{N}{\pi\,(r_i + r_o)} \tag{20.1}$$

and was used to depict FDs.

Two kinds of walking were performed in the experiment. In the first case, we did not give any specific instructions to the participants, so that they walked normally. In the latter case, the participants were instructed to walk with the sound from the electric metronome, whose rhythm is 70 [BPM]. Note that we did not inform which foot to move first.

In the case $N = 1$, we measured the lap time for completing a circuit. In the case $N \geq 3$, we measured the time that each participant passes the measuring point in the circuit and depicted the cumulative plots, which show the evolution of the total number of participants who passed the measuring point. Then, linear regression analysis gives pedestrian flows as the slope of the cumulative plots. The condition $N = 1$ and 3 (the smallest and the second smallest density case) were performed three times so that the maximum and minimum values are plotted by the error bars in Fig. 20.1. In the case $N \geq 6$, the errors obtained from the linear regression analysis is too small to be depicted by error bars.

20.2.2 Experimental Verification of the Effect of Rhythm

Figure 20.1b shows the FDs obtained in our experiment (Normal2 and Rhythm). From the figure, first, we see that the flow is larger in the normal case than the rhythmic case in the low-density regime. Hence, the pace 70 [BPM] is much slower than the normal-walking pace of the participants, and the flow becomes smaller if the participants try to walk with the slow rhythm. Second, linearity of the flows in the low-density regime verified that participants walked with constant step size and pace. Third, we see that the flow decreases linearly in the rhythmic case, whereas, the flow in the normal case is convex downward in the high-density regime. This observation is supported by calculating average second difference of the data between $N = 15$ and 30, which are 0.14 and 0.03 in the normal and rhythmic cases, respectively. We consider that the clear convexity, which was not seen in the FD in [8] (Fig. 20.1b Normal1), is observed because we performed the experiment with a density of more than 2.0 [persons/m]. In [3], these experimental data are compared with (fitted by) the mathematical model. They indicates that linear decreasing of the flow implies that step size of pedestrians becomes smaller. On the other hand, convexity indicates that both step size and pace become smaller. Thus, the walking pace decreases from the influence of the predecessors in the normal case, while it is maintained by rhythm in the rhythmic case. Finally, we observe the crossing of the two plots. In other words, the flow of the rhythmic case exceeds that of the normal case in the high-density regime. Therefore, we have verified that slow rhythm improves the pedestrian flow.

20.3 Simple Model of Egress Process

Egress process is vigorously studied since it is strongly related with evacuation in emergency situation. Many simulations as well as experiments by real pedestrians are performed [5, 11]; however, there are few theoretical research [12]. In this section, we introduce a simple model, which explains the effect of competitive and cooperative behavior near an exit.

We consider a system, which is composed of an exit cell and its Moore neighboring cells as in Fig. 20.2a. Time is discrete in this model. It is assumed that pedestrians come from the outside of the system, and at each five neighboring cell, there is a pedestrian with the probability σ. They try to move to the exit cell with the

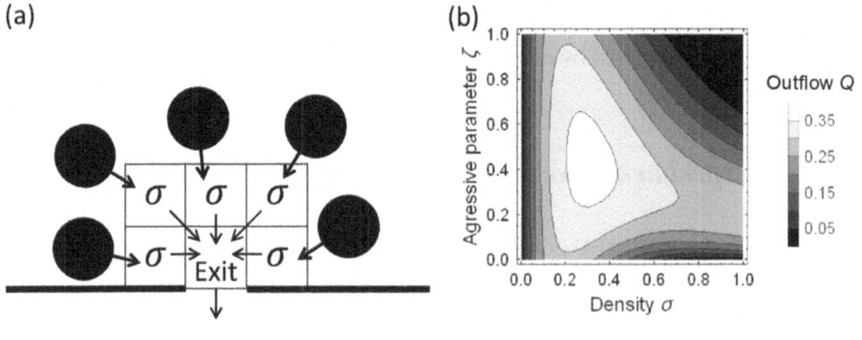

Fig. 20.2 (**a**) Schematic view of a simple egress model. (**b**) Contour plot of outflow Q

probability 1 if it is not occupied by others.[1] The pedestrian at the exit cell get out from the system with the probability 1. We denote the number of the neighboring cells by n. (Here $n = 5$.) Then the probability of m pedestrians trying to move to the exit cell is described by the binomial distribution as follows.

$$b(m) = \binom{n}{m} \sigma^m (1 - \sigma)^{n-m} \tag{20.2}$$

If more than one pedestrian try to move to the exit cell, conflict occur since only one pedestrian can stay at the exit cell. Thus, we introduce frictional function, which is the probability that no one can move to the exit cell.

$$\psi(m) = \begin{cases} 0 & (m = 0, 1) \\ 1 - m\zeta(1 - \zeta)^{m-1} & (m \geq 2) \end{cases}, \tag{20.3}$$

where ζ is the aggressive parameter, which is the probability that the pedestrians trying to move to the exit cell in spite of the conflict among them. The parameter ζ represents the competitiveness of the pedestrians. If ζ is small, pedestrians are in cooperative mood and often give way to each other. By contrast, pedestrians are in competitive mood and collide with each other at the exit when ζ is large. The second term in the second line in Eq. (20.3) corresponds to the situation that one pedestrian aggressively move to the exit cell and the other $m-1$ pedestrians give way to others. By subtracting this term from 1, we calculate the probability that no one can move. Furthermore, Eq. (20.3) indicates that if all the pedestrians give way to others, no one moves to the exit cell. By using Eqs. (20.2) and (20.3), the probability that one pedestrian reaches the exit cell is calculated as

$$r(n) = \sum_{m=1}^{n} (1 - \psi(m))\, b(m). \tag{20.4}$$

We denote the probabilities that the exit cell is vacant and occupied by $P(0)$ and $P(1)$, respectively. Then, the master equations in the stationary state are described as

$$\begin{bmatrix} P(0) \\ P(1) \end{bmatrix} = \begin{bmatrix} 1 - r(n) & 1 \\ r(n) & 0 \end{bmatrix} \begin{bmatrix} P(0) \\ P(1) \end{bmatrix} \tag{20.5}$$

We solve Eq. (20.5) with the normalization condition $P(0) + P(1) = 1$ and obtain the expression of pedestrian outflow.

$$Q(\sigma, \zeta, n) = 1 \cdot P(1) = \frac{r(n)}{1 + r(n)}. \tag{20.6}$$

[1]If we introduce the moving probability from the neighboring cells to the exit cell, which is 1 in this paper, we can represent slow and first pedestrians.

Figure 20.2b is contour plot of the outflow Q in the case $n = 5$. When $\sigma < 0.2$, few conflicts occur, so that the outflow Q is hardly affected by the aggressive parameter ζ. When $\sigma > 0.2$, conflicts occur more often, so that Q achieves maximum against the change of ζ with constant σ. Few pedestrians move to the exit cell in a conflict situation in the small ζ case since they often give way to others. On the contrary, a conflict among aggressive pedestrians is not solved in the large ζ case. Therefore, both small and large ζ diminish the outflow Q. It is interesting that the simple model introduced in this section succeed to show the existence of the optimal strength of giving way to others.

In [4], the effect of competitive and cooperative behavior on evacuation is experimentally studied. It indicates that pedestrians can evacuation faster when they are in cooperative mood if the width of the exit is narrow. This result corresponds to our result in Fig. 20.2. In the experimental evacuation process, it is feasible to consider that the density around the exit is large, i.e., σ is large. Figure 20.2b implies that the outflow achieves maximum at small ζ, i.e., cooperative mood, when density is large.

20.4 Queueing Process

Pedestrian queueing system, which we see at cash registers in super markets, ticket-vending machines in stations, and automated teller machines in banks, is also one of the important themes in the field of pedestrian dynamics for the following reasons. Firstly, pedestrians become stressful when they wait at a queue for a long time. Secondly, long waiting time in one queueing system affects the starting time of other events, for instance, if some passengers take a long time to pass a security check in an airport due to a long waiting queue, the departure time of the flight may delay [13]. Lastly, a long queue prevents smooth movement of pedestrians and encourages forming a jam around it. Thus, we investigate efficient type of pedestrian queueing system in this section.

According to the queueing theory [6], mean waiting time (MWT) in the fork-type queueing system (Fork) is always shorter than that in the parallel-type queueing system (Parallel). However, in the queueing theory, the effect of walking distance from the head of the queue to the service windows is not included. The effect of the distance may significantly influences on the MWT of pedestrians in large Fork such as an immigration inspection floor in an international airport where walking distance is very long. Therefore, we have developed a walking-distance introduced Parallel (D-Parallel, Fig. 20.3a) and a walking-distance introduced Fork (D-Fork, Fig. 20.3b). We show that MWT becomes shorter in D-Parallel than D-Fork when sufficiently many pedestrians are waiting in the queue.

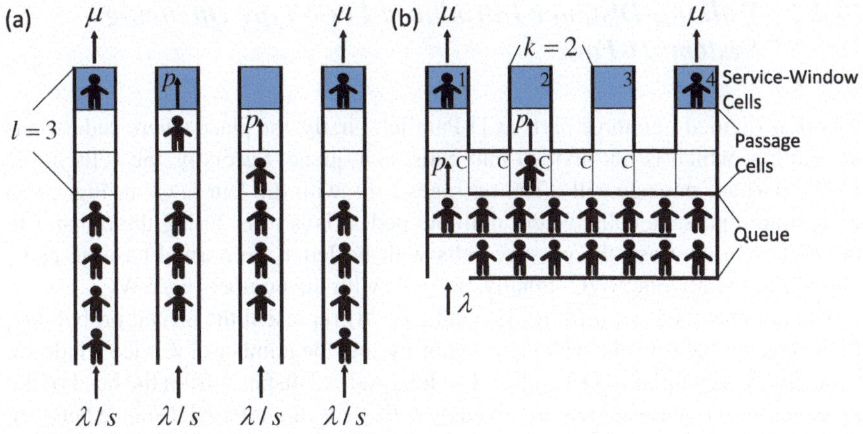

Fig. 20.3 Schematic view of walking-distance introduced queueing model. (**a**) D-Parallel. (**b**) D-Fork

20.4.1 Walking-Distance Introduced Parallel-Type Queueing System: D-Parallel

D-Parallel (Fig. 20.3a) is divided into three parts, which are queues, passage cells, and service-window cells (SWCs). The number of SWCs are denoted by s. The SWCs have two states: vacant and occupied. Note that each cell contains only one pedestrian at most. Time is discrete in the model. A pedestrian arrives at the queueing system (each queue) with the probability λ (λ/s). When a SWC becomes vacant state, a pedestrian at the head of the queue decides to proceed to the SWC, and it becomes occupied state. The pedestrian walks passage cells to the SWC for l cells with probability p and starts receiving the service.[2] It finishes with the probability μ and the pedestrian leaves the system. At the same time the SWC becomes vacant state again. The walking effect delays the start of service and affects MWT. Note that the size of the SWCs is considered as one cell in our model, so that $l \geq 1$ is satisfied in D-Parallel. Besides, we focus on the situation that $p > \lambda, \mu$, which is natural for queueing situation.

[2]The parameter p controls the walking velocity of pedestrians. If we set p to small value, we can consider slow pedestrians, who are greatly affected by long walking distance. In this paper, it is fixed to $p = 1$ in the simulation, so that fast pedestrians are considered.

20.4.2 Walking-Distance Introduced Fork-Type Queueing System: D-Fork

D-Fork is divided into three parts as D-Parallel. Firstly, the place where pedestrians are waiting, which is not divided into cells, is a queue. Secondly, the cells in the middle part are passage cells. The passages cells with the letter "C" in Fig. 20.3b are common passage cells, where multiple pedestrians pass during the transfer to the SWCs. In contrast, the passage cells with no letter are normal passage cells, which lead to only one SWC. Finally, the cells with the numbers are SWCs.

The parameters λ, μ, p ($\in (0, 1]$), and s ($\in \mathbb{N}$) represent the arrival probability, the service probability, the walking probability, and the number of service windows, respectively, as similar to D-Parallel. The longitudinal distance from the head of the queue to the service windows are given by l. Besides, the interval distance between two service windows is given by k. Figure 20.3b represents the case where $s = 4$, $l = 3$, and $k = 2$.

Outline of the movement of pedestrians and the state transition of the SWCs are as follows (Details are described in Sect. 20.4.3). A pedestrian arrives at the queueing system with the probability λ. When he/she reaches the head of the queue and there is at least one vacant SWC, he/she decides to move to it, and its state changes into occupied state. The pedestrian proceed to the SWC by one cell with the probability p in one time step if his/her proceeding cell is vacant. A service starts when the pedestrian arrives at the SWC, and after it finishes with the probability μ the state of the SWC changes into the vacant state.

20.4.3 Update Rule in Simulation

The simulation of D-Parallel and D-Fork consists of the following five steps per unit time step.

1. If the following three conditions:

   - there is at least one pedestrian in the queue,
   - target SWC of the pedestrian at the head of the queue is not determined,
   - there is at least one SWC in the vacant state,

   are satisfied, the pedestrian at the head of the queue decides to proceed to the nearest vacant SWC, and the state of the SWC becomes occupied. Note that he/she never changes the target SWC even if some other SWCs which are nearer than his/her target become vacant state during his/her walking process.
2. Add one pedestrian to the queue with the probability λ.
3. If the target SWC of the pedestrian at the head of the queue is determined and the first cell of the passage cell is vacant, proceed him/her to the first passage cell with the probability p.

4. Proceed each pedestrian in the passage cells (except the pedestrian who moved in the step 3) by one cell to his/her service windows with the probability p if there is no pedestrian at their proceeding cell.
5. Remove pedestrians at the SWCs (except the pedestrian who reach the SWC in the step 4) and change their states into vacant state with the probability μ.

20.5 Comparison of Mean Waiting Time in D-Parallel and D-Fork

Figure 20.4 show MWT against arrival-service ratio ρ ($= \lambda/(s\mu)$). We clearly observe the crossing of the curves of D-Parallel and D-Fork. This indicates that when the arrival-service ratio ρ is small, i.e., there are not frequent arrival of pedestrians against total service efficiency of the system; we should form D-Fork to decrease MWT. On the contrary, when ρ is large, i.e., pedestrians arrive frequently against total service efficiency of the system, we should form D-Parallel.

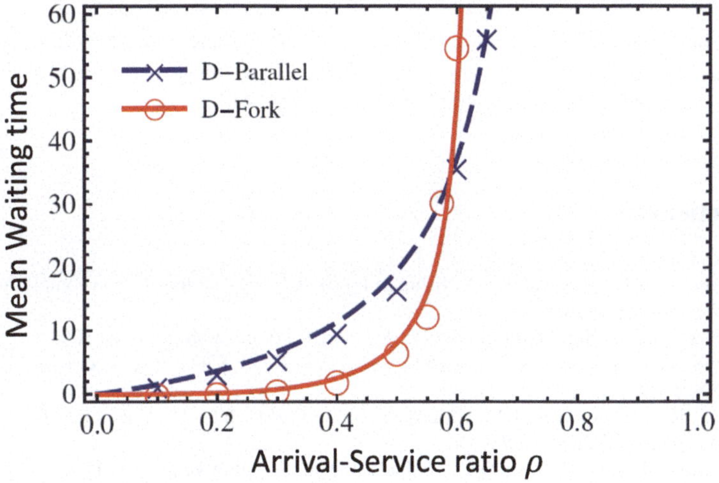

Fig. 20.4 Mean waiting time against arrival-service ratio ρ in the case $\mu = 0.1$ and $p = 1$. The *marker plots* are the result of our simulation and the *solid* and *dashed lines* are result of approximated theoretical analysis in [7]

20.6 Summary

In this paper, we briefly review our result in pedestrian dynamics. We have discovered that slow rhythm improves pedestrian flow in congested situation. Our simple theoretical model well explains the effect of competitive and cooperative behavior at a narrow exit on outflow. Simulation of walking-distance introduced queueing models indicates that the fork-type is not always the suitable type for pedestrian queueing system.

Egress process is studied in more detail. A simple egress model introduced in Sect. 20.3 also helps explaining the effect of an obstacle at a narrow exit [11]. It is also revealed that diminution of local flow enhance the total flow at the exit in the case there are successive bottlenecks [14].

In the near future, combined topics such as the effect of rhythm on the egress process should be also studied. Furthermore, as we have investigated suitable type of queueing system, effective room arrangement and position of exits for egress process is needed to be considered. It is expected that application of our results will diminish "jams" in the real world.

Acknowledgements This work was supported by JSPS KAKENHI Grant Number 24760058.

References

1. Schadschneider A, Chowdhury D, Nishinari K (2010) Stochastic transport in complex systems. Elsevier, Amsterdam/Oxford
2. Helbing D (2001) Rev Mod Phys 73(4):1067 doi:10.1103/RevModPhys.73.1067
3. Yanagisawa D, Tomoeda A, Nishinari K (2012) Phys Rev E 85(1):016111. doi:10.1103/PhysRevE.85.016111
4. Muir HC, Bottomley DM, Marrison C (1996) Int J Aviat Psychol 6(1):57 doi:10.1207/s15327108ijap0601_4
5. Kirchner A, Klüpfel H, Nishinari K, Schadschneider A, Schreckenberg M (2003) Phys A Stat Mech Appl 324(3–4):689 doi:10.1016/S0378-4371(03)00076-1.
6. Bolch G, Greiner S, de Meer H, Trivedi KS (2006) Queueing networks and Markov chains: modeling and performance evaluation with computer science applications, 2nd edn. Wiley, Hoboken.
7. Yanagisawa D, Suma Y, Tomoeda A, Miura A, Ohtsuka K, Nishinari K (2013) Transp Res C Emerg Technol 37:238 doi:10.1016/j.trc.2013.04.008.
8. Seyfried A, Steffen B, Klingsch W, Boltes M (2005) J Stat Mech Theory Exp 2005(10):P10002. doi:10.1088/1742-5468/2005/10/P10002.
9. Jelić A, Appert-Rolland C, Lemercier S, Pettré J (2012) Phys Rev E 85(3):036111. doi:10.1103/PhysRevE.85.036111
10. Styns F, van Noorden L, Moelants D, Leman M (2007) Hum Mov Sci 26:769 doi:10.1016/j.humov.2007.07.007.

11. Yanagisawa D, Kimura A, Tomoeda A, Nishi R, Suma Y, Ohtsuka K, Nishinari K (2009) Phys Rev E 80(3):036110. doi:10.1103/PhysRevE.80.036110.
12. Yanagisawa D, Nishinari K (2007) Phys Rev E 76(6):061117. doi:10.1103/PhysRevE.76.061117.
13. Kazda A, Caves RE (2007) Airport design and operation, 2nd edn. Elsevier, Oxford
14. Ezaki T, Yanagisawa D, Nishinari K (2012) Phys Rev E 86(2):026118. doi:10.1103/PhysRevE.86.026118.

Chapter 21
Qualitative Methods of Validating Evacuation Behaviors

Tomoichi Takahashi

Abstract Multi-agent simulations (MAS) have been used to study the dynamics of social systems. Disaster-related simulation is one of application fields. The simulation is applied to scenarios that are difficult to perform drills in the real world. The results provide useful data such as the amount of time people take to evacuate buildings and how smoothly rescue responders arrive at target points in the buildings. Making use of the simulation results to plan disaster-prevention measure, we need to verify that the simulation results that are reasonable at scenarios that are not confirmed from real data and observations. In this paper, we discuss the standardization process of MAS-based evacuation simulations by examining qualitative differences perceived in our evacuation simulations.

21.1 Introduction

Disasters may occur anytime and anywhere in the world. Disaster prevention methods are planned and drills are conducted to check disaster-related social systems involving damage assessment, response measurement, and evacuation guidance because these help save lives during emergencies. These drills are used to estimate the required safe egress time (RSET) and improve prevention plans for emergency situations. Students at schools and occupants of buildings are encouraged to participate in such drills. However, it is difficult to conduct drills that involve a large number of people in real-world environments, such as the scenario that occurred on September 11, 2001, at the World Trade Center (WTC) buildings in New York City [1].

We learn how people behaved during disasters from media stories and reports published by those in authority. Their actions involve the following general phenomena: they begin evacuation based on their individual circumstances; they communicate with each other and share information about the emergency; people in

T. Takahashi (✉)
Meijo University, Shiogamaguchi Tempaku, Nagoya, Japan
e-mail: ttaka@meijo-u.ac.jp

H. Takayasu et al. (eds.), *Proceedings of the International Conference on Social Modeling and Simulation, plus Econophysics Colloquium 2014*, Springer Proceedings in Complexity, DOI 10.1007/978-3-319-20591-5_21

the vicinity of the area where the emergency occurred adjust their actions according to the shared information. These phenomena reflect features of collective behavior in emergency situations. The simulation of these phenomena includes modeling individual emotions, interactions between humans, and characterizing the behavior of groups of people in a crowd. The simulation system that consists of these components provides a solution to a possible emergency. However, nobody can validate the results of the simulations or guarantee how the system would really work during an emergency.

We believe that evacuation simulation systems can be used not only to estimate the time taken for evacuation but also to check how smoothly rescue responders reach their targets at emergency sites. In this paper, we discuss qualitative standards of validating simulations result that are hard to be checked with experiments in the real world. Section 21.2 describes related works and background scenarios. Section 21.3 shows our evacuation simulations as an example of disaster-related social systems. Conditions of validating simulations qualitatively are discussed in Sect. 21.4. Section 21.5 provides a descriptive summary.

21.2 Related Works

The National Institute of Standards and Technology (NIST) published reports on the WTC egress of September 11, 2001. The report describes the evacuation of Towers 1 and 2, and offers an explanation for the variation in the time taken for evacuation of the two towers despite their layout, size, and number of occupants being almost the same. The NIST report includes a description taken from a simulation of occupant evacuations during the WTC disaster and highlighted several concerns that future simulation systems should address.

The social psychological factors involved in human behavior are related to the validation of crowd simulation models. A substantial amount of data on pedestrian dynamics was presented at the Pedestrian and Evacuation Dynamics conference [2]. Zhang et al. conducted experiments on human bidirectional flows at the laboratory level [3]. Helbing's empirical social forces model simulated interactions among people and resultant behaviors such as the arch-like blocking of an exit and faster-is-slower effect [4]. The results of crowd simulations using these models have been validated with real world data.

Pelechnao et al. proposed a HiDAC model that enables high-density crowd simulation in dynamically changing environments [5]. Their model is based on Helbing's work and is composed of geometrical information and psychological rules with a force model resembling behaviors of real people. Durupinar et al. extend the HiDAC model by specifying agents personalities in order to mimic human behaviors from normal and disaster environments [6]. Guy et al. use Eysenck's three-personality model for crowd simulation and show how personality affects the social behavior of crowds, including faster-in-slow effect [7]. Okaya et al. proposed an information-transfer and sharing model during evacuation and demonstrated how guidance methods can improve evacuation time [8]. While the simulations including human person factors make the simulations more realistic ones, the human related factors and their behaviors pose a problem of validating the simulation results.

Validating the results of social simulations is critical to ensure that they are applicable to real-world cases. Especially, life-threatening applications require data from real-world situations to assure its usefulness [9]. The existing experimental measures often rely on ad-hoc applications, e.g., local crowd densities are measured to verify patterns of human movements in crowd. The conditions may differ from ones obtained from prior cases and experiments. The differences of conditions are such as at what times the simulation starts; daytime or night, and the intentions of people; they aim to the same place or have their own destinations. The conditions are also assumed to be well set as well as the model of social behaviors.

21.3 Agent Based Evacuation Behavior Simulation

Computer simulations allow to examine out-the-box scenarios that are hard to be experimented in the real world. Human evacuation behaviors are examples of such situations and agent based simulations can express the microscopic behaviors of humans. To show the features of human change simulation results, two evacuation simulations are demonstrated, one is the behavior of occupants starting to evacuate followed by an announcement broadcast through public address system (PA) and the other is the action of rescue responders during emergencies. TENDENKO that we have been developed is used to simulate two cases [8].

21.3.1 Evacuation Behaviors According to PA

21.3.1.1 Simulation Background

During emergencies, the authorities activate alarms or announce evacuation instructions to begin the evacuation. According to the GEJE report, only 40 % of evacuees heard the emergency alert warning given over the PA system [10].[1] Of those who heard the warning, 80 % recognized the urgent need for evacuation and the remaining 20 % did not understand the announcement owing to the noise and confusion. In case of the September 11 incident, messages were announced on a limited number of floors of the WTC buildings that were hit by planes. The messages had been pre-prepared for the types of accidents that prompted phased evacuation

[1] The report was based on investigations conducted with 870 people who were from Iwata, Miyagi and Fukishima prefectures. The percentages were different for the three prefectures and the average values are listed in this paper.

in advance, and unfortunately, these messages did not provide proper guidance to occupants of the buildings based on dynamically changing situations. Furthermore, some people on the impacted floors did not hear the announcement.

We believe three components, i.e.,—rate of transmission, content, and method—are explicitly embodied in communication during emergencies. The components used in the simulations are based on existing documents that have been reported in situations of past emergencies, and the behaviors of agents in the MAS are designed to perform similar actions as described in the documents.

21.3.1.2 Simulation Results

Figure 21.1 shows a snapshot of simulating evacuation of 1000 people (with 200 people on every floor) evacuating from a five-story building. This building is a library at our university and has stairs between floors and two exits. One is the main entrance, 3.7 m wide, on the second floor, and the other is an emergency exit on the first floor. Figure 21.2 shows the simulation results of four scenarios (Table 21.1). The simulations were run three times for each scenario. The averages of evacuation rates (percentages of evacuated agents in all agents) were plotted in simulation-time sequence. The first scenario is that the broadcast is heard by all and everyone evacuates instantly after the announcement. The other three scenarios differ from scenario 1 in terms of: the type of evacuees, contents of announcement, and timing of announcement, respectively. Simulations for scenarios 3 and 4 indicate a better evacuation rate than scenarios 1 and 2. Checking the locations of the agents shows that congestions occurred at stairwells in scenarios 1 and 2, and this leads to the low rates of evacuation and the big variations among the simulations.

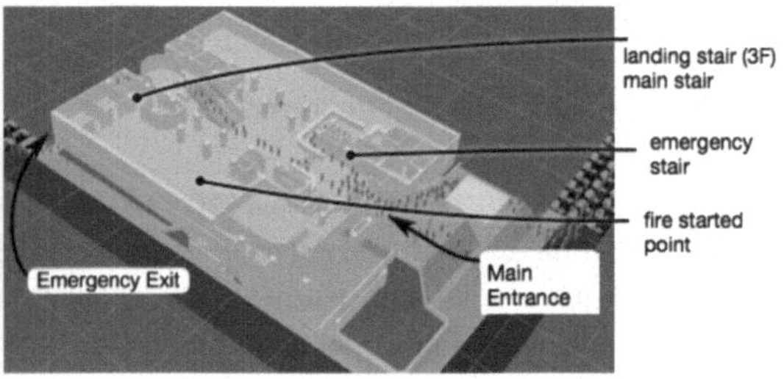

Fig. 21.1 Library facade (*left*) and image of agent behavior on the second floor (*right*)

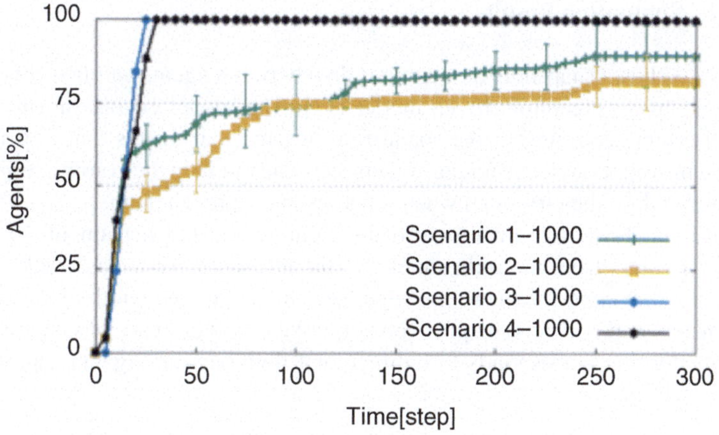

Fig. 21.2 Time change of evacuation rates for scenarios

Table 21.1 Scenario setting for evacuation

Scenario	Evacuation type[a]	Announcement Content[b]	Step
1	A	A	6
2	B	A	6
3	A	B	6
4	A	A	6 (other floors), 20 (4th floor)

[a]Type A and B correspond to instant evacuation and evacuation after jobs, respectively [10]
[b]Content A and B are "evacuate from main entrance" and "for 1, 3, 5th floor evacuate at main entrance, others from emergency exit"

21.3.2 Rescue Responders' Action During Emergencies

21.3.2.1 Simulation Background

The arrival of first responders affects the end time of RSET. We need to check how smoothly rescue responders reach their targets during emergency situations. It is natural for people to swerve when they come close to colliding with one another. Survivors of the WTC attacks considered the counter flow of first responders as both evacuation support and obstacles to their exit.

Zhang's experiment was the counter flow between two groups; the numbers of the groups and the types of agents in the groups are the same, and they compared the agents' movement of their simulation with the experimental data [3]. People's behaviors differ according to who is approached by whom. The mass behavior of pedestrians is thought to affect the arrival time of first responders. To our best knowledge, counter flow between agents and responders has not been tested in experiments.

21.3.2.2 Simulation Results

Figure 21.3 shows snapshots of a counter flow between agents and first responders with/ without a perception-driven model [11]. The model enables agent change their behaviors according to the social role of particular agents that is perceived by visual information; for example, agents step aside to help the coming responders go through, while they try to go when other agents approach them.

Agents on the left room move to the right room, and a team of 10 rescue responders enters from the right room into the left room. They pass each other in the corridor that connects the rooms together. The length and width of the corridor are 10 meters and 3 meters, respectively. Figure 21.3 (a) is a case of no perception-driven model that corresponds to bidirectional flow between agents, and (b) is a

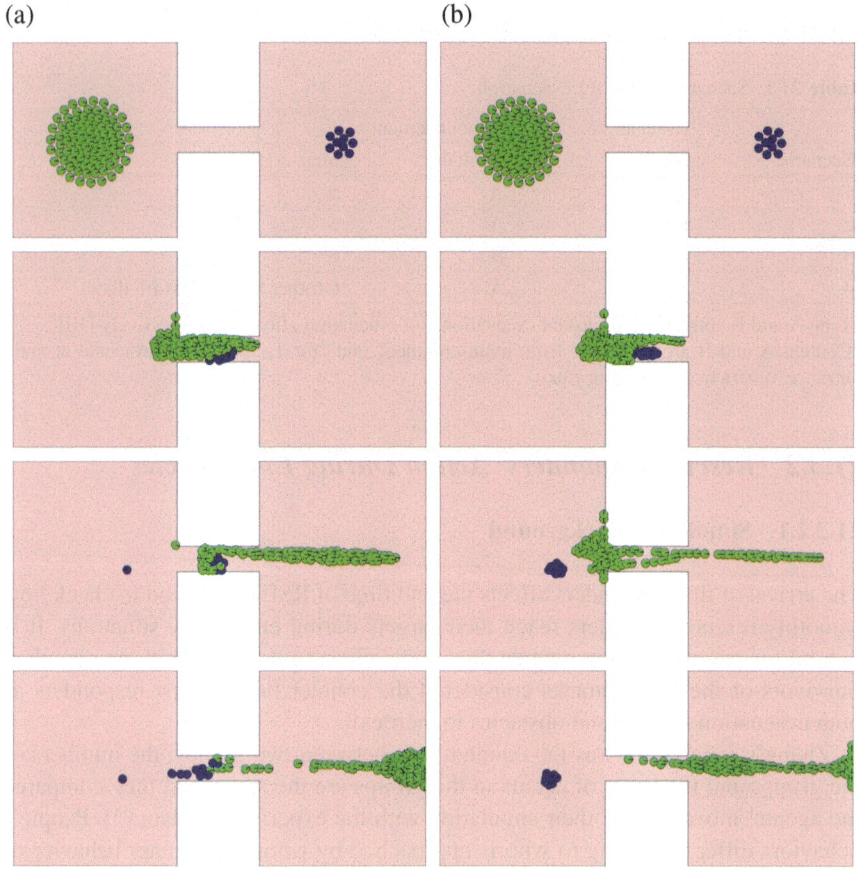

Fig. 21.3 Counterflow movements between 10 rescue responders (*blue*) and 100 occupants. (Figures at the *left* show the initial position, which moves to the right as time proceeds.) (**a**) Without perception, (**b**) with perception

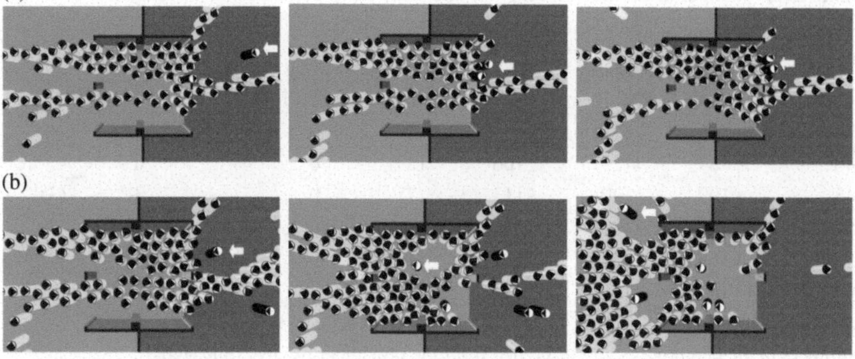

Fig. 21.4 Counter flow between evacuees from the library and rescue responders entering from the *right*; snapshots are 40, 45 and 50 steps from *left* to *right*, one responder is pointed by *filled left arrow*. (**a**) Without perception (responders (*black body*) remain outside), (**b**) with perception (responders move inside against occupants (*light color body*))

case of perception-driven model. By introducing a perception-driven model, the responders move to the other room faster with cooperative behavior from the agents.

Figure 21.4 shows snapshots of the other simulations in the library mentioned in Sect. 21.3.2. The scenario is that 1000 people evacuate from the library and five responders enter from the main entrance to help the injured inside. Figure 21.4a, b are the counter flow of agents and fire responders at the main entrance without and with the perception factor, respectively; they are snapshots at time step 40, 45 and 50 sequentially from left to right. The agents (sector mark on the top of light color body) evacuate from left to right and the responders (triangle sector on the top of black body) enter the library from the right. The marks on the agents' heads indicate the direction of their movements. A responder is pointed with a white arrow. The responder remained at entrance at 50 time step in (a), while the responder entered and went to a directed site in the library in (b).

The left column of Table 21.2 shows the number of agents who evacuated the library and the right column is number of responders who entered. From time steps 45 to 55 time steps, fewer agents evacuate in a simulation with perception-driven model than in the other simulation without the model. However, after all responders entered the building, there is no one who blocks the evacuation at the entrance obstructs the flow of evacuation; more agents successfully evacuate with-perception. This is an interesting finding and presents problems on how to estimate the simulation results and to make use of the finding in making prevention plans for emergencies.

Table 21.2 Number of evacuated occupants and entering responders at main entrance

Time step	Number of Evacuated agents: Without PM		With PM		Entering responders: Without PM	With PM	
35	66	(21)	83	(27)	0	0	
40	89	(23)	106	(23)	0	0	
45	109	(20)	113	(7)	0	3	(3)
50	129	(20)	115	(2)	0	5	(2)
55	153	(24)	129	(14)	0	5	(0)
60	169	(16)	148	(19)	0	5	(0)

PM perception-driven model
Numbers in the parentheses are difference from the previous row (five time steps)

21.4 Validation of ABS Results for Scenarios Containing Human Actions

21.4.1 Validation Problems in Conventional Social Systems

The two simulations provide useful results that contain practical information to building managers and rescue officers. However, following concern makes the managers pose to adopt the simulation results in their policymaking.

1. Although the behaviors driven by perception-driven model seem to be similar to ones reported in the GEJE and WTC accidents and experimenting the counter-flow behavior at unexpected situations is hard to perform in the real world, the model are inadequately-supported from the real-world data so the simulations cannot be applied to other cases.
2. The evacuation times of Towers 1 and 2 on September 11, 2001, varied despite the fact that their layout, size, and number of occupants were almost the same. This fact indicates that there are other factors that should be taken in consideration to explain the difference in evacuation times of Towers 1 and 2. PA is known to change the occupants' actions, and evacuation announcement may be one factor.

The points to be used in the policy making are that the system should be well designed to present the behavior of targets, and the results are guaranteed to be reasonable ones for the scenarios even if they are applied to outside-the-box scenarios. The following points are hypothesized in modeling the social systems.

H1:(whole-part relation) A social system, Σ, may be composed of subsystems, S^i. Every one has some knowledge on phenomena that social systems simulate. The knowledge is implemented in S^i. They are modeled with a finite set of parameters, $\Pi = \{p_1, p_2, \ldots p_n\}$. The parameters, p_i, represent features of agents, environments, interactions among them or others.

H2:(causality of subsystem) The procedure followed in the system is described by formulas or rules. In the case of a discrete-time dynamic system, it can be described as: $\Sigma_{t+1} = F(\Sigma_t, \Pi_t)$.

H3:(validness of subsystem) When subsystems, S^i, are well defined, then the system, Σ, may be well designed and expanding or refining parameters and functions cover more phenomena or make simulations consistent with experimental data or empirical rules.

Social systems involve various factors. These factors are also required to be well defined. Table 21.3 shows subsystems and parameters of evacuation simulations. The simulators in the NIST reports are characterized in physical properties of agents and TENDENKO focuses the representation of mental/social states and information distribution through communication domains based on existing documents [12]. With implementing human actions as agent behaviors, the evacuation simulations and its subsystems are mainly categorized into agent and environment. The parameters of agents are physical factors, mental status, sensing ability and actions of agents. Compared to the simulation systems listed in NIST report,

Table 21.3 Parameters specifying evacuation simulations

Subsystem	Parameters		Simulators in [12]	TENDENKO
Agent	Physical	Age	√	
		Sex	√	
		Impaired/unimpaired	√	
	Metal/social	Mind state		√
		Human relationship		√
		Role		√
	Perception	See	√	√
		Hear		√
	Action	Walk/run	√	√
		Communicate		√
		Help others		
Environment	Map/buildings	2D/3D	√	√
		Elevator		
	Pedestrian dynamics	Occupant behavior	√	√
	Communication channel	Announcement		√
		Information sharing		√

21.4.2 *Qualitative Standard to Simulation Results for Social Scenarios*

Nobody can validate the results of evacuation simulations for emergency situations that have not occurred and affirm that the planning based on the simulation results work well at a possible emergency situation. People evaluate the results of simulation from their personal perspectives. The perspectives may be outside of scopes that the social systems aim to simulate, even though they understand the model of simulation are based on the past cases and do not cover all characteristics of disasters.

In scientific and engineering fields, a principle: guess of model, compute consequence, and compare experiment, has been used to increase the fidelity of simulations [13]. It is difficult that evacuation simulation do follow the principle, because we cannot repeat evacuation drills that many people take part in at the same conditions of the simulations. We propose the following qualitative standards that are necessary to apply such simulations without real-world data:

S1:(consistency with data) Simulation results of Σ or its changes after changing parameters or modifying subsystems are compatible with the past anecdotal reports,

S2:(generation of new findings) The results involve something that are not recognized important before simulations, and that points are reasonable from empirical rule

S3:(accountability of results) The cause of the changes can be explainable from the simulation data systematically.

Table 21.4 shows relevance to the hypotheses in the design of simulation mentioned in Sect. 21.4.1.

The two TENDENKO's simulations in the previous section demonstrate that the simulations with the same size of real environments help to reflect behaviors that would occur in a real situation. The simulations suggest possible solutions that can be used as an alternative of evacuation drills. The possibilities of two simulations are checked using the standards:

- In the case of evacuation behaviors according to PA, Table 21.1 shows that scenarios 2, 3 and 4 differ from scenario 1 in one factor. Scenarios 3 and 4 correspond to phased evacuations that ease congestions through certain evacuation

Table 21.4 Relationship between hypothesis in modeling and standard of estimation

	H1	H2	H3
	Whole-part	Causality	Validness
S1 (consistency with data)	√	√	
S2 (generation of new findings)		√	√
S3 (accountability of results)	√		√

behaviors. The aforementioned three points are satisfied and additionally the advice corresponds to one of advices proposed in the GEJE report.

- In the case of rescue responders' action, S1 standard is satisfied. The perception-driven model makes the simulation real one; however, they do not give any findings to improve the rescue operations of responders. The results do not meet the other two standards S2 and S3.

21.5 Discussions and Summary

We believe that MAS-based evacuation systems can replace evacuation drills that guide people in real environments. During real disasters, people respond to directives and helpful information from authorities, fellow citizens, family, and friends. They behave differently in response to such information and their intentions. Evacuation simulations using various scenarios provide us, especially safety officer, with data for analyzing the qualitative differences of these scenarios.

In this paper, we propose standards to check whether the results of social simulations are effective or not by using two examples of simulations under various conditions. Both results seem to improve disaster prevention plans, however one is ranked as effective one and the other is not. We believe such qualitative standard on effectiveness of MAS is important to be widely used.

Acknowledgement This work was supported by JSPS KAKEN Grant Number 24500186.

References

1. Averill JD, Mileti DS, Peacock RD, Kuligowski ED, Groner NE (2005) Occupant behavior, egress, and emergency communications (NIST NCSTAR 1-7). Technical report. National Institute of Standards and Technology, Gaitherburg, MD
2. Peacock RD, Kuligowski ED, Averill JD (2011) Pedestrian and evacuaion dynamics. Springer, Berlin
3. Zhang J, Klingsch W, Schadschneider A, Seyfried A (2012) Ordering in bidirectional pedestrian flows and its influence on the fundamental diagram. J Stat Mech Theory Exp P02002. doi:10.1088/1742-5468/2012/00/P00000. arXiv:1107.5246 [physics.soc-ph]
4. Helbing D, Farkas I, Vicsek T (2000) Nature 407:487
5. Pelechano N, Allbeck JM, Badler NI (2007) Proceedings of the 2007 ACM SIGGRAPH/eurographics symposium on computer animation, SCA '07. Eurographics Association, Aire-la-Ville, pp 99–108. http://www.portal.acm.org/citation.cfm?id=1272690.1272705
6. Durupinar F, Pelechano N, Allbeck J, Gudukbay U, Badler NI (2011) IEEE Comput Graph Appl 31(3):22. doi:10.1109/MCG.2009.105. http://www.dx.doi.org/10.1109/MCG.2009.105

7. Guy SJ, Kim S, Lin MC, Manocha D (2011) Eurographics/ACM SIGGRAPH symposium on computer animation, pp 43–52
8. Okaya M, Niwa T, Takahashi T (2014) The Autonomous agents and multiagent systems (AAMAS), pp 169–1670
9. Banerjee B, Kraemer L (2010) AAMAS2010, pp 1551–1552
10. C.O.G. of Japan Prevention disaster conference, the great West Japan earthquake and tsunami. Report on evacuation behavior of people (in Japanese). http://www.bousai.go.jp/kaigirep/chousakai/tohokukyokun/7/index.html. Accessed 21 Jan 2014
11. Okaya M, Takahashi T (2012) First International workshop on pattern recognition and crowd analysis
12. Kuligowski ED (2005) NIST SP 1032; workshop on building occupant movement during fire emergencies
13. Feynman RP (1967) The character of physical law. The MIT Press, Cambridge

Chapter 22
Collective Dynamics of Pedestrians with No Fixed Destination

Takayuki Hiraoka, Takashi Shimada, and Nobuyasu Ito

Abstract In order to understand pedestrian dynamics, we construct a model of self-propelled disk particles interacting repulsively with no fixed destination. From molecular dynamics simulations, we find that the model exhibits collective motion and transition from a disordered to a polar-ordered, heterogenous state. Binary scattering study suggests that ordering originates from parallel alignment of particles' velocity after collision. The dependency of alignment tendency on the model parameter agrees well with the behavior of multiparticle systems. We verify that the model reproduces the actual pedestrian phenomena in a straight pathway. Although there is still a gap with empirical findings, especially in high densities, the result implies that pedestrian crowds can spontaneously build up a collective motion even in the situation where they have lost their destinations.

22.1 Introduction

From schooling of fish, flocking of birds, swarming of insects to migration of cells or bacteria, we often observe collective behaviors of biological organisms. One may presume that these fascinating pattern formation in nature is produced by sophisticated information processing mechanism specific to the species, by elaborate interaction between individuals, or by presence of a special individual who

T. Hiraoka (✉) • T. Shimada
Department of Applied Physics, Graduate School of Engineering, The University of Tokyo, 7-3-1 Hongo, Bunkyo-ku, Tokyo 113-8656, Japan

JST CREST, 4-1-8 Honcho, Kawaguchi, Saitama 332-0012, Japan
e-mail: hiraoka@serow.t.u-tokyo.ac.jp; shimada@ap.t.u-tokyo.ac.jp

N. Ito
Department of Applied Physics, Graduate School of Engineering, The University of Tokyo, 7-3-1 Hongo, Bunkyo-ku, Tokyo 113-8656, Japan

JST CREST, 4-1-8 Honcho, Kawaguchi, Saitama 332-0012, Japan

RIKEN AICS, 7-1-26, Minatojima-minami-machi, Chuo-ku, Kobe, Hyogo 650-0047, Japan
e-mail: ito@ap.t.u-tokyo.ac.jp

H. Takayasu et al. (eds.), *Proceedings of the International Conference on Social Modeling and Simulation, plus Econophysics Colloquium 2014*, Springer Proceedings in Complexity, DOI 10.1007/978-3-319-20591-5_22

takes leadership. However, recent studies on self-propelled particle systems revealed that collective motion could arise even in the absence of long-range communication, complex behavioral rules, and global leadership [1–3]. Furthermore, it has been shown that the non-equilibrium character enables the systems to develop long-range order and anomalously large density fluctuations, which are unusual in equilibrium systems [4–8]. Recent studies have found these features are shared not only by biological organisms but also by non-biological systems such as vibrated granular particles, in which explicit alignment with neighbors are absent [9].

One of the important research subject studied under the concept of the self-propelled particle is the pedestrian dynamics. We notice that there are characteristic patterns of crowds in streets, intersections, train stations, airport terminals, concert halls, sport stadiums, political demonstrations, etc. Such patterns in the urban environment spontaneously arise from individuals moving with their own destinations or intentions. Many microscopic models have been proposed to describe pedestrian movement. They can be categorized into two main types: cellular automata models [10, 11], in which time and space is discretized, have an advantage in computational cost, while force-based models, which is inspired by Newtonian mechanics, can simulate realistic trajectories [12, 13]. Excluded volume interaction and repulsion due to social psychological effect play an important role in both types of model.

In this proceedings we aim to establish a kinetic understanding towards the collective dynamics of self-propelled particle systems with repulsive interaction. In order to find out whether the crowd develops collective behavioral order, we construct a simple self-propelled particle model which assumes no constant destination nor explicit alignment interaction. We report the details of the model and the results obtained from the numerical simulations.

22.2 Model and Simulations

Among many pedestrian models that has been previously proposed, the social force model [12] is the one that has been widely recognized. It assumes that each pedestrian follows Newtonian equation of motion, which consist of the sum of self-driving force towards the destination and repulsive forces, namely the exponential "social force" from other pedestrians. In addition to the social force, another literature [14] introduces normal body force and tangential friction that describes physical contacts between people.

In order to construct a simple model and to clarify the physical meaning of crowd dynamics, we start with two assumptions:

- Pedestrians do not have constant destination.
- Pedestrians interact with linear elastic repulsive forces.

It may seem improbable that pedestrians have no destination on their way. However, they can in fact lose or abandon their destinations in extremely dense crowd. The latter point also reflects highly dense conditions where excluded volume effect plays a pivotal role in crowd dynamics.

Let us consider N polar disk particles of equal radius a moving on a two-dimensional continuous surface. The polarity of each disk is defined by an unit vector $\hat{\mathbf{e}}(\psi_i) = \cos\psi_i\hat{\mathbf{x}} + \sin\psi_i\hat{\mathbf{y}}$. The equation of motion is given by

$$\frac{d\mathbf{v}_i}{dt} = \alpha\hat{\mathbf{e}}(\psi_i) - \beta\mathbf{v}_i + \sum_j \mathbf{f}_{ij}, \tag{22.1}$$

where $\mathbf{r}_i$ denotes the position of particle i, and the direction of their velocity is θ_i, i.e., $\mathbf{v}_i = v_i\,(\cos\theta_i\hat{\mathbf{x}} + \sin\theta_i\hat{\mathbf{y}})$.

Particle i drives itself with a self-propulsion force of constant magnitude α along its polarity axis while the velocity is damped by a drag force of coefficient β. The dynamics of the polarity is overdamped by a torque proportional to the angular deviation from the direction of the velocity, as

$$\frac{d\psi_i}{dt} = \gamma\,(\theta_i - \psi_i)\,, \tag{22.2}$$

where γ is the damping coefficient.

We assume that the interaction between particles i and j is given by a steric repulsive force with a linear elasticity, i.e., $\mathbf{f}_{ij} = -k\left(2a - r_{ij}\right)(\mathbf{r}_i - \mathbf{r}_j)/r_{ij}$ if $r_{ij} = |\mathbf{r}_i - \mathbf{r}_j| < 2a$ and $\mathbf{f}_{ij} = 0$ otherwise. Note that the momentum is conserved by the interaction itself. Without loss of generality, we set length unit $2a = 1$ and time unit $\beta^{-1} = 1$ and obtain rescaled equations of motion. The model is then governed by three time scales: α^{-1} is the time for a free particle to travel its own diameter, γ^{-1} is the angular relaxation time of polarity, and $k^{-1/2}$ is the elastic time scale of collision.

The magnitude of self-propulsion force and elastic modulus are fixed as $\alpha = 1$ and $k = 100$, respectively. Under such choice of parameter values, particles penetrate their neighbors by at most $\sim 1\,\%$ of their diameter. Therefore the elasticity is large enough to avoid unrealistic situation where particles in contact pass through each other. Although the empirical value of elasticity of human body is not known, Helbing et al. [14] estimates it as $k = 1.2 \times 10^5\,\mathrm{kg\,s^{-2}}$. Given that the mass of a pedestrian is 80 kg and $\beta^{-1} \simeq 0.5\,\mathrm{s}$, the scaled elasticity will be $k \sim 10^2$, which is consistent with the above value.

22.3 Spontaneous Ordering with Periodic Boundary Conditions

We performed molecular dynamics simulations with $N = 10{,}000$ particles on a square plane of size L with periodic boundaries. Initial configurations are assigned randomly in terms of particles' position and their direction of polarity. The overlaps between particles have been reduced by evolving the system only by the interaction forces for a sufficient time prior to each simulation. The two control parameters of the simulations are the angular damping coefficient γ and the packing fraction

$$\rho = \frac{\pi N a^2}{4L^2}. \tag{22.3}$$

In the region where damping is weak, the system exhibits polar ordering and clustering as shown in Fig. 22.1. To characterize the collective motion, we employ as the order parameter the global polarization

$$\phi = \frac{1}{N}\left|\sum_{i=1}^{N} \hat{\mathbf{e}}(\psi_i)\right|, \tag{22.4}$$

whose value is finite in the phase with polar order, and goes to zero in a globally disordered state.

For a fixed packing fraction, the growth of the order parameter is slow when damping coefficient is small ($\gamma \ll 1$). It is because each particle tends to keep its polarity to the same direction as given in the initial random state. As the damping parameter increases, the speed of polarity alignment becomes faster. However, increasing the damping parameter further slows down the development of the order. Above a certain value of γ, no collective motion takes place so that the system remains disordered and isotropic. Close to this phase boundary, the time until the system builds up a polar order exceeds the computationally feasible time, which makes us difficult to identify the exact transition point. Therefore, we carried out multiple (typically 16) runs with different initial configurations for each set of control parameters, ρ and γ, and categorized the corresponding point in the parameter space as polar-ordered phase if ϕ grows larger than 0.5 for at least one realization. Obtained phase diagram is shown in Fig. 22.2.

22.4 Binary Scattering Study

In this section, we give a simple explanation to understand the mechanism that underlies the characteristic ordering behavior shown in previous section. Let us limit our discussion only to the binary particle collision process [15]. Here we assume the system is dilute enough ($\rho \to 0$) so that only the uncorrelated, binary collisions take place, and both the velocity and the polarity are fully relaxed before each collision.

If the damping is weak, the polarities of two particles remain unchanged, so the directions of motion are temporally changed by the collision but eventually restored to the original direction. Here the relative angle between the velocities does not change before and after the collision. By contrast, if the damping is strong, the polarities rotates themselves quickly to align to the directions of motion, so the particles moves as if they have exchanged their momentum. Here again the absolute

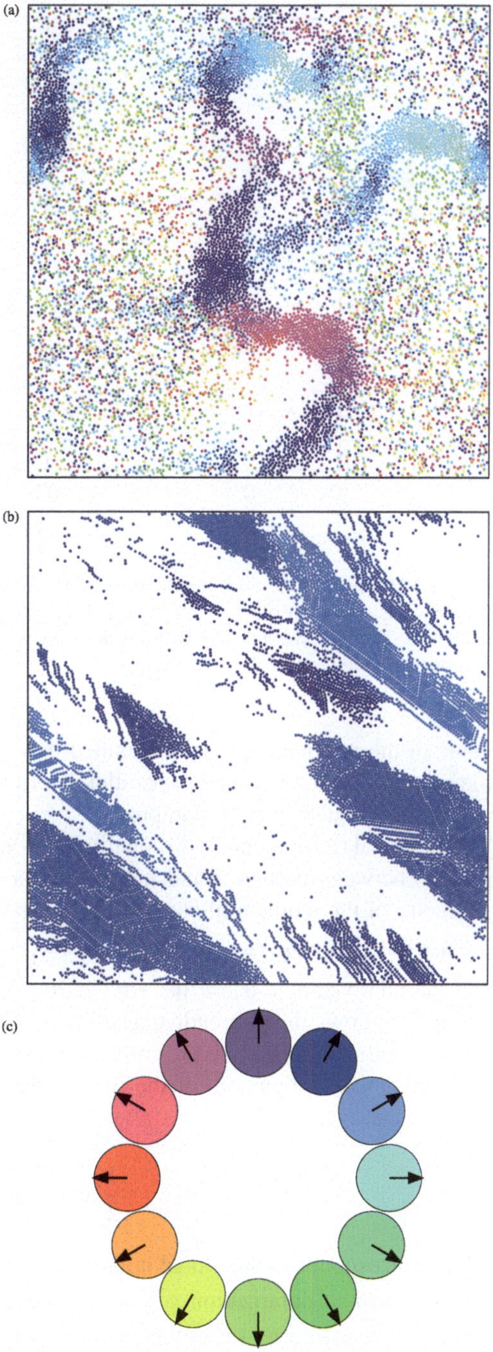

Fig. 22.1 Typical snapshots of a system in periodic boundary boxes (**a**) at onset of the collective motion, and (**b**) at fully ordered state. The packing fraction $\rho = 0.3$ and the damping parameter $\gamma = 20$. The *color* denote the polarity of each particle, as shown in (**c**)

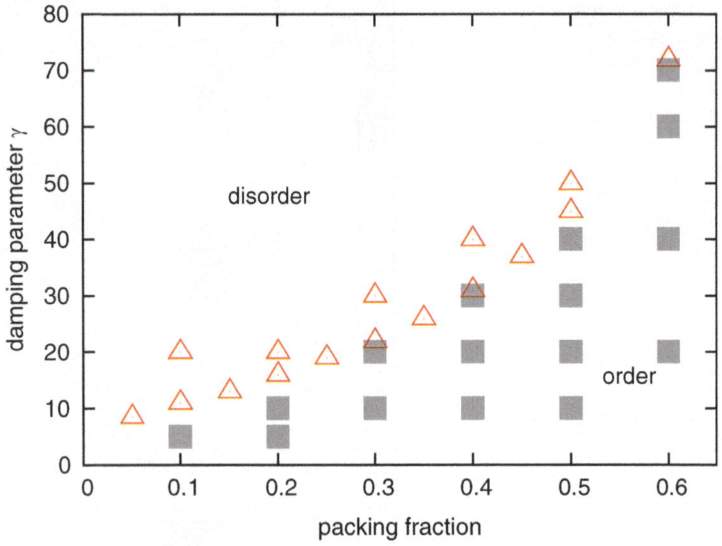

Fig. 22.2 The ρ-γ phase diagram obtained from molecular dynamics simulations. *Gray squares* denote states that polar order is observed and *red triangles* indicate the phase where the system remains disordered. We suppose that disordered phase stretches to the upper left domain, where we does not have numerical results yet

value of the relative angle is maintained. For an intermediate damping parameter, the motion of two bodies align parallel due to the competing effect of the collision and the subsequent angular damping.

Numerical results support this qualitative conjecture. Consider a binary scattering process between particle i and j. Since we assume the rotational invariance, the geometry of the moment of contact is fully specified by two scalar parameters: the impact parameter $b_{ij} = \sqrt{r_{ij}^2 - \mathbf{r}_{ij} \cdot (\mathbf{v}_i - \mathbf{v}_j)/v_{ij}} \in [0, 1]$, where $v_{ij} = |\mathbf{v}_i - \mathbf{v}_j|$ and the relative angle $\theta_{ij} = |\theta_i - \theta_j| \in [0, \pi]$, as shown in Fig. 22.3. The impact parameter shows the perpendicular offset of the two bodies' center of mass from head on collision. If $b_{ij} = 0$ the collision is head on whereas it is a miss if $b_{ij} > 1$.

Instantaneous alignment of the two particles are characterized by two-particle polarization

$$\phi^{(2)} = \frac{1}{2} \left| \hat{\mathbf{e}}(\psi_i) + \hat{\mathbf{e}}(\psi_j) \right|, \tag{22.5}$$

which corresponds to the global polarization [Eq. (22.4)] with $N = 2$. We measure the two-particle polarization $\phi^{(2)}_{\text{out}}$ at an adequate time for the polarities and the velocities to relax after the collision, and compare it to the polarization $\phi^{(2)}_{\text{in}}$ before the collision. The increment,

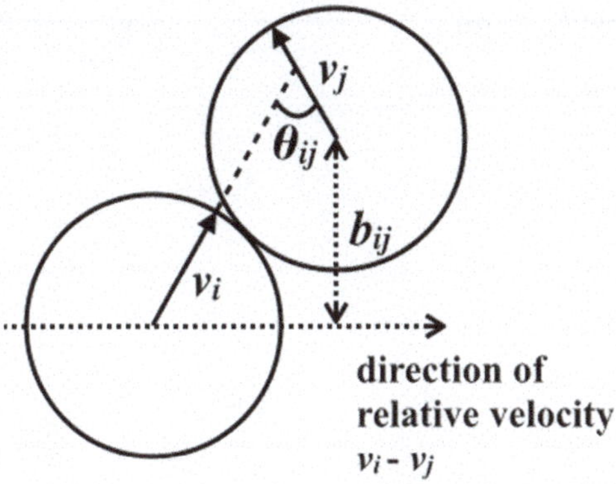

Fig. 22.3 Schematic view of collision geometry. The geometry is defined by the relative angle θ_{ij} and impact parameter b_{ij}. We assume that the two particles are fully relaxed in terms of velocity and polarity before each collision, therefore $\mathbf{v}_i/|\mathbf{v}_i| = \hat{\mathbf{e}}(\psi_i)$

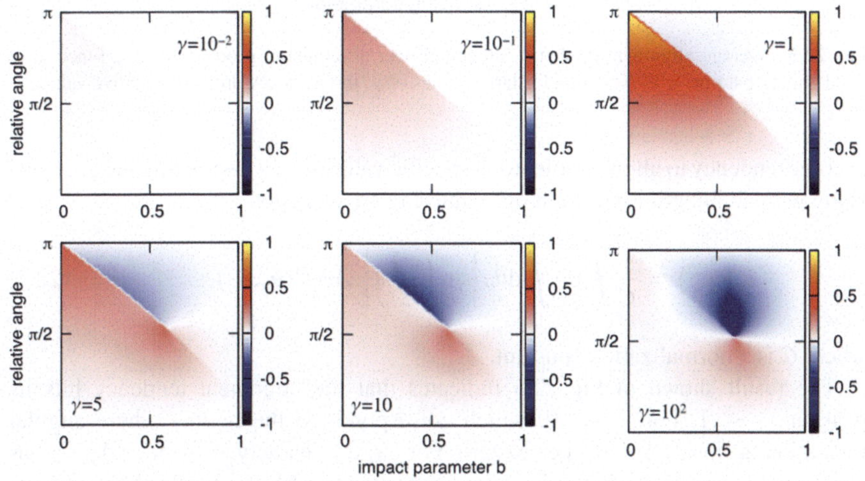

Fig. 22.4 The two-particle polarization increment $\Delta\phi^{(2)}$ as a function of relative angle θ_{ij} and impact parameter b_{ij}. γ is varied as denoted in the figure. For the collision geometry in the *red* region, the collision makes particles align to each other ($\Delta\phi^{(2)}(b, \theta) > 0$), while in the *blue* region it result in antiparallel alignment ($\Delta\phi^{(2)}(b, \theta) < 0$)

$$\Delta\phi^{(2)} = \phi^{(2)}_{\text{out}} - \phi^{(2)}_{\text{in}}, \tag{22.6}$$

indicates the magnitude of parallel alignment caused by the binary scattering process. Figure 22.4 depicts $\Delta\phi^{(2)}$ as a function of the collision geometry (b_{ij}, θ_{ij}).

Assuming that the multiparticle system is homogenous and isotropic, the probability that the two particles collide in the relative angle of θ_{ij} is proportional to v_{ij} and that impact parameters b will be equally distributed. Therefore we can obtain

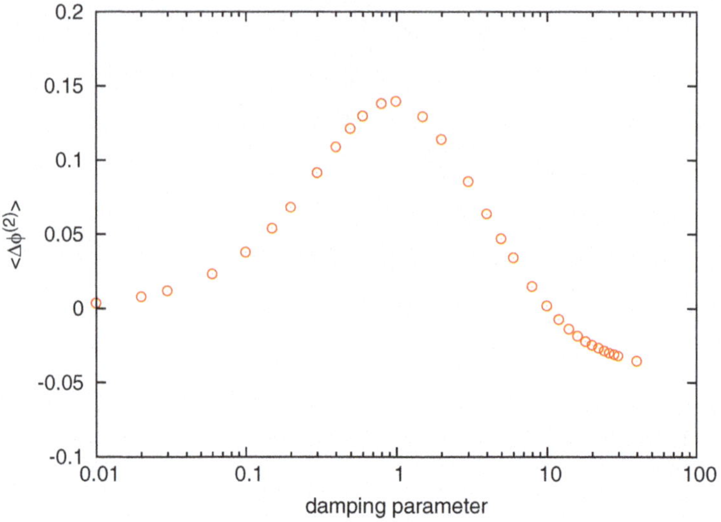

Fig. 22.5 Average alignment tendency integrated over all collision geometries, as a function of the damping parameter γ. Note that for large values of γ, the tendency drops to negative values

average tendency to align parallel for a specific value of γ by estimating the expected value with an integration weight of "scattering cross section"

$$\langle \Delta\phi^{(2)} \rangle = \frac{1}{C} \int \mathrm{d}b \int \mathrm{d}\theta_{ij} \left| \sin\left(\frac{\theta_{ij}}{2}\right) \right| \Delta\phi^{(2)}(b_{ij}, \theta_{ij}), \tag{22.7}$$

where C is a normalization constant.

The result shown in Fig. 22.5 indicates that the alignment tendency hits its peak at $\gamma \sim 1$. For $\gamma \to 0$, which corresponds to the regime where angular relaxation is slow, $\langle \Delta\phi^{(2)} \rangle$ goes to zero. For large γ, namely $\gamma \to \infty$, $\langle \Delta\phi^{(2)} \rangle$ has a negative value. This picture have two consistency with the result obtained from the multiparticle simulations; (1) The ordering in many-body system is the fastest in the parameter region that maximize the value of $\langle \Delta\phi^{(2)} \rangle$; (2) The transition in the dilute system occurs at $\gamma \sim 10$, where $\langle \Delta\phi^{(2)} \rangle$ changes its sign. These points imply that at least for the dilute limit $\rho \to 0$, the onset of collective motion arise from the repetition of binary collision process.

22.5 Flow in a Pipe

In order to validate the correspondence of the model with actual pedestrian phenomena, we performed multiparticle simulations in a "pipe", i.e., a rectangular area with periodic boundaries in the longitudinal direction and fixed repulsive

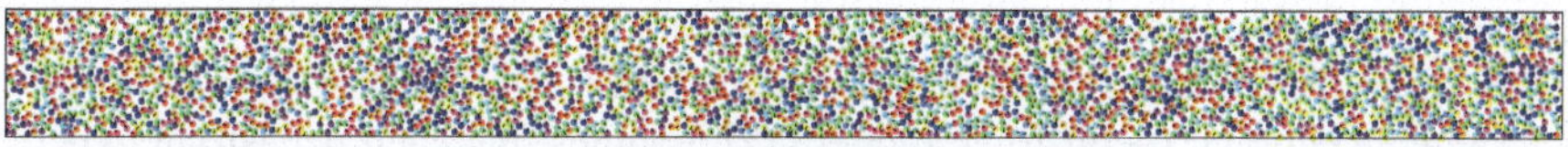

(a) snapshot at $t = 0$

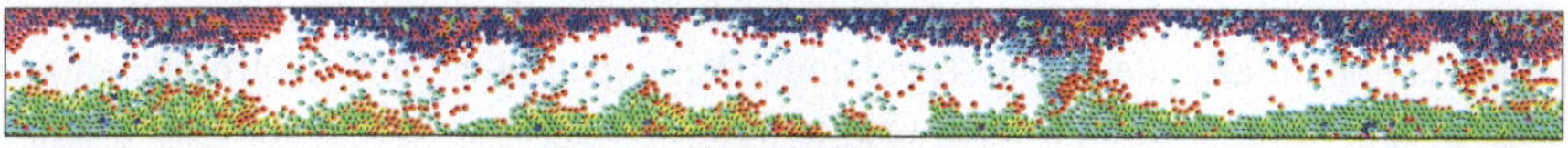

(b) $t = 100$

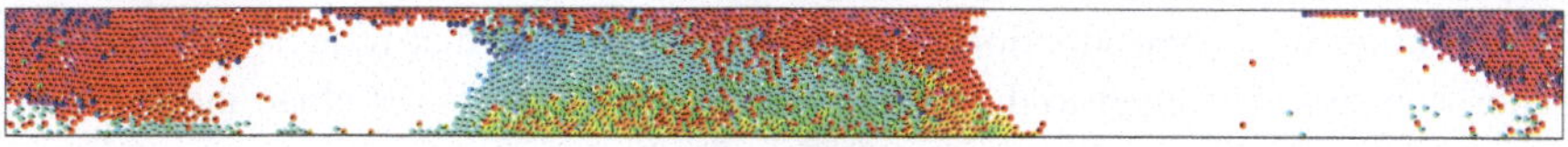

(c) $t = 300$

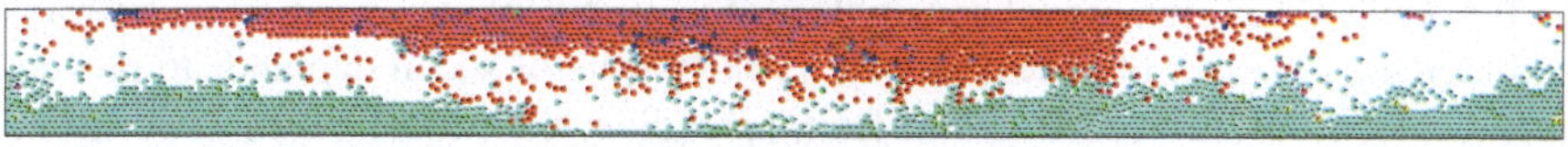

(d) $t = 500$

Fig. 22.6 Snapshots of the simulations carried out under "pipe" condition. The system consist of $N = 3200$ particles and is periodic for x-direction while bounded by elastic slipping walls for y-direction. The width of the pipe is 20.0, $\rho = 0.5$, $\gamma = 0.01$. At $t = 300$, three-lane structure is observed. (**a**) snapshot at $t = 0$, (**b**) $t = 100$, (**c**) $t = 300$, (**d**) $t = 500$

boundaries in the lateral direction. The interaction between the particles and the repulsive boundaries is assumed to be similar to the particle-particle interaction, that is, the interaction potential is elastic and frictionless. Starting from random initial condition, the system develops into two lanes of particle flow moving in opposite directions for certain sets of parameters (Fig. 22.6). Three-lane structure is also observed in a transient state. These results indicate that the model can reproduce the lane formation, which is one of the basic self-organization phenomena observed in pedestrian crowd [12]. The stability of lane structures and their dependency to the width of the pipe are subject of future investigation.

22.6 Summary and Discussion

In order to close the gap between theory on self-propelled particles and pedestrian dynamics study, we proposed a self-propelled particle model with repulsive interaction, and examined its collective behavior through many-body simulations. Binary scattering studies demonstrate the microscopic mechanism underlying the transition from a disordered to a polar-ordered phase.

Our model assumes that the collision process itself is elastic, i.e., no dissipation is taken into account. Nevertheless, the effective inelasticity introduced by the angular damping allows the formation of clusters, or "flocks," similar to that of granular gases. Of course, this argument is not exact because the many-body correlation cannot be ignored once local clusters are formed. Still, it provides a qualitative and, to some extent, quantitative explanation. We look forward to a further discussion on many-body effect and on cluster-cluster interaction.

The conventional social force model [12] assumes that each pedestrian experiences the social force from other pedestrians in his/her eyesight, which is described as an exponentially decaying repulsion. We expect that any isotropic, short-ranged repulsive potential, including exponential one, would not deviate the overall property of the system from the results we obtained with linear elastic repulsion. On the other hand, introduction of anisotropic potential that reflects the fact that pedestrians react stronger to the situation in front of them is not clear and yet to be discovered.

It is known that high crowd density leads to a turbulent movement of pedestrians and increases the risk of crowd disaster [16]. In spite of social demands to prevent such accidents from occurring and from spreading, their mechanism is yet to be uncovered, since experiments cannot be carried out due to ethical reasons, and observational data are hardly available. Previous pedestrian models assume that every agent is aware of its own destination and keeps driving itself until it reaches to that point. However, there are circumstances when pedestrians are not so conscious of where they are heading to. In fact we scanned the footage from the crowd disaster happened in Germany in 2010 and found that people sometimes behave as if they have lost or abandoned their initial destination in extremely dense crowd.

To this end, we verified that our model, which has no fixed destination, could display bidirectional lanes similar to what is observed in pedestrian flows in a straight pathway. Here, the damping parameter γ in the model can be regarded as the quickness of one's reaction to a contact with neighbor walkers. However, the phase diagram shows that in higher density, the order develops for a broader range of the parameter, which does not meet with the empirical facts. By improving the model we expect that our result can lead to a further understanding on the mechanism of crowd disasters.

References

1. Vicsek T, Czirók A, Ben-Jacob E, Cohen I, Shochet O (1995) Phys Rev Lett 75:1226
2. Czirók A, Stanley HE, Vicsek T (1997) J Phys A Math Gen 30:1375
3. Grégoire G, Chaté H, Tu Y (2003) Phys D Nonlinear Phenom 181:157
4. Toner J, Tu Y (1995) Phys Rev Lett 75:4326
5. Toner J, Tu Y (1998) Phys Rev E 58:4828
6. Simha RA, Ramaswamy S (2002) Phys Rev Lett 89:058101
7. Ramaswamy S, Simha RA, Toner J (2003) Europhys Lett 62:196
8. Chaté H, Ginelli F, Montagne R (2006) Phys Rev Lett 96:180602
9. Narayan V, Ramaswamy S, Menon N (2007) Science 317:105
10. Burstedde C, Klauck K, Schadschneider A, Zittartz J (2001) Physica A 295:507

11. Kirchner A, Nishinari K, Schadschneider A (2003) Phys Rev E 67:056122
12. Helbing D, Molnár P (1995) Phys Rev E 51:4282
13. Yu WJ, Chen R, Dong LY, Dai SQ (2005) Phys Rev E 72:026112
14. Helbing D, Farkas I, Vicsek T (2000) Nature 407:487
15. Hanke T, Weber CA, Frey E (2013) Phys Rev E 88:052309
16. Helbing D, Johansson A, Al-Abideen H (2007) Phys Rev E 75:046109

Chapter 23
Traffic Simulation of Kobe-City

Yuta Asano, Nobuyasu Ito, Hajime Inaoka, Tetsuo Imai, and Takeshi Uchitane

Abstract A traffic simulation of Kobe-city was carried out. In order to simulate an actual traffic flow, a road network was constructed utilizing a high-quality digital map data, and an origin-destination information of vehicles was estimated by a geographical population distribution data. The result obtained in this way was incompatible with the traffic census data due to the differences between the simulation and actual traffic, such as routing, OD information and so on. In order to improve the reproducibility of the traffic flow, the parameter search whose adjustable parameter was the speed limit of the road was conducted. This adjustment showed that reproducibility improves. Further improvement of the reproducibility needs to reconsideration of the routing algorithm.

23.1 Introduction

Cars and trains play major roles in today's urban mobility. According to a person trip survey of Keihanshin area [1], the percentages of car, two wheels, bus, railway, and walk in people's means of transportation are 33 %, 22 %, 3 %, 18 %, and 24 %, respectively. The percentage of the vehicle makes up a large fraction in the mobility. Therefore, it is not too much to say that people is living in automobile-dependent

Y. Asano (✉)
RIKEN Advanced Institute for Computational Science, Kobe, Japan
e-mail: yuta.asano@riken.jp

N. Ito
RIKEN Advanced Institute for Computational Science, Kobe, Japan

Department of Applied Physics, School of Engineering, The University of Tokyo, Tokyo, Japan

CREST, JST, Saitama, Japan
e-mail: ito@ap.t.u-tokyo.ac.jp

H. Inaoka • T. Imai • T. Uchitane
RIKEN Advanced Institute for Computational Science, Kobe, Japan

CREST, JST, Saitama, Japan
e-mail: hajime.inaoka@riken.jp; tetsuo.imai@riken.jp; takeshi.uchitane@riken.jp

H. Takayasu et al. (eds.), *Proceedings of the International Conference on Social Modeling and Simulation, plus Econophysics Colloquium 2014*, Springer Proceedings in Complexity, DOI 10.1007/978-3-319-20591-5_23

community. On the other hand, social problems such as the traffic jam are becoming a remarkable matter. These problems pose disadvantage to daily life, economics, global warming and other environmental problems. However, taking into account of risks and cost of a social experiment, it is difficult to conduct the experiment to resolve such problems.

A virtual social experiment which utilize a traffic simulation becomes a way to resolve these problems, because today's computer has a performance to execute a large scale simulation in a practical time. In this study, we focused on a traffic of Kobe-city as a specific example, and the traffic simulation of Kobe-city was carried out as a first step to resolve the social problems. Our purpose is to reproduce the actual traffic of Kobe-city as far as possible. When an actual traffic in Kobe-city is reproduced by the simulation, it will be possible to use optimization of a traffic system. In addition, this simulation will also be a basis for a simulation about a flow of people. Because, as mentioned in the beginning, traffic flow plays an important role in the people's means of transportation.

To simulate the actual traffic of Kobe-city, the road network and the traffic volume of Kobe-city have to be reproduced. Although almost all these kind of the simulations have used the OpenStreetMap (OSM) [2] for convenience, the OSM is not necessary sufficient in our purpose. Therefore, to attain the former, a high-quality digital map data was used. In order to realize the latter, we have to find an appropriate scenario and optimal parameters. In this study, only an attendance of the people was assumed as a traffic demand, and the speed limit of the road was chosen as an adjustable parameter. The results are compared with the traffic census data [3], and we discuss the variation of reproducibility by changing the adjustable parameter. We conclude this paper by commenting on the possibility of our parameter search.

23.2 Method

Although various traffic simulators exist, the SUMO (Simulation of Urban Mobility) [4] was adapted because it is fast enough to execute a large-scale traffic simulation whose road network size and the number of vehicles are large in a practical time. A method of the traffic simulation utilizing the SUMO consists of constructing a road network and estimating an origin-destination (OD) information of the vehicles. We are planning on carrying out the simulation of whole Kobe-city finally, a simulation in an area of mesh code of 523501 was conducted first to establish a methodology which reproduce the actual traffic flow. This is because, as far as we know, there are no general methods to do data assimilation with the actual traffic flow in the large-scale road network.

23.2.1 *Road Structure*

The purpose of this study is to reproduce the actual traffic of Kobe-city. For this reason, as mentioned in the previous section, utilizing the OSM was avoided because the OSM has more than a little difference from an actual road structure. For example, the number of traffic lanes and structure of the traffic intersection are different from actuals (see Fig. 23.1), and unrealistic isolated traffic lanes exist. In order to construct the actual road network of Kobe-city, a high-quality digital map data provided from Zenrin Co. Ltd. was used. The detailed data of the road network was converted to a format of an input data of the *NETCONVERT* [5] which is attachment program of the SUMO by our own script code of the Ruby. The resultant road network of mesh code of 523501 is depicted in Fig. 23.2.

23.2.2 *OD Information*

Because there was no available data about the OD information of Kobe-city, this was estimated by a geographical population distribution data estimated by NTT DoCoMo, Inc. from its mobile phone position data. In the geographical population distribution data, Kobe-city was divided into the meshes of 500 × 500 m, and the population of every 4 h for each mesh in a normal weekday was described.

We assumed that the distribution of the OD is proportional to the population distribution, and people is in a home in the mid-night (00:00 ∼ 04:00), and in a workplace in the daytime (12:00 ∼ 16:00). The *ActivityGen* [6] which is an

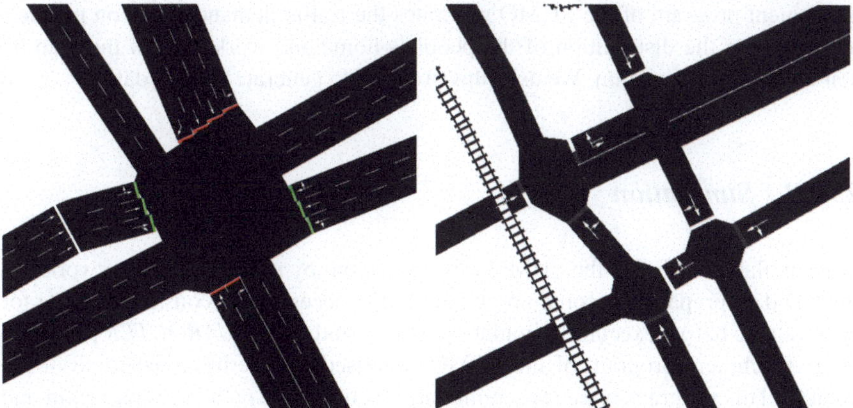

Fig. 23.1 Comparison of the road network of traffic intersection at Naka-koen-minami between Zenrin's and OSM's data. *Left panel* and *right panel* show road network constructed using Zenrin's data and OSM's data, respectively

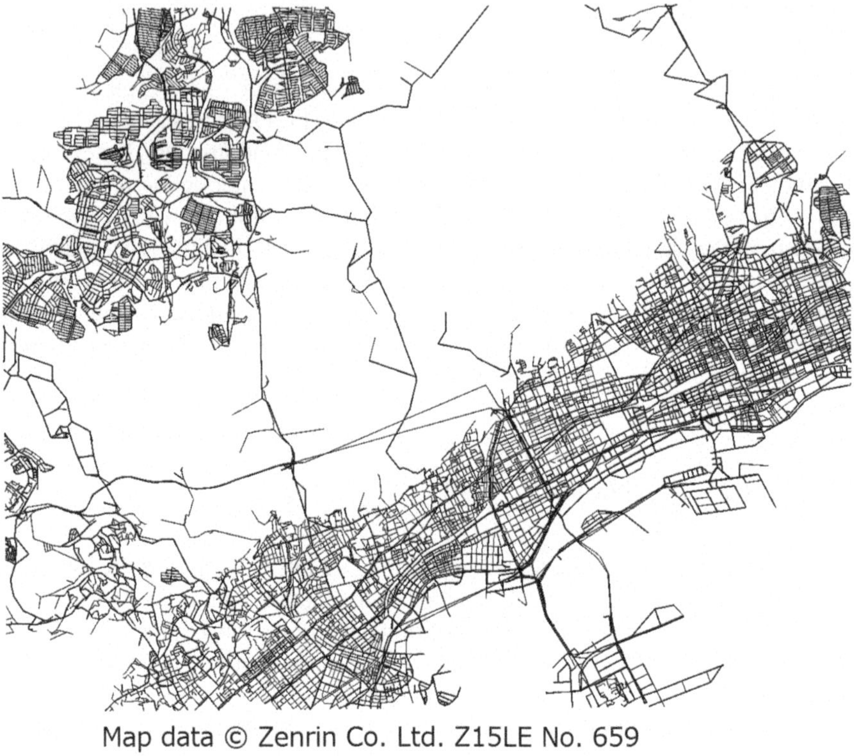

Fig. 23.2 The road network of mesh code of 523501

attachment program of the SUMO generates the traffic demand based on people's activity from the distribution of the people's home and workplace in the map by some statistical algorithm. We used this program to generate the OD data.

23.2.3 Simulation

The simulation utilizing the SUMO was carried out by using the road network and the OD data prepared by foregoing method. It is necessary to construct a route for each vehicle before executing simulation. In this study, the *DUAROUTER* [7] which is an attachment program of the SUMO was used to determine the route of the vehicle. This program searches a minimal traveling time path between origin and destination on the map by the Dijkstra method [8].

Since the route cannot be changed dynamically in the simulation of the SUMO, a glidlock results from the congested traffic. Therefore, iterative computation which repeats searching the route of each vehicle and simulation to run each vehicle was performed to find a route whose traveling time is minimum. The *DOUAITERATE* [9] which is attachment program of the SUMO was used to this end. This program does iterative computation so that the travel time of the vehicles attain minimum. In this study, the number of iteration was set to five times.

A scenario of the traffic simulation in this study is as follows. Only attendance was assumed as a traffic demand. The OD (the origin is the home of people, and the destination is the workplace of people.) was estimated from the geographical population distribution data as mentioned above. The total number of vehicle N is one of the adjustable parameters in this treatment. We set it to 100,000 per one day by reference to the traffic survey [10]. In addition, incoming traffic and outgoing traffic from out of the map of 523501 were ignored. The measurement of the traffic volume of the road was performed in a period of 7200–10,800 s. We assumed for the first 2 h the preparation period to set up the vehicle on the map.

The results obtained in this way are compared with the traffic census data of Kobe-city [3] in the next section. We emphasize that the data assimilation of the traffic flow is challenging because there are many free parameters and the response to the parameter change is nonlinear.

23.3 Results And Discussions

To make a comparison between the simulation result and the traffic census data of Kobe-city [3], the number of vehicles which passed a point where the traffic census measurement was performed [11] was counted in the simulation. The points where the traffic census measurement was performed were categorized into three types. One is on the national route, another is on the prefectural road, and the other is on the city road. The results is depicted in Fig. 23.3. As seen from Fig. 23.3, the overall traffic volume of the simulation is less than that of the traffic census data in each kind of point.

From this result, we considered that the OD model and/or the route of the vehicle were needed to improve. However, it is difficult to improve the former because there is no available data about the OD information, as mentioned before. In addition, even if the distribution of the OD was changed, the result differed only slightly. Therefore, we focused on the improvement of the latter. The simplest way to change the route of the vehicle is varying the speed limit of the road. Because we used the Dijkstra algorithm whose weight of the road is traveling time of the vehicle. In fact, the result when the speed limit of the thin road is changed from 60 km/h to 3.6 km/h is depicted in Fig. 23.4. The difference between Figs. 23.3 and 23.4 is subtle. However, in Fig. 23.4, the green points near 300 in abscissa axis are approaching the diagonal line compared with Fig. 23.3. Therefore, we expected that the data assimilation is attainable by the adjustment of the speed limit of the road.

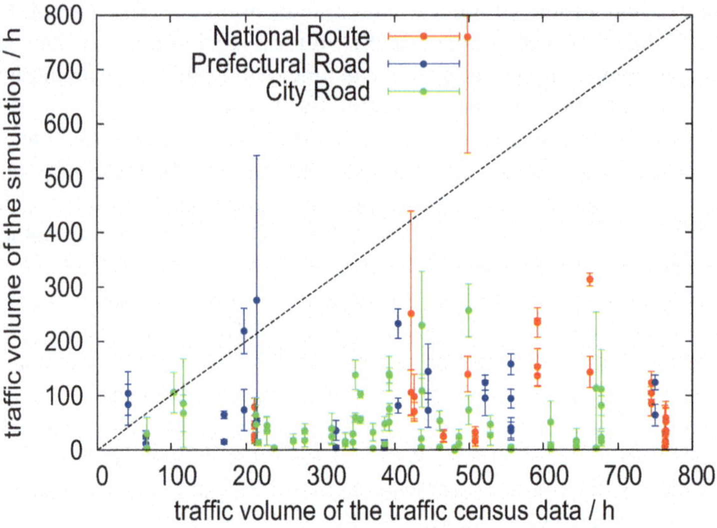

Fig. 23.3 The traffic volume obtained by the simulation is plotted against that of the traffic census data [3]. The results of the national route, the prefectural road, and the city road are depicted by *red*, *blue*, and *green*, respectively. The *error bars* show the standard deviation. The speed limits of each roads are set to the legal speed in Japan

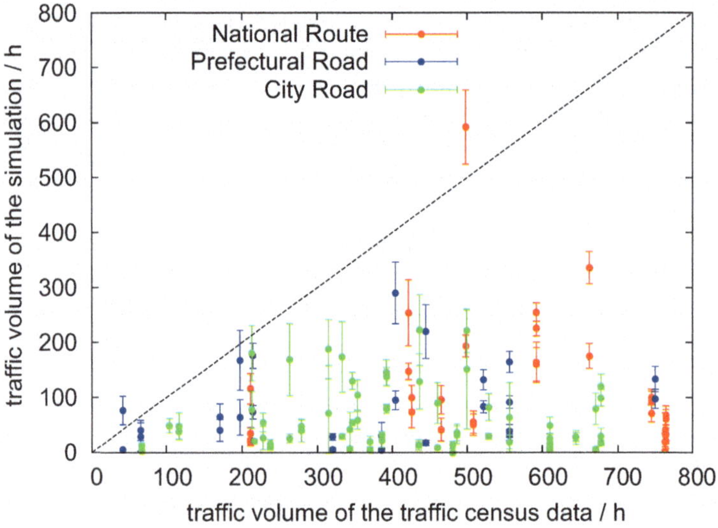

Fig. 23.4 The traffic volume obtained by the simulation is plotted against that of the traffic census data [3]. The results of the national route, the prefectural road, and the city road are depicted by *red*, *blue*, and *green*, respectively. The *error bars* show the standard deviation. The speed limits of each roads except for the thin road are set to the legal speed in Japan. The speed limit of the thin road is set to 3.6 km/h

To find a optimal set of the speed limit of the road, a loss function was defined as

$$F_\alpha = \sum_{i=1}^{N_\alpha} \left(\frac{f_{i\,\alpha}^{\mathrm{sim}} - f_{i\,\alpha}^{\mathrm{exp}}}{f_{i\,\alpha}^{\mathrm{exp}}} \right)^2 , \tag{23.1}$$

where $f_{i\,\alpha}^{\mathrm{sim}}$ and $f_{i\,\alpha}^{\mathrm{exp}}$ denote the traffic volume obtained by the simulation and the traffic census data at the ith measuring point on the road whose type is α. N_α represents the total number of measuring point belonging to the road whose type is α. In the area of the mesh code of 523501, $N_{\text{national route}}$, $N_{\text{prefectural road}}$, and $N_{\text{city road}}$ are 39, 26, and 64, respectively. As mentioned above, the speed limit of the road was chosen as an adjustable parameter, and the loss function was calculated to each parameter set.

The loss function of the national route, the prefectural road, and the city road are plotted against the speed limit of the national route in Figs. 23.5, 23.6, and 23.7, respectively. Figure 23.5 shows that the loss function decrease with the increasing the speed limit of the national route, and decrease with the decreasing the speed limit of the thin road. The former is nothing special, because the vehicles which pass the national route are increased by increasing the speed limit of the national route. The latter indicates that the current routing algorithm makes thin road readily traversable to the vehicle. Figure 23.6 shows that the loss function decrease with the decreasing the speed limit of the national route when the speed limit of the thin road is greater than 5 m/s. Otherwise the loss function is independent of the speed limit

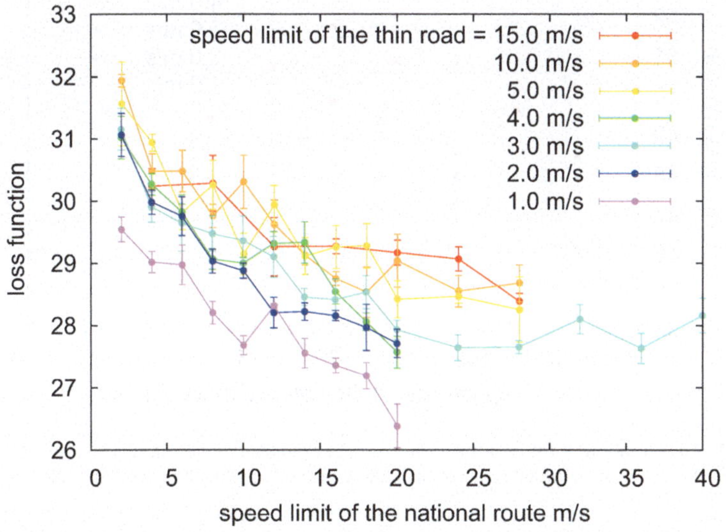

Fig. 23.5 The variation of the loss function $F_{\text{national route}}$, evaluated using Eq. (23.1), against a change in the speed limit of the national route at each value of the speed limit of the thin road

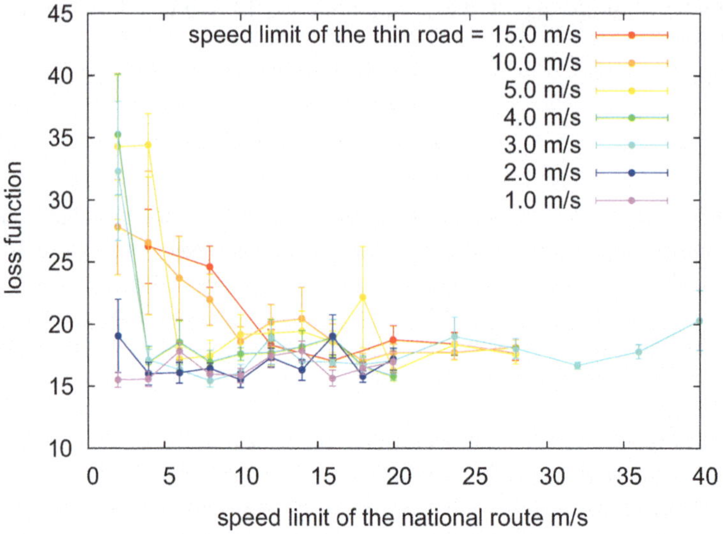

Fig. 23.6 The variation of the loss function $F_{\text{prefectural road}}$, evaluated using Eq. (23.1), against a change in the speed limit of the national route at each value of the speed limit of the thin road

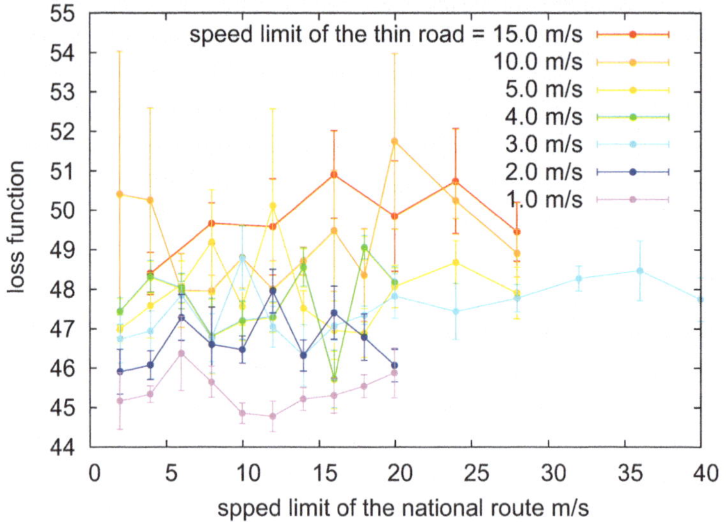

Fig. 23.7 The variation of the loss function $F_{\text{city road}}$, evaluated using Eq. (23.1), against a change in the speed limit of the national route at each value of the speed limit of the thin road

of the national route. Figure 23.7 shows that the loss function is independent of the speed limit of the national route, and slightly decrease with decreasing the speed limit of the thin road.

As seen from Figs. 23.5, 23.6, and 23.7, the value of the loss function is changed by varying the speed limit of the road. In these figures, the minimum values of the $F_{\text{national route}}/N_{\text{national route}}$, $F_{\text{prefectural road}}/N_{\text{prefercral road}}$, and $F_{\text{city road}}/N_{\text{city road}}$ are $0.68(1), 0.60(2)$, and $0.70(1)$, respectively. The number in parentheses indicates the accuracy of the last digit. To do the same adjustment with the speed limit of the other roads, the reproducibility will be better.

23.4 Summary

A traffic simulation of Kobe-city was performed using the SUMO whose map was constructed by high-quality digital map data provided by Zenrin Co. Ltd. Because there was no available data for the OD information, it was estimated from a geographical population distribution data. The traffic volume obtained by the simulation was compared with that of the traffic census data, and the result of the simulation is less than that of the traffic census data. In order to improve the reproducibility of the traffic census data, a parameter search whose adjustable parameter was chosen as the speed limit of the road was conducted. We found that this adjustment improved the reproducibility of the traffic census data. The reproducibility will be better by adjusting all of the road.

Further improvement of the reproducibility will need to reconsideration of the routing algorithm. Specific example is that, in the current routing algorithm, the national route and the thin road are indistinguishable. However, in real world, a driver should distinguish the individuality of the road.

Acknowledgements The systematic simulations in this study were assisted by OACIS [12, 13]. This research is supported by the Center of Innovation Program from Japan Science and Technology Agency, JST.

References

1. http://www.kkr.mlit.go.jp/plan/pt/research_pt/h22/pt_h22.pdf
2. http://www.openstreetmap.org
3. http://www.city.kobe.lg.jp/information/data/statistics/traffic/keikaku/img/22census_w.xls
4. Krajzewicz D, Erdmann J, Behrisch M, Bieker L (2012) Int J Adv Syst Meas 5:128
5. http://www.sumo.dlr.de/wiki/NETCONVERT

6. http://www.sumo.dlr.de/wiki/Demand/Activity_based_Demand_Generation
7. http://www.sumo.dlr.de/wiki/DUAROUTER
8. Dijkstra EW (1959) Numer Math 1:269
9. http://www.sumo.dlr.de/wiki/Demand/Dynamic_User_Assignment
10. http://www.city.kobe.lg.jp/information/project/urban/sogokotsu/img/1-4.genkyo-kadai_0807.pdf
11. http://www.city.kobe.lg.jp/information/data/statistics/traffic/keikaku/img/22kasyo.pdf
12. Murase Y, Uchitane T, Ito N (2014) Phys Procedia 57:73
13. https://www.github.com/crest-cassia/oacis

Chapter 24
MOSAIIC: City-Level Agent-Based Traffic Simulation Adapted to Emergency Situations

Guillaume Czura, Patrick Taillandier, Pierrick Tranouez, and Éric Daudé

Abstract In this paper, we present MOSAIIC, an agent-based model to simulate the road traffic of a city in the context of a catastrophic event. Whether natural (cyclone, earthquake, flood) or human (industrial accident) in origin, catastrophic situations modify both infrastructures (buildings, road networks) and human behaviors, which can have a huge impact on human safety. Because the heterogeneities of human behaviors, of land-uses and of network topology have a great impact on the traffic flows, the agent-based modeling is particularly adapted to this subject. In this paper, we focus on the new traffic model itself: the way geographical data is used to build a network, the various behaviors of our agents, from the individual to the collective level.

24.1 Introduction

Nowadays, traffic simulations are often used by urban planners to make decisions concerning road infrastructures. Many models have been developed these last years. These models are grouped according to their levels of representation: macroscopic [1], mesoscopic [2], microscopic [3] and nanoscopic [4].

A modeling approach that is particularly well-fitted for micro-simulation is agent-based modeling. It allows to consider the heterogeneity of driver behaviors and to take into account the global impact of local processes.

Such approach is increasingly used as many frameworks allowing urban planners to easily build their own scenarios (MATSIM [5], SUMO [6], AgentPolis [7]) are

G. Czura (✉) • P. Taillandier
UMR CNRS IDEES 6266, University of Rouen, Mont-Saint-Aignan, France
e-mail: guillaume.czura@gmail.com; patrick.taillandier@univ-rouen.fr

P. Tranouez
EA LITIS 4108, University of Rouen, Saint-Étienne-du-Rouvray, France
e-mail: pierrick.tranouez@univ-rouen.fr

É. Daudé (✉)
CNRS, UMIFRE 20 CSH, CNRS-MAE, Delhi, India
e-mail: eric.daude@univ-rouen.fr

H. Takayasu et al. (eds.), *Proceedings of the International Conference on Social Modeling and Simulation, plus Econophysics Colloquium 2014*, Springer Proceedings in Complexity, DOI 10.1007/978-3-319-20591-5_24

developed. However, if these frameworks are often well-adapted to traffic in normal condition, very few tools allow the simulation of uncommon events such as natural or technological hazards. Actually, in this context, being able to simulate the traffic in a realistic way while taking into account the road infrastructure (crossing, traffic signals...), the properties of the cars (length, max speed...) and the personality of the drivers (tendency to respect the norms) is mandatory. Most other frameworks work at a higher level, supposing regularities and statistical behaviors. But in a disaster individual decisions can lead to important collective consequences. Two drivers can leave their car, thus blocking hundred behind them. Drivers may react to things they see, fleeing by taking one-way streets in reverse, creating a jam in the road leading to this street. Individual-based modeling and micro-simulation are the only way to incorporate those possibilities, not just origin destination matrices and shortest path algorithms. For modelers without high level programming skills, adapting these platforms to specific application contexts is out of reach as they require to write code in JAVA or C++. As a result, many simulators are still developed from scratch or with a generic platform (e.g. [8–10]).

In this paper, we propose a new generic model dedicated to traffic simulation based on the work of [10–12] called MOSAIIC. This model, which have been implemented using the GAMA modeling platform [13], is easily tunable through a specific modeling language. Moreover, this model manages road infrastructures and traffic signals, input from real geographical data, as well as a detailed implementation of cars and drivers: choice of destination, acceleration and deceleration according to the surrounding traffic and the regulation, lanes changing, crossroads crossing etc. In addition, it allows to take into account the personality of each driver: respect of norms (traffic light, right of way, speed limits...), the management of tailgating.

The paper is organized as follows: Sect. 24.2 is dedicated to the presentation of the generic MOSAIIC Agent-based model. Section 24.3 concludes and presents some perspectives.

24.2 The MOSAIIC Agent-Based Traffic Model

As stated in the previous section, we chose to implement the model with the GAMA platform. The GAMA platform provides modelers—who quite often are not developers—with tools to develop highly complex models. In particular, it offers a complete modeling language (GAML: GAma Modeling Language) and an integrated development environment that allows modelers to quickly and easily build models. Indeed, the GAML language is as simple to use and to understand as the Netlogo modeling language [14] and does not require high level programming skills. In addition, GAMA provides different features that can be used by modelers to develop traffic models. In particular, GAMA allows to simply load GIS data (shapefiles, OSM data...), to define graphs from polyline geometries, to compute shortest paths and to move agents on a polyline networks. At last, it integrates an extension dedicated to fine-scale traffic simulation [15].

24.2.1 Structure of the Network

A key issue for our model is to be versatile enough to be usable with most of classic road GIS data, in particular OSM[1] data. We choose then to use a classic format for the roads and nodes (Fig. 24.1). Each road is a polyline composed of road sections (segments). Each road has a target node and a source node. Each node knows all its input and output roads. A road is considered as directed. For bidirectional roads, 2 roads have to be defined corresponding to both directions. Note that for some GIS data, only one road is defined for bidirectional roads, and the nodes are not explicitly defined. In this case, it is very easy, using the GAML language, to create the reverse roads and the corresponding nodes (it only requires few lines of GAML).

A road can be composed of several lanes (Fig. 24.2). The vehicles are able to change at any time its lane and even use a lane of the reverse road. In this case, the vehicle "cross" the road (for example going from Road 2 to Road 1 in the Fig. 24.1). Legal speed is another property of the modeled road. Note that even if the user of the model has no information about these values for some of the roads (the OSM data are often incomplete), it is very easy using the GAML language to fill in the missing values by a default values. It is also possible to change these values dynamically

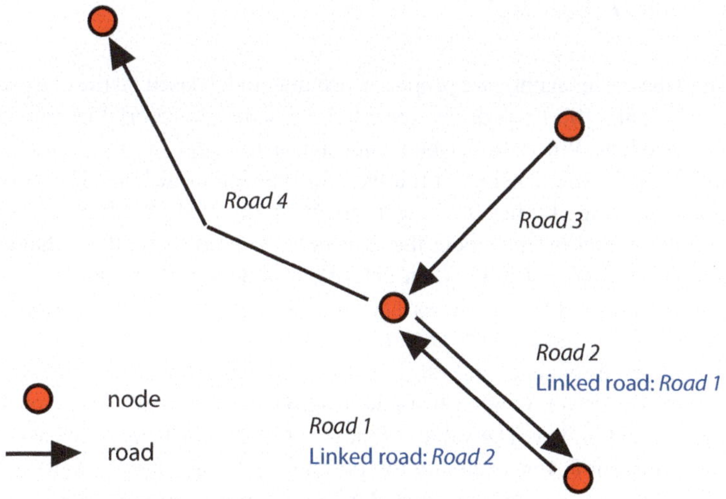

Fig. 24.1 Roads and nodes description in the model

[1]OSM: OpenStreetMap.

Fig. 24.2 Roads and lanes description in the model

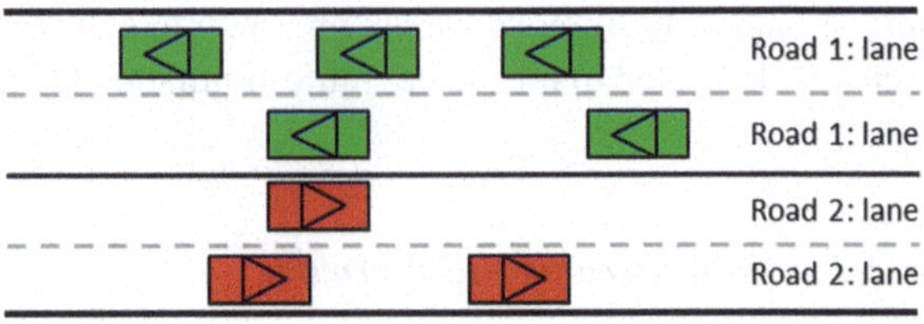

during the simulation (for example, to model that after an accident, a lane of a road is closed or that the speed of a road is decreased by the authorities).

In order to give the modelers the possibility to simply add dynamics to these infrastructures (e.g. to add a deterioration dynamic to roads), we chose to represent all the road infrastructures (road, traffic signals) as agents.

For each roads, a list of predefined variables is defined. Some of them are linked to the road properties:

- *lanes*: number of lanes.
- *maxspeed*: maximum authorized speed on the road.

In the same way, for each nodes, a list of predefined variables is defined. Amongst them, the most important is the list of stop signals, and for each stop, the list of roads concerned by it.

The complete list of variables for roads and nodes can be founded in [15].

24.2.2 *Driver Agents*

Concerning the driver agents, we propose a driving model based on the one proposed by Tranouez et al. [10]. Each driver agent has a planned trajectory that consists in a succession of edges. When the driver agent enters a new edge, it first chooses its lane according to the traffic density, with a bias for the rightmost lane. The movement on an edge is inspired by the Intelligent Driver Model [16]. A difference with our driving model is that in our model the drivers have the possibility to change their lane at any time (and not only when entering a new edge). In addition, we have defined more variables for the driver agents in order to give more possibilities for the modelers to tune the driver behavior.

The driver agents have several variables that will define the car properties and the personality of the driver, ranging from the length of the vehicle to the probabilities of respecting right of way. The values of these variables can be modified at any time during the simulation. For example, the probability to take a reverse road can be increased if the driver is stuck for several minutes behind a slow vehicle.

24.2.3 Dynamics of the Model

One step of the simulation represents 1 s. The dynamics of the model is based six consecutive steps:

1. Each road agent computes the potential traffic jams
2. Each traffic signal computes its new state.
3. New drivers arrive in the simulation
4. Drivers that do not have a path to reach their destination (or that should recompute them owing to changes in their context) compute it.
5. Drivers drive toward their final target. Note that the driving step is asynchronous. agents move one after the other. The order of activation of the driver Agents depend on their distance to the end of their current road: the drivers closer to the road end are activated first.
6. Drivers that reach their final target are removed from the simulation

24.2.3.1 Traffic Jam Management

Each road has the capability to compute the traffic jams on it. A traffic jam is defined as the presence on the road of at least *number_threshold* drivers of which the speed is inferior to *speed_threshold*. The *number_threshold* variable depends on the capacity of the road (will be lower for a small road than for a long road) and the *speed_threshold* variable on the *max_speed* on the road. A traffic jam becomes real for drivers if it exists for at least *time_threshold*.

24.2.3.2 Traffic Signal Update

Each traffic signal (only traffic light in our model) update its state counter.

24.2.3.3 New Driver Arrival

According to the current time and the data provided, a certain number of drivers are created. We assume that the model user has data (scenario) concerning the number of drivers departing at each period of time (for instance, there are often more drivers departing at 8AM than at 11PM). These drivers are located on one of the nodes according to the scenario data. Indeed, we assume here that the user has also data concerning the Origin and Destination of drivers. According to this data, a probability to use each node as an origin is computed and used to define the initial location of each new driver. In the same way, a probability to use each node as a destination according to a given origin is computed and used to define the final target of each new driver.

24.2.3.4 Computation of the Path

A driver can compute the path between its current location and its final target using a graph structure (each road will be an edge of the graph). In order to do so, the driver will use its own weights concerning for the edge. We defined four profiles of driver (i.e. four types of weight):

- minimize the travel distance
- minimize the travel time
- minimize the travel time and favorise roads with many lanes
- minimize the travel time and avoid traffic lights

In addition, when a driver perceived that its path will cross at least one known uncommon event (traffic jam, blocked roads), it will test a the *proba_avoid_event* probability to define if it will try to avoid it or not. If it tries to avoid it, it has two specific behaviors that will depends on the test of the *proba_know_map* probability. If the driver knows the map, it will compute a new shortest path taking into account all the information it has concerning uncommon events. In the other case, it will just try to choose through heuristics roads without uncommon events that will allow it to move closer to its final target.

24.2.3.5 Driving Step

The driving action of the driver agents work as follow: while the agent has the time to move, it first defines the speed he tries to reach based on different variables. Then, the agent moves toward the current target and computes the remaining time (Fig. 24.3). More specifically, each driver has a remaining time which is initially set to 1 s. Remaining time decreases after it drives, and it can continue to drive until remaining time becomes 0 or it has to stop at the intersection.

During the movement, the agents can change lanes (see below). If the agent reaches its final target, it stops; if it reaches its current target (that is not the final target), it tests if it can cross the intersection to reach the next road of the current path. If it is possible, it defines its new target and continues to move. The function that defines if the agent crosses or not the intersection to continue to move works as follow (Fig. 24.4): first, it updates its known uncommon events by adding all the uncommon events it perceives (the ones at a distance lower than its *perception_distance).* If its current path crosses an uncommon event, it will apply its path computation action. After that, it tests if the road is blocked by a driver at the intersection (if the road is blocked, the agent does not cross the intersection). Then, if there is at least one stop signal at the intersection (traffic signal, stop sign. . .), for each of these signals, the agent tests its probability to respect or not the signal (note that the agent has a specific probability to respect each type of signals). If there is no stopping signal or if the agent does not respect it, the agent checks if there is at least one vehicle coming from a right (or left if the agent drives on the left side) road at a distance lower than its security distance (i.e. minimal distance to the closest vehicle

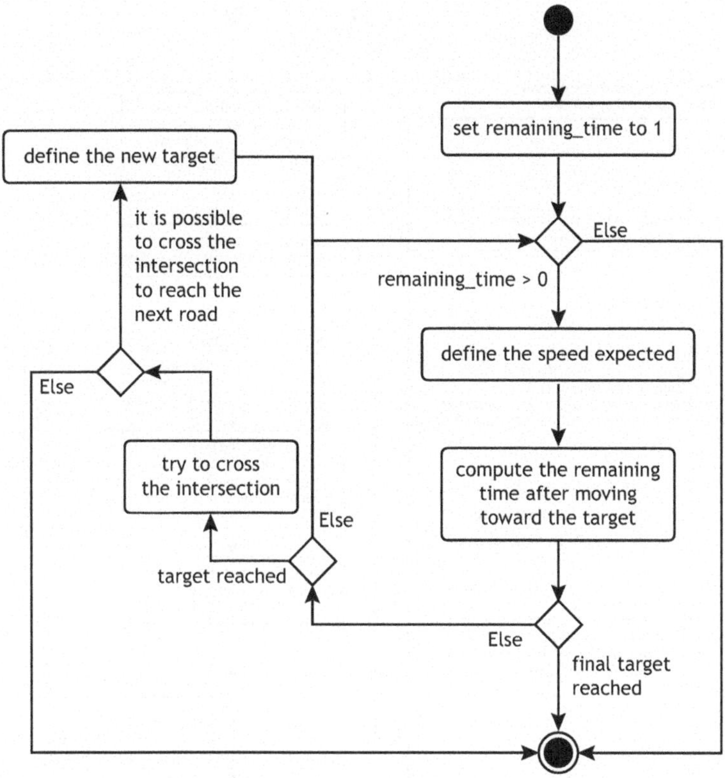

Fig. 24.3 Agent driving action algorithm

from which the agent feels safe). If there is one, it tests its probability to respect this priority. If there is no vehicle from the right roads or if it chooses to do not respect the right priority, it tests if it is possible to cross the intersection to its target road without blocking the intersection (i.e. if there is enough space in the target road). If it can cross the intersection, it crosses it; otherwise, it tests its probability to block the node: if the agent decides nevertheless to cross the intersection, then the perpendicular roads will be blocked at the intersection level (these roads will be unblocked when the agent is going to move).

Concerning the movement of the driver agents on the current road, the agent moves from a section of the road (i.e. segment composing the polyline) to another section according to the remaining time and to the maximal distance that the agent can moves (Fig. 24.4). For each road section, the agent first computes the initial remaining distance it can travel according the remaining time and its speed (i.e. max distance it can travel if there is no other vehicle). Then, the agent computes its security distance (i.e. minimal distance to the next vehicle from which the agent feels safe) according to its speed and its *security_distance_coeff*. While its remaining distance is not null, the agent computes the maximal distance it can travel

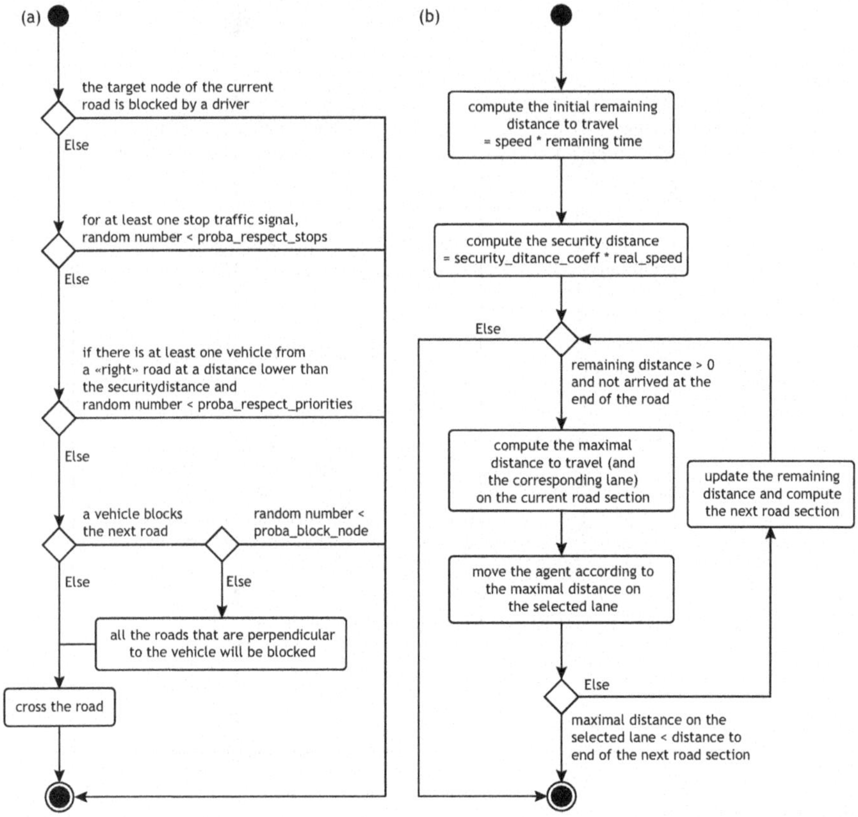

Fig. 24.4 (**a**) Crossing of an intersection (case where the driver drives on the right side of the road) and (**b**) Move on the current road algorithms

(and the corresponding lane), then it moves according to this distance (and updates its current lane if necessary). If the agent is not blocked by another vehicle and can reach the end of the road section, it updates its current road section and continues to move. The agent changes lanes if it computes it could go further in its time slot on another lane.

Figure 24.5 shows a snapshot of a simulation carried out for the agglomeration of Rouen (France).

24.3 Conclusion

In this paper, we presented a new generic traffic model. From this model, which was implemented with the GAMA platform, traffic simulations with a detailed representation of the driver operational behaviors can be built. In particular, it

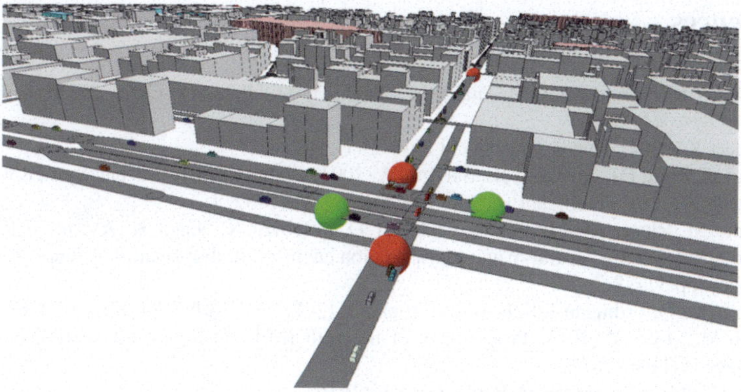

Fig. 24.5 MOSAIIC traffic model applied to Rouen agglomeration, France

models the road infrastructures and traffic signals, the lane changes of the drivers and their respect of norms. In comparison to the use of existing traffic simulation frameworks, the advantage of our model is to enable modelers to easily define models adapted to their application context.

We plan to enrich the model in order to make the driver agents more cognitive, in particular concerning their choice of path and their adaptation to the their current context in emergency situation. For this, we plan to give the driver agents a BDI architecture that can be based on [17, 18].

If the model is already capable to simulate tens of thousands of driver agents, we plan to improve its efficiency by using High Performance Computing and in particular distribution on GPU to enable large scale simulations with millions of driver agents.

At last, we plan as well to develop new tools to help people to prepare their data. The goal will be to offer the possibility from incomplete OSM data to automatically fill the missing attributes, and to create a consistent network (with its infrastructure and traffic signals). A particular attention will be brought on traffic signals and traffic lights.

Acknowledgements This research was founded by the project MOSAIIC, funded by the Region Haute-Normandie, the Large-Scale Research Network TERA and the Technological Risk Control Network (TRCN).

References

1. Kotsialos A, Papageorgioy M, Diakaki C, Pavlis Y, Middleham F (2002) IEEE Trans Intell Transp Syst 3(4):282
2. Waldeer K (2004) Transp Theory Stat Phys 33(1):31
3. Miller JE, Hunt DJ, Abraham JE, Salvini PA (2004) Comput Environ Urban Syst 28:9
4. Daiheng N (2003) Intelligent vehicles symposium, pp 47–52
5. Balmer M, Rieser M, Meister K, Charypar D, Lefebvre N, Nagel K, Axhausen K (2009) Multi-agent systems for traffic and transportation engineering. Information Science Reference, Hershey, pp 57–78
6. Krajzewicz D, Erdmann J, Behrisch M, Bieker L (2012) Int J Adv Syst Meas 5(3&4):128
7. Jakob M, Moler Z (2013) Proceedings of the 16th IEEE intelligent transportation systems conference (ITSC 2013)
8. Horn M (2002) Transp Res A Policy Pract 36(2):167
9. Banos A, Marilleau N, Team M (2012) ECSS
10. Tranouez P, Daudé E, Langlois P (2012) J Nonlinear Syst Appl 3(2):98
11. Daudé É, Provitolo D, Dubos-Paillard E, Gaillard D, Eliot E, Langlois P, Zimmermann E, Saint-Gérand T 2009) From system complexity to emergent properties. Springer, Berlin/Heidelberg, pp 165–178
12. Daudé É, Tranouez P, Langlois P (2009) The 3rd international conference on complex systems and applications, pp 116–121
13. Grignard A, Taillandier P, Gaudou B, Vo D, Huynh N, Drogoul A (2013) PRIMA 2013: principles and practice of multi-agent systems. Lecture Notes in computer science, vol 8291. Springer, Berlin, pp 117–131
14. Wilensky U (1999) Netlogo. Technical report. Center for Connected Learning and Computer-Based Modeling, Northwestern University, Evanston, IL. http://www.ccl.northwestern.edu/netlogo/
15. Taillandier P (2014) Eighth international workshop on agents in traffic and transportation (ATT)
16. Kesting A, Treiber M, Helbing D (2007) J Transp Res Board 1999:86
17. Taillandier P, Therond O, Gaudou B (2012) International environmental modelling and software society (iEMSs)
18. Le VM, Gaudou B, Taillandier P, Vo DA (2013) Advanced methods and technologies for agent and multi-agent systems (KES-AMSTA). Frontiers in artificial intelligence and applications, vol 252. IOS Press, Amsterdam, pp 395–403

Chapter 25
GUI for Agent Based Modeling

Tadashi Kurata, Hiroshi Deguchi, and Manabu Ichikawa

Abstract In this paper, we discuss how to build a model by SOARS VisualShell intuitively and explain its architecture. SOARS (Agent based simulation modeling language) SOARS Project (http://www.soars.jp), Tanuma et al. (Post-proceedings of AESCS04. Springer, Japan, pp 49–56, 2004) and Tanuma and Deguchi (Inst Electron Inf Commun Eng D J90-D(9):2415–2422, 2007) is a programming language to model social phenomena by agent-based simulation. We aim to make SOARS a simulation description language by which a domain expert can simulate social interactions occurred in the real world by ones conceptual model intuitively. Therefore, a support tool for realizing and achieving specialized concepts is necessary for a domain expert to build and run a simulation model based on his/her only domain knowledge without possessing complex programming skill, and SOARS VisualShell is an application to support such intuitive modeling by SOARS.

25.1 Background

In this paper, we focus on the agent-based simulation modeling of social phenomena. By agent-based simulation, we can model a system composed of agents who make decisions autonomously, and simulate the interactions between them. On the other hand, big data analysis is becoming more important in IoT era, for experts in divergent fields, and it is becoming challenging yet promising to construct an agent-based simulation model by applying the big data [3]. We consider it important to design IoT or IoE, which handle the interrelationship among autonomous agents such as real person and things on the internet, by using agent-based simulation. As we have already implemented Pub/Sub(the standard protocol for IoT) library for SOARS, it is possible to communicate the agent of IoT through the broker [1].

T. Kurata (✉) • H. Deguchi • M. Ichikawa
Tokyo Institute of Technology, 4259 Nagatsuta-cho, Midori-ku, Yokohama, Kanagawa, 226-8503, Japan
e-mail: kurata12@cs.dis.titech.ac.jp; deguchi@dis.titech.ac.jp; ichikawa@dis.titech.ac.jp

H. Takayasu et al. (eds.), *Proceedings of the International Conference on Social Modeling and Simulation, plus Econophysics Colloquium 2014*, Springer Proceedings in Complexity, DOI 10.1007/978-3-319-20591-5_25

For agent-based simulation modeling, domain experts need to design and model the complex interactions among agents in the real world. However, it is hard work for domain experts who have no programming experiences to construct such agent-based simulation models since many such models are built by programming languages. To solve this problem, it is necessary to construct an application for experts without programming skills to build models, same as using a CAD / CAM [4, 11] based on their domain knowledge.

In order to fulfill this purpose, a Modeling GUI which could naturally realize the agent concept and their interactions that expresses is essential and necessary. Modeling GUI provides a support environment for constructing models by a domain specific language under the specific model frame. For example, Stella [10] is the domain specific language under the system dynamics model frame, and its Modeling GUI allows users to construct stock-flow type models visually and intuitively while only mathematical knowledge about the system dynamics is required. Another domain specific language is Matlab [6] whose Modeling GUI is designed to construct control models visually and intuitively under the control model frame, and only mathematical knowledge on the control theory is necessary.

However, in terms of agent-based modeling, although there are modeling GUIs which can model the agent over two-dimensional cells intuitively, i.e. NetLogo [7], few of them could express agents social interactions intuitively, and there is no Modeling GUI to express agents social role interaction neither.

25.2 Objectives

SOARS, a domain specific language under agent-based simulation model frame, was developed by Deguchi Laboratory of Tokyo Institute of Technology since 2004 [9, 12, 13] and continued to evolve. Since its simulation engine is a text-based programming language, programming skills by the text editor are required in order to construct models. As a result, domain experts without programming skills may face difficulties of constructing agent-based simulation models by using it. Therefore, SOARS VisualShell was developed as the Modeling GUI for SOARS. It enables domain experts who have few programming experiences to construct agent-based simulation models only based on their domain knowledge.

Similar to Stella, Matlab and Scratch [5, 8], SOARS VidualShell should be designed with an intuitive user interface, by achieving two objectives. The first one is to prevent the occurrence of syntactical bugs. In this way, the user could identify the semantic bug, i.e. the model design bug, when the simulation does not behave in accordance with the original intention. The other one is for domain experts to realize SOARS specific model concepts intuitively, such as social role interactions.

This paper is organized as follows. In Sect. 25.3, we will explain the design of SOARS VisualShell developed to achieve these two objectives, and conclusions and future work are discussed in Sects. 25.4 and 25.5 respectively.

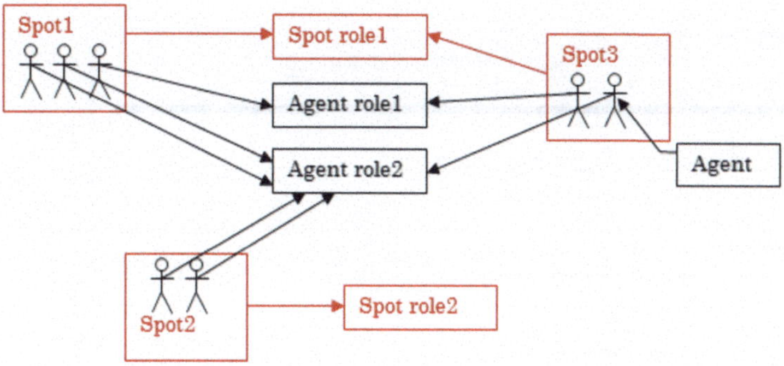

Fig. 25.1 Agent, spot and role in SOARS

25.3 Design

25.3.1 Architecture of SOARS

SOARS is composed of agents/spots as entity with their associated roles. Agents can move between any spot on the space. Agents and spots have variables, such as string, numeric, array, hash table variables and so on. A role has instructions as conditions and actions. Agent and spot can select any role to execute corresponding instructions with their variables in order to advance the simulation progresses, as shown in Fig. 25.1.

Furthermore, SOARS has the stage concept to control the instruction execution order in a role. One loop of a simulation is divided into one or more stages, which are executed in a specific order. While each stage, it is possible to define parallel executable instructions of which the execution order does not matter. By imposing such restrictions, all instructions will be executed in the correct order.

However, as its program is based on text, programming with the text editor is required, as shown in Fig. 25.2.

25.3.2 Design Concept of SOARS VisualShell

General computing languages, such as C, C++, Java, Fortran, and so on, are not domain specific languages, and the programming freedom degree is high and the description is fine-grained. As a result, their corresponding GUI is complicated to handle due to the high freedom degree, and a text editor is required for programming, such as Eclipse, Microsoft Visual Studio and so on.

On the other hand, the programming freedom of domain specific language, such as Stella, Matlab, SOARS, and so on, is low and the description is coarse-grained.

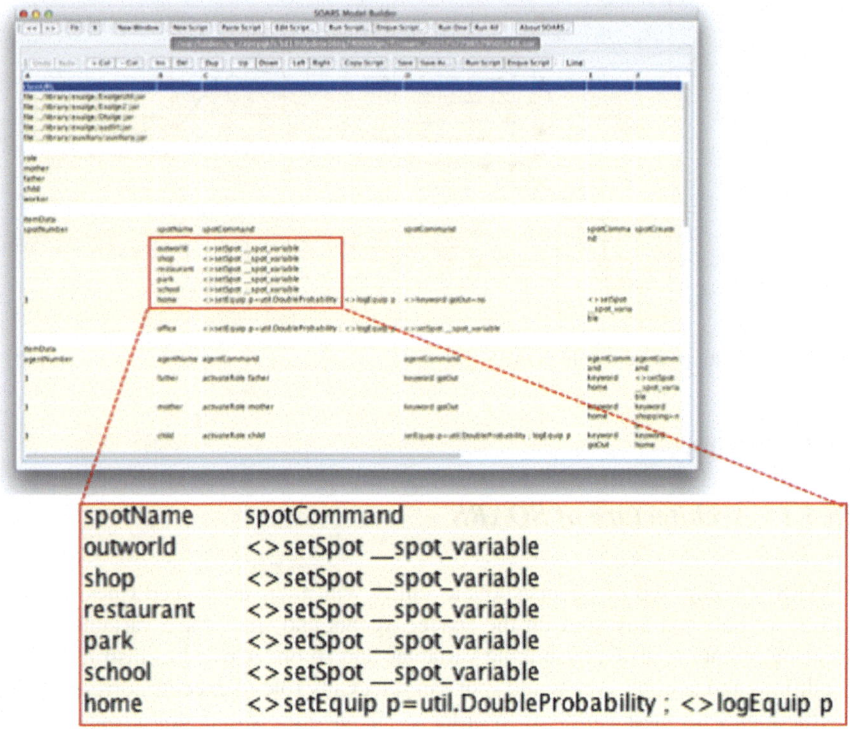

Fig. 25.2 SOARS programming in text base. It requires to type the program in each cell without syntactical error check

Table 25.1 Comparison among programming languages, model frame and modeling GUI

Programming language	DSL[a]	Model frame	Modeling GUI
General computing language[b]	No	None	None
Stella	Yes	System dynamics	Stella GUI
Matlab	Yes	Control theory	Matlab GUI
SOARS	Yes	Agent-based simulation	SOARS VisualShell

[a] Domain specific language
[b] C, C++, Java, Fortran and so on

As a result, the GUI is not complicated to handle due to the low freedom degree, and it is possible to construct the model with Modeling GUI.

The comparison among above-mentioned programming languages is shown in Table 25.1.

In this study, we construct SOARS VisualShell, the Modeling GUI for SOARS by Java.

25.3.3 *User Interface of SOARS VisualShell*

SOARS VisualShell is the Modeling GUI for model construction by SOARS, and can manipulate agents, spots, roles and stages of SOARS visually and intuitively. Since SOARS VisualShell can define every element to create SOARS text base program, SOARS VisualShell can describe every SOARS program as the simulation development environment. SOARS VisualShell holds the visual description language in XML format to describe elements designed for SOARS, such as agents, spots, roles and stages. There are two features of SOARS VisualShell. One is the interactive interface which does not require users to remember instructions and formats, while the other is the automatical SOARS text-based program generator.

Furthermore, SOARS VisualShell requires only mouse operations for input and output, and the keyboard input is only required to set the value of variables. Syntactical bugs are prevented as the check of input strings is done automatically. Therefore, while knowing only the agent/spot concept of SOARS, it is possible to use this GUI intuitively without holding complex programming skill. SOARS VisualShell does not require users to handle syntactical bugs. In SOARS VisualShell, agents, spots and roles are represented as icons, as shown in Fig. 25.3, and the data structure is shown in Fig. 25.4. Its user interface visualizes the structure of SOARS intuitively. Through the interface, users are able to create a new agent, spot and role only by drag and drop from the icon menu. By clicking the start button, users can create a SOARS text base program automatically and launch the simulation more easily.

There is the editing tool Dia Diagram Editor(UML editor) such as SOARS VisualShell. Dia Diagram Editor is a document generation tool to edit UML diagram [2]. On the other hand, SOARS VisualShell is a tool for automatically generating the source code of the agent-based simulation actually works. SOARS VisualShell can build the model without syntactical bugs and run it.

25.3.3.1 Agent/Spot Edit

In SOARS VisualShell, agents and spots are represented as icons. By double clicking the icon, users can define the agent/spot name, the number of agent/spot and variables, such as string, numeric, array, hash table and so on, as shown in Fig. 25.5. It is possible to specify the number of agent/spot of up to several billion. SOARS VisualShell is actually used to infection simulation model building of huge city of about 300,000.

This user interface requires only mouse operations to select a variable type, and the keyboard input is only used for setting an agent/spot definition, a variable name and the initial value of variables.

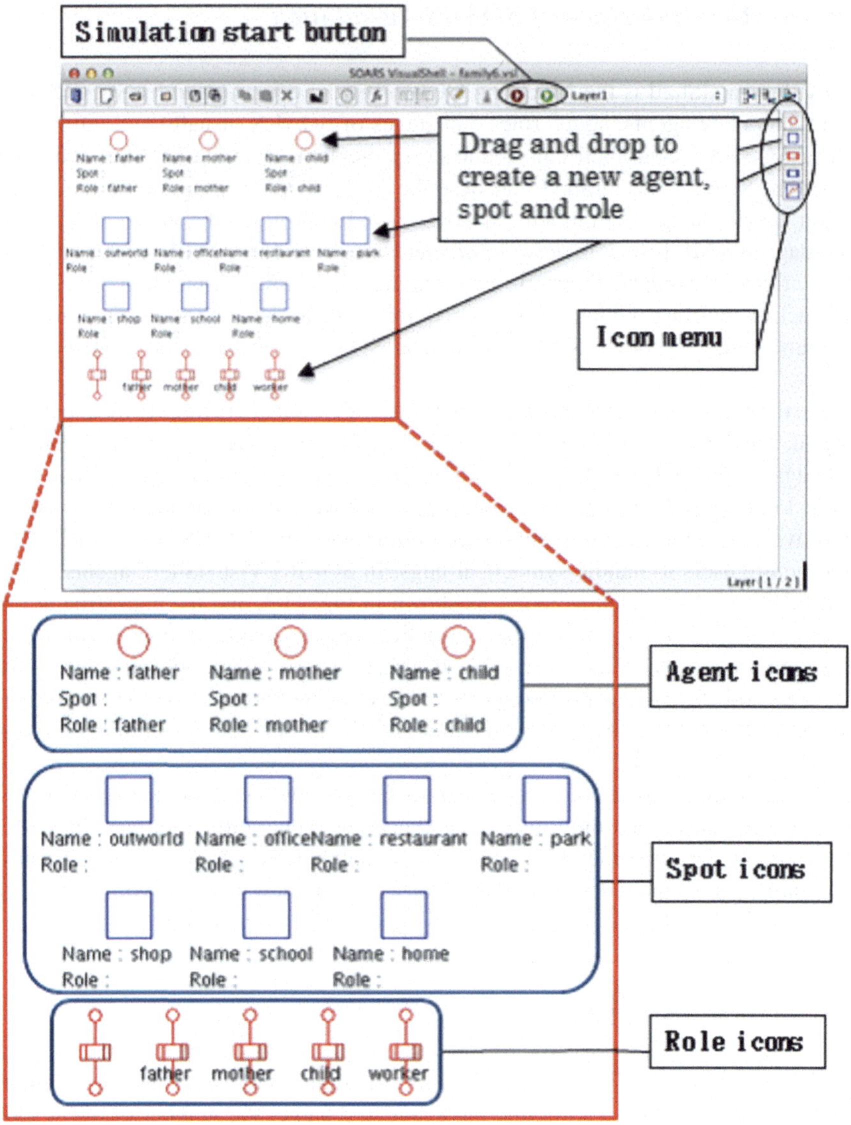

Fig. 25.3 SOARS VisualShell

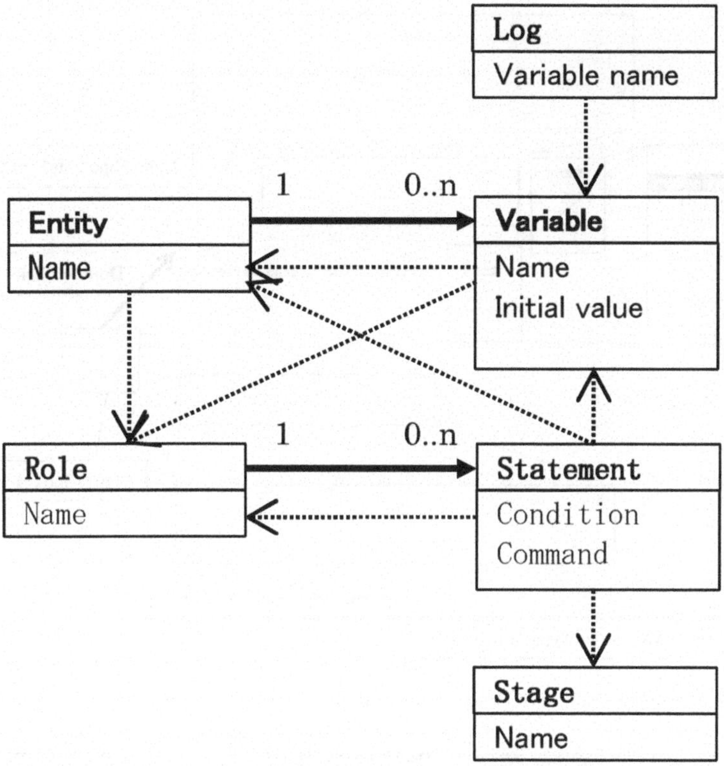

Fig. 25.4 Data structure of SOARS VisualShell

25.3.3.2 Role Edit

In SOARS VisualShell, roles are represented as icons, as shown in Fig. 25.6. By double clicking the icon, it is possible to define the role name, conditions and actions executed on each stage. The conditions and actions are defined as instructions.

This user interface is in spreadsheet format, and a stage and associated instructions should be set to each cell. Besides, it requires only mouse operations to select a stage, a condition type, an action type, an instruction and each instructions arguments, and the keyboard input is only necessary for inputting the role name.

25.3.3.3 Stage Edit

In SOARS VisualShell, stages are defined in the GUI, as shown in Fig. 25.7. This user interface requires the keyboard input for only inputting the stage name, and it requires only mouse operations to set the stage order.

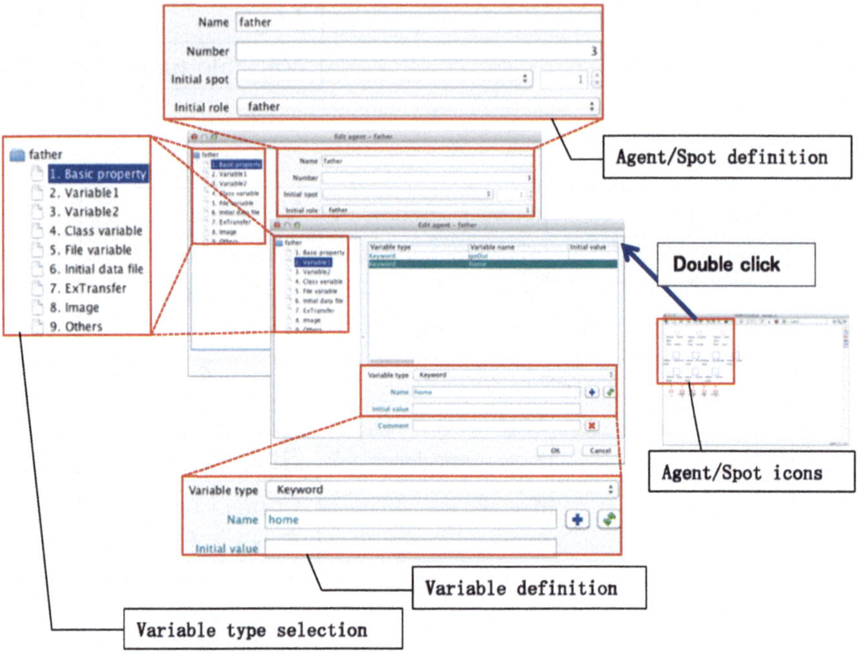

Fig. 25.5 Agent/Spot edit in SOARS VisualShell

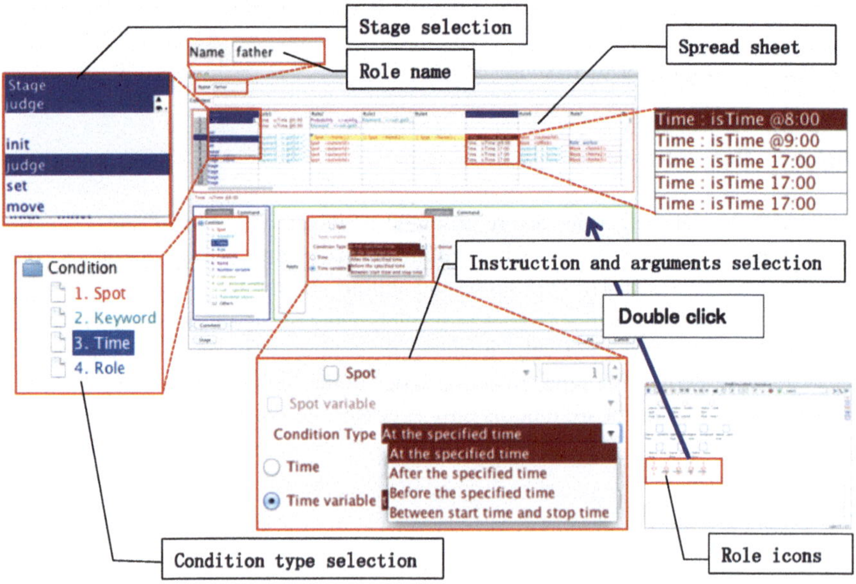

Fig. 25.6 Role edit in SOARS VisualShell

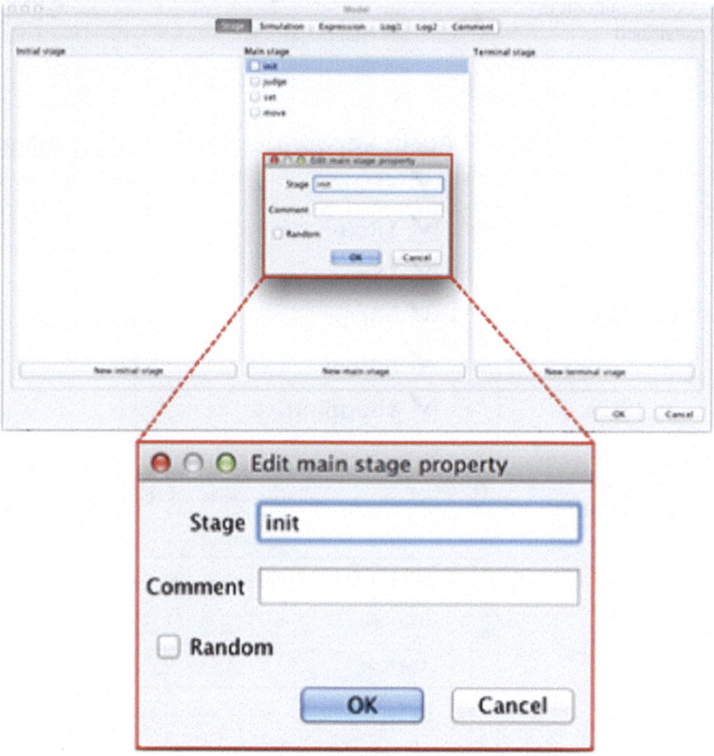

Fig. 25.7 Stage edit in SOARS VisualShell

25.3.3.4 Log Output Specification

In SOARS, the value of agent/spot variables is recorded in a log file during the simulation running. In SOARS VisualShell, the logged variables are defined in the GUI, as shown in Fig. 25.8. This user interface requires only mouse operations to select the logged variables from the list.

25.3.3.5 Simulation Condition Specification

In SOARS, it is necessary to set simulation conditions, such as the simulation start time, step time, stop time and so on. In SOARS VisualShell, they are defined in the GUI, as shown in Fig. 25.9. This user interface requires only mouse operations to select numeric values and keyboard operations for direct value input.

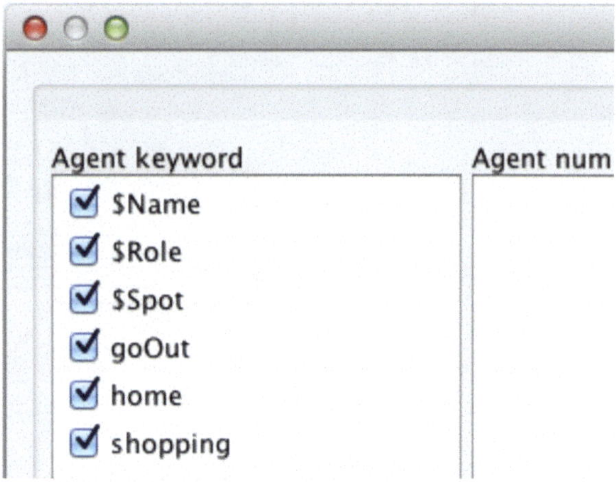

Fig. 25.8 Log output edit in SOARS VisualShell

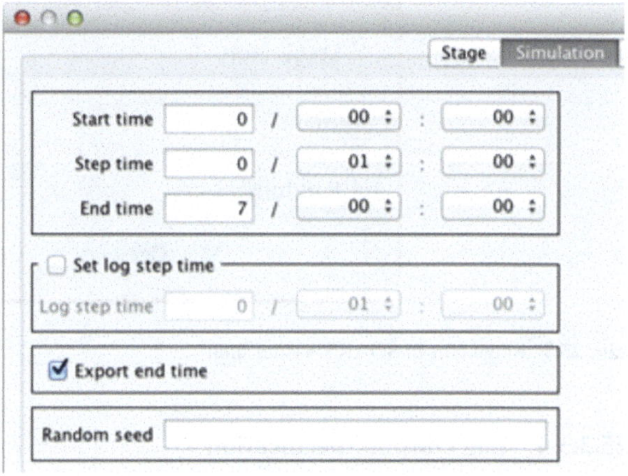

Fig. 25.9 Simulation condition edit in SOARS VisualShell

25.4 Conclusion

In SOARS VisualShell, as the concepts necessary to construct a model are expressed systematically, it is possible for the domain expert to construct a model only by selecting those elements sequentially. In other words, it is possible to construct a model merely by mouse operations. Though the string input is done via keyboard, the checking of input strings is done automatically. Therefore, it is possible to construct a model intuitively and syntactical bugs are avoided naturally. When the simulation goes against ones original intention, domain experts can identify the causes as semantic bugs, i.e. model design bugs, immediately. In addition, users can get familiar with the SOARS VisualShell operations within a very short period on any Java-installed PC (Windows, Mac, Linux).

In order to promote SOARS and for education purposes, SOARS Project holds SOARS Workshop every year [9]. By attending the intense tutorials with the assistant of experienced SOARS programmers, anyone can construct an agent-based simulation model by SOARS through the SOARS VisualShell within a short period. SOARS VisualShell is available in the humanities universities(Waseda University, Tokyo Institute of Technology, and so on).

In this paper, we have shown that this GUI, which is designed for a domain specific language of agent-based modeling, SOARS, is effective in modeling. More specifically, since the syntactical bug never occurs by using this GUI, users can devote their effort to resolving the semantic bugs only. In addition the user can manipulate this GUI to realize the model concept intuitively while programming skills are not required.

25.5 Future Work

Agent-based modeling is expected to develop systems composed of agents in the real world in IOT era. In future, SOARS is not only expected to model autonomous agents abstracted from the real world, but also to become an agent-based designing language which could utilize big data collected from sensors and enable motion control of actuator from the real world more intuitively [1]. In addition, SOARS VisualShell is expected to evolve to enable visual description of models.

References

1. Deguchi H (2014) Multiagent simulation: 2. Social and organizational ICT architecture design in the IoE Era—from social simulation to real world OS -. Inf Process Society of Japan 55(6):589–548 (in Japanese)
2. Dia Diagram Editor (2004) http://sourceforge.net/projects/dia-installer/
3. Karnouskos S, Tariq MMJ, (2009) Using multi-agent systems to simulate dynamic infrastructures populated with large numbers of web service enabled devices. In: International symposium on autonomous decentralized systems, 2009 (ISADS 2009), pp 23–25
4. Kirk JA, Anand DK, Anjanappa M, Uppal R (1986) Implementation of a flexible manufacturing protocol. In: Proceedings of 2nd IASTED international conference. Los Angeles, CA, USA, p 71
5. Maloney J et al (2010) The scratch programming language and environment. ACM Trans Comput Educ 10(4), Article 16 (11/2010):15
6. Matlab (late 1970s) http://www.mathworks.com/
7. NetLogo (1999) https://ccl.northwestern.edu/netlogo/
8. Resnick M et al (2009) Scratch: programming for all. Commun ACM 52(11):60–67
9. SOARS Project (2004) http://www.soars.jp
10. Stella (1987) http://www.iseesystems.com/

11. Sutherland I (1963) Sketchpad: a man-machine graphical communication system. In: Proceedings of the 1963 spring joint computer conference. Spartan Books, Baltimore, MD, pp 45–53
12. Tanuma H, Deguchi H (2007) Development of agent-based simulation language: SOARS. Inst Electron Inf Commun Eng D J90-D(9):2415–2422
13. Tanuma H, Deguchi H, Shimizu T (2004) SOARS: spot oriented agent role simulator: design and implementation. In: Post-proceedings of AESCS04. Springer, Japan, pp 49–56

Part V
Social Media

Chapter 26
Emotional Changes in Japanese Blog Space Resulting from the 3.11 Earthquake

Yukie Sano, Hideki Takayasu, and Misako Takayasu

Abstract We quantified the emotional changes observed in social media after major disasters, focusing especially on the Japanese blog space after the Great East Japan Earthquake in 2011. We checked the appearances of Japanese adjectives and found that special emotion adjectives, such as 'impatient', and 'frustrating' which involve the want to help others but the person has no means to and feels frustrated, occur with considerably increasing frequency. To visualize social mood, we drew a co-occurrence network of adjectives showing a major topological change at the site of the quake. Measuring emotional changes after an emergency has been difficult, but, our research has the potential to achieve it.

26.1 Introduction

Social data collected on a large scale has helped to promote our understanding of society and ourselves during this century [1–3]. Drastic changes in information technology enable us to collect various types of communication data, such as those on mobile phone and face-to-face contacts. Hidden patterns of human activities are uncovered, and these results are beginning to used to solve real social problems.

Web-based data attract many scientists particularly because it is easy to collect data on a large scale for discussion in scientific papers and such data reflect various real social phenomena such as political, market, financial, and daily events.

Y. Sano (✉)
University of Tsukuba, Ibaraki, Japan
e-mail: sano@sk.tsukuba.ac.jp

H. Takayasu
Sony Computer Science Laboratories Inc., Tokyo, Japan

Meiji University, Kanagawa, Japan
e-mail: takayasu@csl.sony.co.jp

M. Takayasu
Tokyo Institute of Technology, Kanagawa, Japan
e-mail: takayasu@dis.titech.ac.jp

H. Takayasu et al. (eds.), *Proceedings of the International Conference on Social Modeling and Simulation, plus Econophysics Colloquium 2014*, Springer Proceedings in Complexity, DOI 10.1007/978-3-319-20591-5_26

Examples include Facebook—the world's largest social networking service [4–6]—Google and Yahoo! search engines [7–9], Wikipedia—a multilingual online encyclopaedia [10–14]—YouTube—an online movie site [15]—and Twitter—a microblogging platform on which uses can post messages up to 140 characters [16–22].

Since various large amounts of social media data are available, studies regarding social media have been greatly increasing substantially. In particular, new applications are demanding and attracted, many researches have focused on the connections between real-world phenomena and social media. For example, Mandel et al. focused on the emotional changes in 65,000 tweets during half the month when hurricane Irene approached the U.S. and found that the level of concern had increased depending on region and gender [23]. However, little is known about long-term emotional changes in social media.

There are many ways to detect emotions in texts. Some examples are Linguistic Inquiry and Word Count (LIWC) [24], Positive and Negative Affect Schedule (PANAS) [25], Affective Norms for English Words (ANEW) [26], and Point Of Mood State (POMS) [27, 28]. Furthermore, Emoticons (combination of 'emotion' and 'icon') such as ':)' and WordNet a thesaurus have been used to detect emotions [29–32]. However, it is known that resulting polarity (i.e., positive or negative) is different depending on the methods[33]. Additionally, most of the methods are English, therefore, there is a not strict dictionary to detect emotions from Japanese texts.

In this paper, we use more than 3.2 billion Japanese blog posts for 6 years since 1 November 2007 as typical Japanese texts. Here we simply use whole Japanese adjectives to detect emotions. Our observation periods include the Great East Japan earthquake in 2011 which is said to have changed social mood qualitatively. First, we describe our data and method in Sect. 26.2. Next we compare the relative frequencies of adjectives before and after the quake and draw co-occurrence networks for the adjectives to visualize social emotions in Sect. 26.3. Summary and discussion are in Sect. 26.4.

26.2 Data and Methods

26.2.1 Data

We studied Japanese blog data from 1 November 2007 to 31 October 2013 (2192 days) using an Internet service called 'Kuchikomi@kakaricho'[1] to collect data. This fee-charging service provides an original web-site and an application programming interface (API). The API returns the daily number of blogs in which any given target word appears in a given period. The daily number of blogs containing a blog post

[1]http://kakaricho.jp (Accessed: 11 March 2015).

which includes that the target word occurring more than once is counted. Thus, if one blog post includes a target word multiple times, the API counts it as one. The API also provides a spam filter with three levels—weak, middle and strong—depending on the desired spam detection accuracy. Here, we use the middle level of spam filter to collect data. The full API database contains more than 3.2 billion blog posts from 38 million accounts.

We search for adjectives listed in the MeCab[2]—a Japanese morphological analyzer—dictionary with original surface forms using the API. Because there are various ways of conjugating forms in Japanese, we search for just their original form for simplicity; there are 1741 adjectives in total. Subsequently, we summarize these adjectives' time series in the case that they have the same pronunciations and meanings. Because Japanese uses three different character sets—Hiragana, Katakana and Kanji (Chinese characters)—instead of an alphabet, people tend to use words with different surface forms that have the same pronunciations and meanings. This procedure leads to 839 adjectives. Finally, we removed extremely low- (less than 10 times per day) or high-frequency (more than 100,000 times per day) adjectives, resulting in a total of 550 adjectives' time series.

26.2.2 *z*-Test for the Quality of Two Proportions

In this research, we apply a z-test for the quality of two proportions to determine whether adjective i's occurrence $x_i(t)$ differs at different times T_0 and T_1. The z-score of statistic z_i is calculated as follows:

$$z_i = \frac{\left| \frac{x_i(T_0)}{X(T_0)} - \frac{x_i(T_1)}{X(T_1)} \right|}{\sqrt{X_i\,(1 - X_i) \left[\frac{1}{x_i(T_0)} + \frac{1}{x_i(T_1)} \right]}} \tag{26.1}$$

where $X(t)$ is the total number of blogs at time t and X_i is calculated as follows:

$$X_i = \frac{x_i(T_0) + x_i(T_1)}{X(T_0) + X(T_1)} \tag{26.2}$$

The null hypothesis is that z_i follows a standard normal distribution. Therefore, we can calculate the p-value to check the proportions of adjectives in different time periods.

[2]https://code.google.com/p/mecab/ (Accessed: 11 March 2015).

26.2.3 Partial Correlation Coefficient

To visualize the change in social emotions at the quake, we construct adjective co-occurrence networks for the pre- and post-quake periods. First, we calculate the Pearson's linear correlation coefficient r_{ij} between the frequency $y_i(t)$ and $y_j(t)$ of adjectives i and j respectively, with a difference of normalized time series,

$$y_i(t) = \frac{x_i(t+1)}{X(t+1)} - \frac{x_i(t)}{X(t)}. \tag{26.3}$$

Pearson's linear correlation coefficient r_{ij} between adjectives i and j is calculated as follows:

$$r_{ij} = \frac{\sum_t (y_i(t) - \langle y_i \rangle)(y_j(t) - \langle y_j \rangle)}{\sigma_i \sigma_j} \tag{26.4}$$

where σ_i and σ_j are the standard deviations of $y_i(t)$ and $y_j(t)$, respectively. Here, we use three weeks before and after the quake for the comparison, with $T = 21$ data points. If adjectives i and j have a positive correlation at the 0.01 significant level, then the value of r_{ij} is greater than 0.5487 for $T = 21$.

To extract more essential links from the co-occurrence network, we use the partial correlation coefficient r_{ij}^k as follows:

$$r_{ij}^k = \frac{r_{ij} - r_{ik} \cdot r_{jk}}{\sqrt{1 - r_{ik}^2}\sqrt{1 - r_{jk}^2}}. \tag{26.5}$$

r_{ij}^k is the partial correlation between i and j under the fixed condition of k. Partial correlation is equivalent to the correlation between residuals $y_i(t)$ and $y_j(t)$ after the removing correlation between each r_{ik} and r_{jk}.

Normally, we can calculate the partial correlation D_{ij} that is removed by all other adjectives' effects by calculating an inverse correlation matrix. However, we cannot calculate it that way in this case, because we have only $T = 21$ data points for each time series and there are 550 adjectives (samples). Therefore, we calculate the partial correlation D_{ij} for each pair of adjectives i and j by removing k's effect and checking the following condition.

$$D_{ij} = \begin{cases} \max r_{ij}^k & (\text{if } \forall k, r_{ij}^k \geq D_c) \\ 0 & (\text{otherwise}), \end{cases} \tag{26.6}$$

where $D_c = 0.5613$ at the 0.01 significance level of partial correlation. To stress the significant partial correlation coefficient, we use maximum value of r_{ij}^k in this research.

26.3 Results

26.3.1 Adjectives in Emergency and Normal Periods

We focus on relative changes in adjectives before and after the quake by calculating z_i, as shown in Eq. (26.1). Here, we define the pre-quake period T_0 as three weeks before the quake, from 18 February to 10 March in 2011. Similarly, we define the post-quake period T_1 as three weeks after the quake, from March 12 to April 1 in 2011.

According to the results of the calculation of z_i for all 550 possible adjectives, 72 adjectives increased and 74 adjectives decreased significantly at the 0.01 significant level. Tables 26.1 and 26.2 show top 10 increased and decreased adjectives respectively. Figure 26.1 shows examples of increased and decreased adjectives.

Table 26.1 Adjectives that increased significantly ($p < 0.01$) after the quake

#	Word	Original form (Japanese)	$x_i(T_1)/x_i(T_0)$
1	Impatient	はがゆい, 歯がゆい, 歯痒い	5.91
2	Lonely	こころぼそい, 心細い	5.48
3	Precious	とうとい, 貴い, 尊い	4.84
4	Sorry	こころぐるしい, 心苦しい	4.36
5	Frustrating	もどかしい	4.13
6	Cannot stand the situation	いたたまれない	4.07
7	Heartless	こころない, 心無い	3.57
8	Shameless	あさましい, 浅ましい	3.50
9	Barely afford	ままならない	3.42
10	Base	さもしい	3.40

Table 26.2 Adjectives that decreased significantly ($p < 0.01$) after the quake

#	Word	Original form (Japanese)	$x_i(T_1)/x_i(T_0)$
1	Brand new	めあたらしい, 目新しい	0.29
2	A little earlier	ひと足早い	0.31
3	Itchy	かゆい, 痒い	0.52
4	Tawdry	やすっぽい, 安っぽい	0.56
5	Cannot wait	まちどおしい, 待ち遠しい, 待遠しい	0.58
6	Salty-sweet	あまからい, 甘辛い	0.59
7	Celebration	めでたい, メデタい, 目出度い	0.62
8	Bitter	しぶい, 渋い	0.63
9	Spicy	こうばしい, 香ばしい, 芳ばしい	0.63
10	Tough	てごわい, 手ごわい	0.63

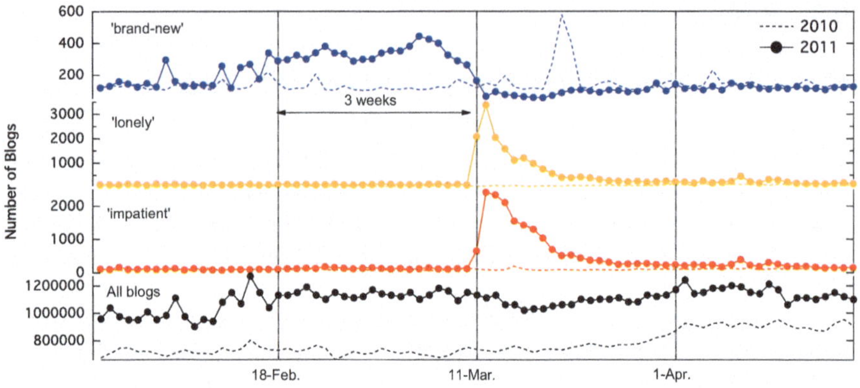

Fig. 26.1 Daily number of blogs including 'brand-new', 'lonely', and 'impatient' from the *top*. Daily number of whole blogs $X(t)$ is shown in the *bottom*. *Solid lines* are in 2011 and *dashed lines* are in 2010 for comparison. Note that sudden increase around 20 March 2010 for 'brand-new' is caused by spam blogs, because we confirm that it diminished when we search with the strong level of spam filter

We found that adjectives such as 'impatient' which express users' feelings of frustration have increased considerably according to Table 26.1 (#1, 4, 5, 6). Even these adjectives have different surface forms, albeit similar meanings, e.g. the frustrated feeling of wanting to help others but being unable to do so.

The usage of words such as 'heartless' and 'shameless' have also increased significantly, according to Table 26.1 (#7, 8, 10). Some people behaved selfishly, buying food and bottles of water under despite the serve shortage conditions after the quake. Therefore, blog posts included complaints about these behaviors with the increased use of these adjectives.

On the other hand, 'brand-new', 'earlier', and 'cannot wait' in Table 26.2 (#1, 2, 5) decreased significantly. These words are often used for positive meanings in expectation of new seasons such as spring and goods such as movies. In fact, many companies canceled or postponed their releases of new product and events. For example, the ceremony of Kyushu Shinkansen opening was canceled, iPad2 (a popular tablet device) release was postponed,[3] and many scheduled movie releases were canceled or postponed.[4] As a result, people lost many chances to use such words.

Furthermore, we found that adjectives related to taste such as 'salty-sweet' and 'bitter' decreased significantly according to Table 26.2 (#6, 8, 9). These decreases of words may reflect the so-called 'self-restraint mood' that people stop to have parties outside such as annual cherry blossom viewing party and to go restaurant. Consequently, the words regarding taste, e.g. restaurant and cook reviews, could

[3]http://gigazine.net/news/20110316_apple_delays_ipad2_launch/(Accessed: 11 March 2015).

[4]http://www.cinematoday.jp/page/N0031042 (Accessed: 11 March 2015).

decrease. These decreased adjectives seem to be more related to social activities rather than emotions, compared to increased adjectives. Therefore, estimating economic situations by using the adjectives is the crucially interesting future topic.

26.3.2 Adjective Changes in the Co-occurrence Network

We constructed co-occurrence network of adjectives during the pre-quake period T_0 and post-quake period T_1. The node size indicates the relative frequency of words compared with the entire period and the color corresponds to the community it belongs to. The community is decided by the modularity Q as follows:

$$Q = \frac{1}{2M} \sum_{i,j=1}^{N} \left(A_{ij} - \frac{k_i k_j}{2M} \right) \delta(c_i, c_j), \tag{26.7}$$

where $N = 550$ and M are total number of nodes and links in network respectively. A_{ij} is the weight of the link between node i and j. Here the weight is correlation coefficient r_{ij} calculated in Eq. (26.4). $k_i = \sum_j A_{ij}$ is the sum of the weights linked to node i, and c_i is the community which i belongs to. $\delta(c_i, c_j)$ is 1 if i and j belong to same community ($c_i = c_j$), otherwise 0. In this paper, we maximize Q for undirected weighted network by a software named Gephi[5] (version 0.8.2) to detect community with the algorithm introduced by Blondel et al.[34].

There are 3354 links among 550 nodes during T_0 using the normal correlation coefficient r_{ij} (Fig. 26.2, left) and 61 links using the partial correlation coefficient D_{ij} (Fig. 26.2, right). As expected, the node size is nearly the same because there was no major news before the quake T_0 and there are no special properties in these networks. There are 14 communities in the network. The largest community shares 16.36 % and the second shares 14.55 %.

In contrast, there are two major communities post-quake period T_1. There are 6125 links among the 550 nodes in T_1 by calculating r_{ij} (Fig. 26.3, top) and 73 links are found by calculating the partial correlation coefficient (Fig. 26.3, bottom). There are five communities in the network. The largest community shares 26.57 % and the second shares 21.59 %. Thus, more nodes categorized into the same communities than T_0 period. This confirms that one community corresponds to the adjectives that increased significantly as shown in the previous section.

[5]http://gephi.org (Accessed: 11 March 2015).

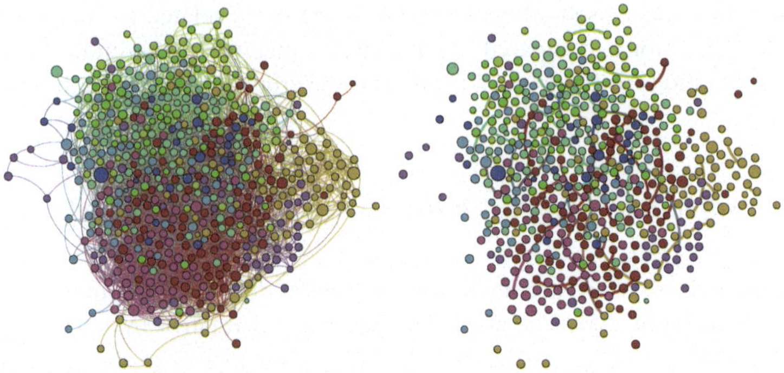

Fig. 26.2 Correlation networks consisting of adjectives before the quake T_0. (*Left*) links are drawn on the basis of the correlation coefficient from Eq. (26.4). (*Right*) links are drawn on the basis of the partial correlation coefficient from Eq. (26.6). Nodes are colored by their community and sized by their relative appearances $\langle x_i(T_0)\rangle/\langle x_i\rangle$, where $\langle x_i\rangle$ is the mean in the entire period and $\langle x_j(T_0)\rangle$ is the mean in the period T_0

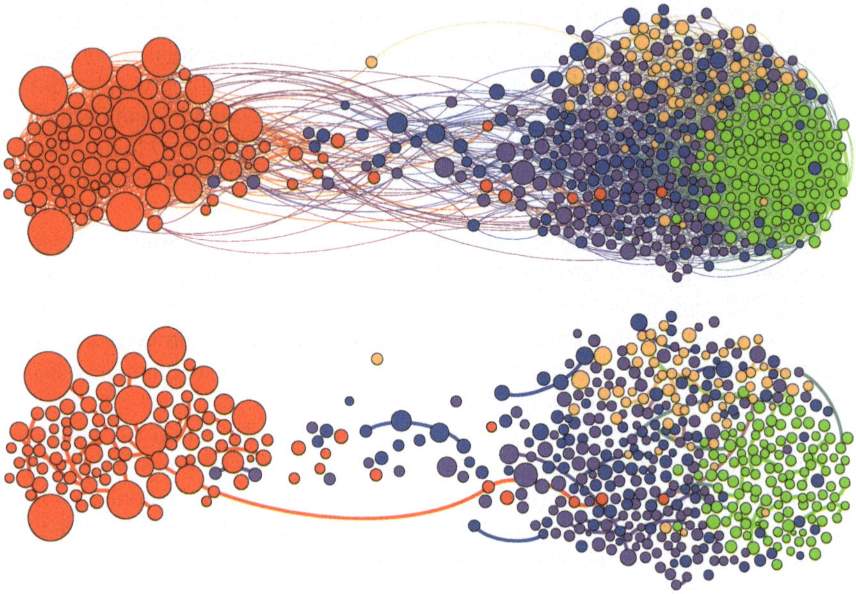

Fig. 26.3 Correlation networks consisting of adjectives after the quake T_1. (*Top*) links are drawn on the basis of the correlation coefficient from Eq. (26.4). (*Bottom*) links are drawn on the basis of the partial correlation coefficient from Eq. (26.6). There is one large grouping that consists of increased frequency nodes soon after the quake (*red*). Nodes are colored by their community and sized by their relative appearances $\langle x_i(T_1)\rangle/\langle x_i\rangle$, where $\langle x_i(T_1)\rangle$ is the mean in the T_1 period

26.4 Summary and Discussion

We observed social emotions in the Japanese blog space during emergency periods, especially before and after the Great East Japan Earthquake. Here, we focus on the relative changes of adjectives. We found that the feelings such as wanting to help others but having no means and the feeling of frustration increased considerably. Thus, adjectives such as ‘impatient’, ‘sorry’, and ‘frustrating’ increased.

To visualize the change in social mood during the quake, we constructed adjective co-occurrence networks. We found that there is a clear topological difference between the pre- and post-quake periods. We found one large community in the post-quake networks with increased adjectives. This result suggests that people tended to share similar emotions post-quake period.

Our results were derived from a limited case study of Japanese social media during the 3.11 Earthquake. However, our results still have novelty and potential, because no one could record small messages from normal people accurately during emergencies until the advent of social media.

Given that during the quake, rumors and false information are said to have spread via mobile phones and social media [35], investigating social emotions during emergency periods has the potential to provide useful warnings regarding false information diffusion. Because psychologists have observed, on the basis of experiments with limited numbers of subjects, that rumors are more likely to diffuse in an emergency situation [36] and when people feel anxious [37].

We expect that data assimilation—a combination of agent-based simulation and real data analysis—will assist us in preventing potential secondary disasters such as false information diffusion and riots during emergencies. The Internet era has even said to foster ‘digital wildfires in a hyperconnected world [38]’ similar to the spread of real wildfires. Our research sheds light on this universal problem and could issue warning signals of potential digital wildfires. We hope that our results can be applied to prevent such problems in the future.

Acknowledgements The authors thank Hottolink, Inc. for providing the data. This study is partially supported by the Grant-in-Aid for Scientific Research (C), Grant Number 24540395 and (B), Grant Number 26310207.

References

1. Watts DJ (2007) Nature 445(7127):489
2. Lazer D, Pentland A, Adamic LA, Aral S, Barabási AL, Brewer D, Christakis N, Contractor N, Fowler JH, Gutmann M, Jebara T, King G, Macy MW, Roy D, Alstyne MV (2009) Science 323:721
3. Pentland AS (2013) Sci Am 309(4):78
4. Onnela JP, Reed-Tsochas F (2010) In: Proc Natl Acad Sci USA 107(43):18375
5. Adamic LA, Lento T (2012) In: Proceedings of the sixth international AAAI conference on weblogs and social media, pp 371–374
6. Kramer A (2012) In: Proceedings of the 2012 ACM conference on human factors in computing systems, pp 767–770
7. Ettredge M, Gerdes J, Karuga G (2005) Commun ACM 48(11):87
8. Ginsberg J, Mohebbi MH, Patel RS, Brammer L, Smolinski MS, Brilliant L (2009) Nature 457(7232):1012
9. Goel S, Hofman J, Lahaie S, Pennock D, Watts DJ (2010) Proc Natl Acad Sci USA 107(41):17486
10. Leskovec J, Huttenlocher D, Kleinberg J (2010) In: Proceedings of the fourth ACM international conference on web search and data mining, pp 98–105
11. Ratkiewicz J, Fortunato S, Flammini A, Menczer F (2010) Phys Rev Lett 104:158701
12. Yasseri T, Sumi R, Kertész J (2012) PLoS One 7(1):e30091
13. Moat HS, Curme C, Avakian A, Kenett DY, Stanley HE, Preis T (2013) Sci Rep 3:1801
14. Mestyán M, Yasseri T, Kertész J (2013) PLoS One 8(8):e71226
15. Crane R, Sornette D (2008) Proc Natl Acad Sci USA 105(41):15649
16. González-Bailón S, Borge-Holthoefer J, Rivero A, Moreno Y (2011) Sci Rep 1:197
17. Mathiesen J, Angheluta L, Ahlgren PTH, Jensen MH (2013) Proc Natl Acad Sci USA 110(43):17259
18. Chew C, Eysenbach G (2010) PLoS One 5(11):e14118
19. Sakaki T, Okazaki M, Matsuo Y (2010) In: Proceedings of the 19th international conference on World Wide Web. ACM, New York, pp 851–860
20. Sasahara K, Hirata Y, Toyoda M, Kitsuregawa M, Aihara K (2013) PLoS One 8(4):e61823
21. Starbird K, Palen L, Hughes A, Vieweg S (2010) In: Proceedings of the 2010 ACM conference on computer supported cooperative work, pp 241–250
22. Toriumi F, Sakaki T, Shinoda K, Kazama K, Kurihara S, Noda I (2013) In: Proceedings of the 22nd international conference on World Wide Web companion, pp 1025–1028
23. Mandel B, Culotta A, Boulahanis J, Stark D, Lewis B, Rodrigue J (2012) In: Proceedings of the second workshop on language in social media, pp 27–36
24. Golder SA, Macy MW (2011) Science 333(6051):1878
25. Watson D, Clark LA, Tellegen A (1988) J Pers Soc Psychol 54(6):1063
26. Kim E, Gilbert S (2009) Detecting sadness in 140 characters: sentiment analysis and mourning Michael Jackson on Twitter. URL http://www.webecologyproject.org/
27. McNaiŗ DM, Lorr M, Droppleman LF (1971) Manual for the profile of mood states. Educational and Industrial Testing Service, San Diego
28. Bollen J, Mao H, Zeng X (2011) J Comput Sci 2:1
29. Lampos V, Lansdall-Welfare T, Araya R, Cristianini N (2013) Analysing mood patterns in the United Kingdom through Twitter content. arXiv:1304.5507v1
30. Hannak A, Anderson E, Barrett LF, Lehmann S, Mislove A, Riedewald M (2012) In: Proceedings of sixth international AAAI conference on weblogs and social media, pp 479–482
31. Bertrand KZ, Bialik M, Virdee K, Gros A, Bar-Yam Y (2013) Sentiment in New York city: a high resolution spatial and temporal view. arXiv:1308.5010v1
32. Zhao J, Dong L, Wu J, Xu K (2012) In: Proceedings of the 18th ACM SIGKDD international conference on knowledge discovery and data mining, p 1528

33. Gonçalves P, Benevenuto F, Cha M (2013) PANAS-t: a pychometric scale for measuring sentiments on Twitter. arXiv:1308.1857v1
34. Blondel VD, Guillaume JL, Lambiotte R, Lefebvre E (2008) J Stat Mech Theory Exp 2008(10):P10008
35. Takayasu M, Sato K, Sano Y, Yamada K, Miura W, Takayasu H (2015) PLoS One 10(4):e0121443
36. Allport GW, Postman L (1948) The psychology of rumor. Henry Holt, New York
37. Walker CJ, Beckerle CA (1987) J Soc Behav Pers 2(3):353
38. World Economic Forum (2013) Global Risks 2013

Chapter 27
Modeling of Enjyo via Process of Consensus Formation on SNS

Takao Komine, Kosetsu Ikeda, Yoichi Ochiai, Keiichi Zempo, and Hiroshi Itsumura

Abstract “The pen is mightier than the sword” said in previous times, the role of dispatching information was given to the people with the special trainee, the people called “Mass-Communication”. However, it has become possible for everyone to dispatch the information on social society with the advent of the Web. Accordingly, Enjyo is often observed on Social Networking Services. Enjyo is a phenomenon that leads the tragedy to individual/company who sends the promotion information via a process of consensus formation as the result of many SNS users. In this research, we analyze the value of reputation on social media in some cases with the purpose of modeling Enjyo. In this study, we tried to numerically analyze and model some cases of Enjyo as well as to classify them by using the data on SNS. For achieving our purpose, we proposed a method of measuring a state of Enjyo and applied the case study method for analysis. With this method, the process of the analysis is likely to be influenced by one’s subjective interpretation or assessment. Therefore, we also tried to facilitate its efficiency and accuracy with random sampling. As a result, several patterns of Enjyo were identified. Moreover, one of the cases of appropriate Enjyo extinction was observed.

T. Komine (✉) • K. Ikeda
Information and Media Studies, Graduate School of Library, University of Tsukuba, Tsukuba, Ibaraki 305-8550, Japan
e-mail: tkomine@slis.tsukuba.ac.jp

Y. Ochiai
Graduate School of Interdisciplinary Information Studies, The University of Tokyo, Bunkyo-ku, Tokyo 113-0033, Japan

K. Zempo
Faculty of Engineering, Systems and Information, University of Tsukuba, Tsukuba, Ibaraki 305-8573, Japan

H. Itsumura
Faculty of Library, Information and Media Science, University of Tsukuba, Tsukuba, Ibaraki 305-8550, Japan
e-mail: hits@slis.tsukuba.ac.jp

H. Takayasu et al. (eds.), *Proceedings of the International Conference on Social Modeling and Simulation, plus Econophysics Colloquium 2014*, Springer Proceedings in Complexity, DOI 10.1007/978-3-319-20591-5_27

27.1 Introduction

Jeff Zucker, the president of CNN Worldwide said in 2014, “Everyone has a blog. We’re all journalists now [1]”. This symbolizes the status quo that every single person can dispatch information to the world by SNS. Conventionally the mass media have played a role to transmit information. But with the advent of the Internet, information can easily be disseminated widely to the people by any individuals via SNS. Professional journalists are well-trained, and understand the huge impact that might be inflicted upon the society through dispatching information. The number of people and organizations in the field of journalism tends to be limited because they need to have logistics channels (e.g. newspaper, magazine, radio, TV, etc.). Therefore the range of information dissemination is relatively small, although the interpretation of information can differ depending on each media. The information is usually cross-referred among the media by the eyes of the general public, thus it is unlikely that the subjective partiality of the transmitters (the mass media, in this case) gives a strong influence on the transmitted information itself (of course it often happens that the media sort out the information; which information is more important than the other, and which should be reported or not). However, as SNS has enabled the general public to widely disseminate the information today, each one of us can play the similar role of mass media. The purposes of the information dispatch by the general public are, in most cases, to bring up some topics for communication between friends, and to simply inform friends and acquaintances of some news. But in some cases, the information is disseminated with the intention of manipulating the story under the subjective partiality of the individual. It is also possible for anybody to add his/her own opinions or beliefs to the information and to transmit it to the public. When information is transmitted in such a way, the original information and facts will inevitably change and be more or less distorted. Conventionally, it took a few months for the public opinion to be formed for certain news through the interaction between the mass media and the general public. But today, the public opinion via SNS only needs a few days or hours to be molded.[1] It sometimes results in a party’s business bankruptcy, losing jobs, or arrest on a crime. This kind of social phenomenon is called “Enjyo”.

On the other hand, there are possible cases in which the promotion via SNS would succeed, and there are some researches which predict the “Hit” by observing the reactions of the people on SNS [2]. To induce the Hit, it is known that not only the quality of the content and the conventional promotion, but also the direct communications (the effects on potential consumers from the player) and the indirect communications (e.g. word-of-mouth, buzz) are important. Moreover, the effectiveness of word-of-mouth is well known and it is researched widely as the method of the marketing [3–7]. This is why people dispatch their activities on SNS.

[1]This speed is astounding for the author who has been working as a journalist, editor, and writer for more than 30 years in magazine industry.

Nevertheless, there is only a fine line between Enjyo and Hit and the dispatch on SNS can sometimes lead to the tragedies.

The purpose of this study is in the quantification and modeling of some cases of Enjyo on SNS. The result will hopefully contribute to developing the tools to predict Enjyo as well as identify the appropriateness of Enjyo for each case. In this research, we numerically analyzed and modeled five cases of Enjyo with an attempt of classifying them by using the data on SNS.

27.2 Principle of Modeling

Some individuals or groups disseminate a piece of information which can cause a sensational event. Most of the time they simply have an intention to gain favorable and positive attention from others. However, it is not always the case. Unfortunate consequences, such as business bankruptcy, job loss, and arrest on a crime, are often produced through some inflammatory arguments or malicious disclosure by unspecified individuals. We call this phenomenon "Enjyo." Also in this article, we use the term "Informing" to describe the process of forming public opinion after transforming its value, being affected by the great amount of unspecified mixed opinions via SNS.

In this study, we adopted the case study approach by using the data collected from SNS, and the details will be described as follows. An operator has categorized each opinion posted on Twitter, Internet discussion boards, or blogs into "positive," "neutral," and "negative" opinions toward the person who triggered the sensational event X. If an operator k categorizes an opinion of an individual i about an event X into "positive,"

$$\epsilon_k(\mathcal{T}_i(X)) = 1 \,, \tag{27.1}$$

If "neutral,"

$$\epsilon_k(\mathcal{T}_i(X)) = 0 \,, \tag{27.2}$$

If "negative,"

$$\epsilon_k(\mathcal{T}_i(X)) = -1 \,, \tag{27.3}$$

where ϵ_k a coding function. By permuting the sequence of options t with the timestamps, we define the public opinion, $Y(t)$, at the time of t as,

$$Y(t) = \sum_{j=1}^{t} \epsilon_k(s_j) \,, \tag{27.4}$$

where $s_t(X)$ denotes the $\mathscr{T}_i(X)$ sorted by timestamps. Moreover, we define the Enjyo as,

$$\lim_{t \to \infty} Y(t) < 0 . \qquad (27.5)$$

Furthermore, if we consider applicative use, it is difficult to find a solution of ϵ_k for every $s_t(X)$, because the analysis on the typical case requires a tremendous number of operators who solve ϵ_k in real time, if we aim to create the forecast system of Enjyo. Thus, we test to compare the results with those in random sampling for i. This random sampling would simulate if the $Y(t)$ by reduced data has the same tendency with $Y(t)$ of all of data or not. Furthermore, the random sampling will be applied for k because of the influence on ϵ_k given by the subjective ideas of operators.

27.3 Analysis of Enjyo

27.3.1 Data Description

To analyze Enjyo, five cases were selected from Twitter and summary sites of the online forum (such as 2-channel), and $Y(t)$ was evaluated by using the formula described in Sect. 26.2. On collecting the samples, we tried to choose each case that could demonstrate the distinctive behavior of the curve $Y(t)$. Also in our intuition, those cases were on a very fine line; whether they would be put in a state of glory or in Enjyo.

27.3.2 Case Study

27.3.2.1 Case 1: Crowd Funding Platform Business

Case 1 is the instance of the service of supporting needy students with scholarship fund collected through crowd-funding. The data were collected from Togetter (the collecting service from the twitter for specific keywords or interests of those who create the thread) from 8:09 to 11:38 on Dec. 26th, 2012. Although the number of tweet was 10053, we used 9676 of them by ignoring clearly meaningless tweets. This service was started by a famous public figure, and became a topic of conversation for a period of time through SNS. In the beginning, the idea was supported by many people, but slowly it was heading for Enjyo. It was due to the disclosure of the fact that the first student chosen as a scholarship recipient was not actually a needy student. This was hunted down by the student's real name as well as her portrait photos revealed on SNS. (According to some reported information, the

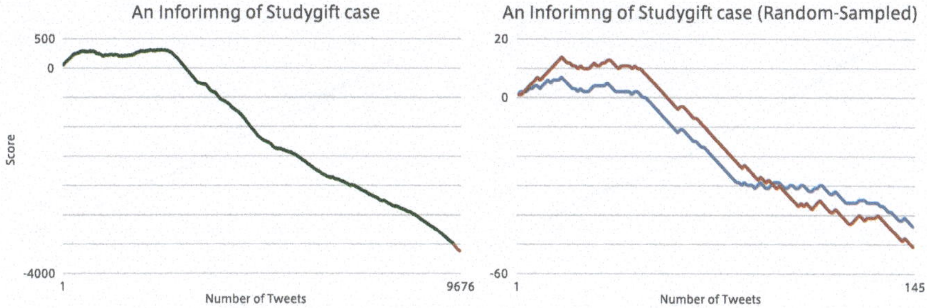

Fig. 27.1 Behavior of $Y(t)$ in case 1 (*left graph*: without sampling, *right graph* (*red*): 145 individual's tweets were random sampled, *right graph* (*blue*): 145 operators were random sampled)

student possessed a smartphone and PC, and went abroad several times.) Figure 27.1 shows the curves of $Y(t)$. Although the value of $Y(t)$ increases at the beginning, it begins to decrease around $t = 2000$, eventually ending up in a state of Enjyo.

27.3.2.2 Case 2: Iceman in Freezer

As Case 2, we selected an event in which juvenile mischievous behavior on SNS resulted in getting a series of negative responses from anonymous people. In this case, Enjyo broke out due to a photo of a teenager inside a convenience store's freezer which was uploaded by his friend onto a site of SNS. Presumably those boys uploaded the photo just for showing off among their friends, but they received such severe bashing because the photo was open to the public. Unfortunately, some SNS users, who were absolute strangers identified the boys by tracking down their accounts. Eventually criminal papers were filed with prosecutors against those teenagers on suspicion of forcible obstruction of business. The data were collected from the two threads on 2-channel which were focused on the Iceman news and were active on 26th and 27th July, 2013. Although the number of posts was 2000, we used 1926 of them by ignoring clearly meaningless tweets. Figure 27.2 shows the behavior of $Y(t)$ in this case. The sharp and continuing decline clearly indicates the occurrence of Enjyo.

27.3.2.3 Case 3: Exposure of Bureaucrat

Case 3 is an instance that a fast-track bureaucrat repeatedly made harsh remarks on his pseudonymous blog. This civil servant was tracked down by the users of SNS outraged over his derogatory and discriminating remarks on the blog, and was eventually faced the consequence of being subjected to disciplinary action. The data were collected from in 2-channel which was focused on the bureaucrat news and was active from 17:55 to 19:17 on 24th Sep, 2013. Although the number of posts

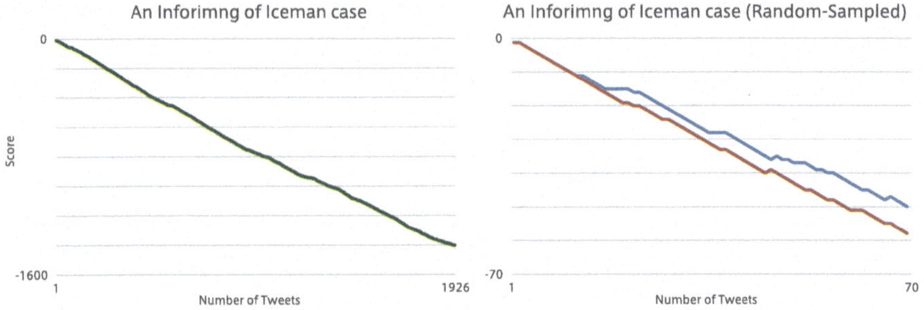

Fig. 27.2 Behavior of $Y(t)$ in case 2 (*left graph*: without sampling, *right graph* (*red*): 70 individual's posts were random sampled, *right graph* (*blue*): 70 operators were random sampled)

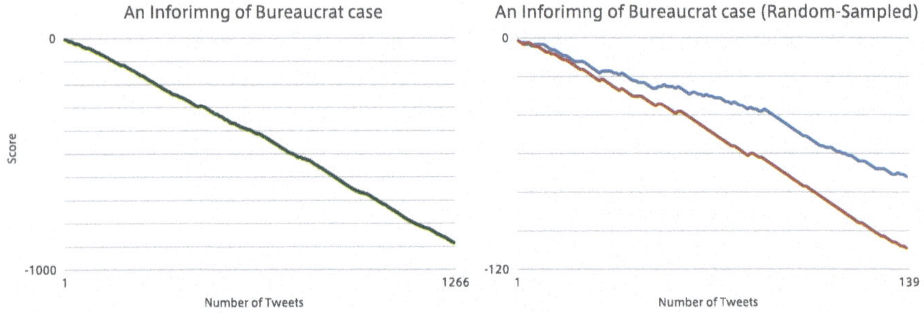

Fig. 27.3 Behavior of $Y(t)$ in case 3 (*left graph*: without sampling, *right graph* (*red*): 139 individual's posts were random sampled, *right graph* (*blue*): 139 operators were random sampled)

was 1163, we used 1266 of them by ignoring clearly meaningless types. Figure 27.3 shows the ongoing state of Enjyo, which is similar to Case 2.

27.3.2.4 Case 4: Rebroadcast of Popular Feature Film

The case of a rerun of a blockbuster feature film is selected as Case 4. The data were collected from on Togetter, which was focused on the rebroadcast and was active from 13th to 20th Jan, 2014. Although the number of tweet was 3847, we use 3701 of them by ignoring clearly meaningless tweets. The behavior of $Y(t)$ is shown in Fig. 27.4. This demonstrates a different pattern of $Y(t)$ with a curve, not declining toward the end, which means Enjyo didn't occur in this case.

27.3.2.5 Case 5: Anger to Headhunting

Case 5 presents the event that a well-known CEO of IT company showed his furious anger through his blog toward an employee who was headhunted and left the company. This was in the news, and Enjyo broke out on SNS. However, the curve

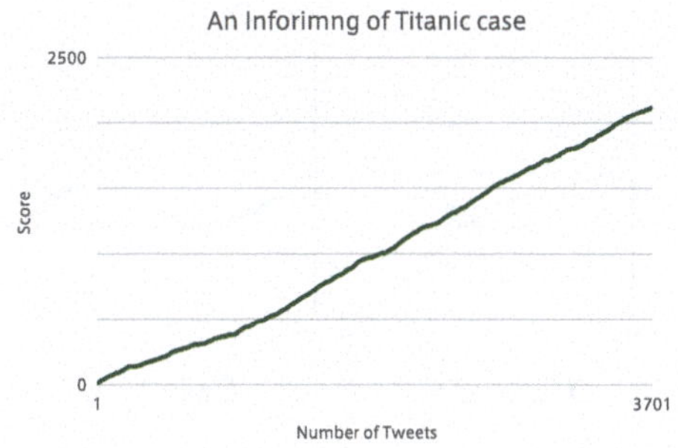

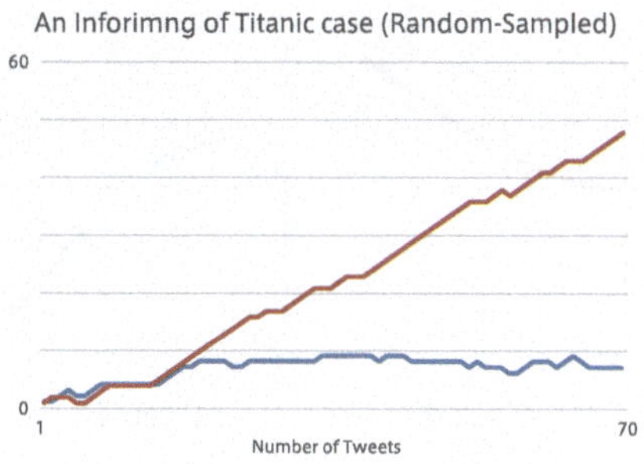

Fig. 27.4 Behavior of $Y(t)$ in case 4 (*left graph*: without sampling, *right graph* (*red*): 70 individual's tweets were random sampled, *right graph* (*blue*): 70 operators were random sampled)

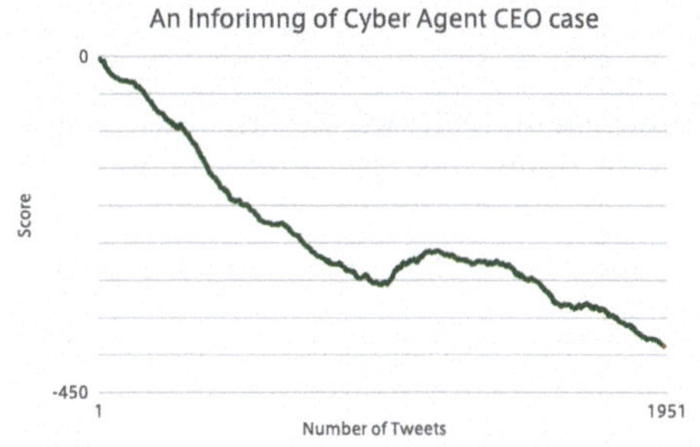

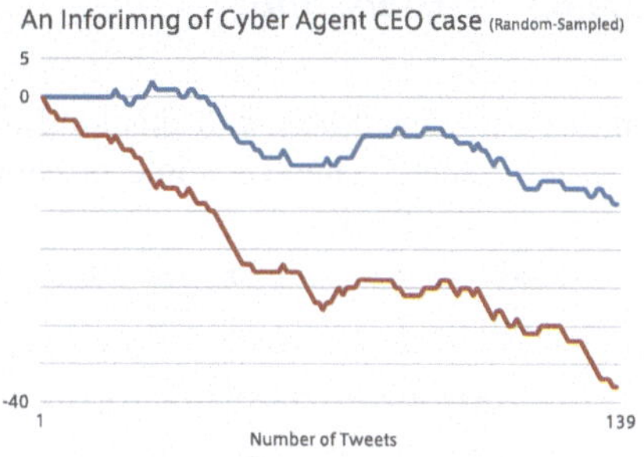

Fig. 27.5 Behavior of $Y(t)$ in case 5 (*left graph*: without sampling, *right graph* (*red*): 139 individual's posts were random sampled, *right graph* (*blue*): 139 operators were random sampled)

of $Y(t)$ in Fig. 27.5 indicates a slight rise in the middle (around $t = 1000$). The data were collected from on Twitter with the name of the company as the searching word from 2nd to 11th Oct, 2014. Although the number of tweet was 2007, we used 1951 of them by ignoring clearly meaningless tweets.

In this case, the curve of $Y(t)$ seems to go straight downhill, but it slightly rises in the middle around $t = 1000$. Since membership fees are required to read the blog posts in question, non-members are accessible only to the headlines. Enjyo broke out instantly because those headlines were flashy. However, there were quite a few readers who criticized the CEO but at the same time commented "I can't read the entire texts." This implies that Enjyo in this case was ignited by those who didn't know much about the news but wanted to enjoy the state of Enjyo. On the other hand, the subscribed members who had access to the entire texts commented that there was no need to attack the CEO. Those favorable comments were posted at the point around $t = 1000$. From this analysis, an assumption can be made that the transmitters of the information on Case 5 intentionally caused Enjyo. Furthermore, there is a possibility that they might have had a foresight on how Enjyo would fade out by making use of the information gaps in advance.

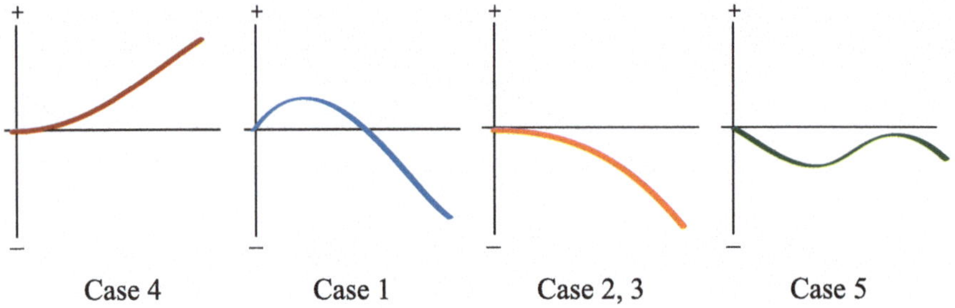

Fig. 27.6 Classes of each cases

27.3.3 Discussion

In this study, we might call this kind of movement described in Case 5 as "Enjyo announcement." There is some possibility of doing this case Enjyo impress for information diffusion.

Based on the above, we classified five cases into four patterns below (Fig. 27.6);

1. $Y(t)$ continues to rise (Case 4),
2. $Y(t)$ temporarily rises but Enjyo breaks out through disclosure of hidden information (Case 1),
3. $Y(t)$ goes straight downhill and Enjyo never stops (Case 2, 3),
4. $Y(t)$ drops in the beginning but starts rising when some measures are taken (Case 5)

Just as in the fourth pattern, it will be essential to take measures against Enjyo. Moreover, to achieve this signification for calculating $Y(t)$ in crowd, the similarity of behavior was observed (the similarity of upper and lower graphs in Figs. 27.1, 27.2, 27.3, 27.4, and 27.5).

27.3.4 Future Works

To reveal the motivations of each user on Informing, we are planning some experiments in the laboratory. Let X denotes an event that can make news, and let X_χ an opinion or interpretation of an individual or a group. We define X_χ as,

$$X_\chi = \mathscr{T}_\chi (X) , \tag{27.6}$$

where the operator $\mathscr{T}_\chi$ denotes information transmission which includes the subjective of χ. Informing occurs due to the intervention of a person or a group during the process of information transmission. Let it assume that there are two senders

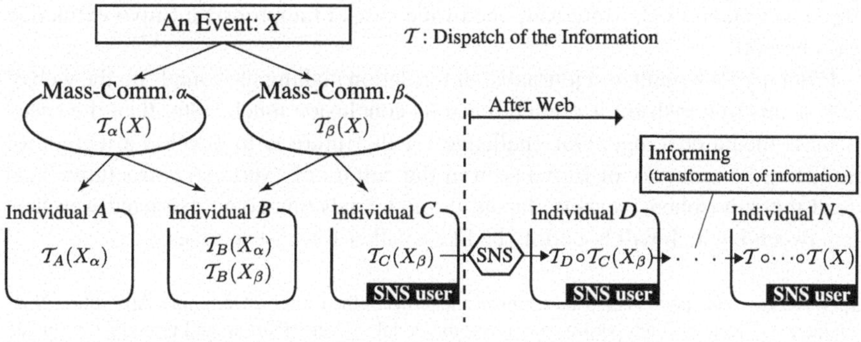

Fig. 27.7 Schematic diagram of "Informing"

of information (mass media α and β), and three individuals (A, B, and C) forming public opinion. The individual A receives the information only from the mass media α, the individual C only from the mass media β, the individual B from both α and β. The opinions and interpretations of each person can be expressed by the following formula;

$$X_A = \mathcal{T}_A(X_\alpha) = \mathcal{T}_A \circ \mathcal{T}_\alpha(X) ,$$
$$X_C = \mathcal{T}_C(X_\beta) = \mathcal{T}_C \circ \mathcal{T}_\beta(X) ,$$
$$X_B = \mathcal{T}_B(X_\alpha + X_\beta) = \mathcal{T}_B \circ \mathcal{T}_\alpha(X) \oplus \mathcal{T}_B \circ \mathcal{T}_\beta(X) . \tag{27.7}$$

The public opinion would be average of the formula. However, since anybody can easily disseminate information today, Informing of the information happens. Naturally, the opinions and interpretations of the individual N should be formed after passing through the subjective eyes of a number of people, which can be expressed by the following formula;

$$X_N = \mathcal{T} \circ \cdots \circ \mathcal{T}(X) . \tag{27.8}$$

Figure 27.7 shows the conceptual diagram of "Informing."

27.4 Conclusion

In this study, we tried to numerically analyze and model some cases of Enjyo as well as to classify them by using the data on SNS. For achieving our purpose, we proposed a method of measuring a state of Enjyo and applied the case study method for analysis. With this method, the process of the analysis is likely to be influenced by one's subjective interpretation or assessment. Therefore, we also tried to facilitate its efficiency and accuracy with random sampling. As a result, several patterns of

Enjyo were identified. Moreover, one of the cases of appropriate Enjyo extinction was observed.

From my viewpoint as a journalist, information making the rounds on the web or SNS seems to transform itself and reach its conclusion much faster than the speed of mass media coverage. Our challenge for the future is to develop a method of predicting an outbreak of Enjyo seen in the culture of "vertically-structured" and "read-the-atmosphere" society, Japan, through analyzing some extracted web data (e.g. tweets), which will hopefully be like weather forecast services.

References

1. Asahi.com (2015) http://www.asahi.com/articles/ASG9V04HFG9TUHBI03C.html. Accessed 31 Mar 2015
2. Ishii A et al (2012) New J Phys. doi:10.1088/1367-2630/14/6/063018
3. Leskovec J, Adamic L, Huberman B (2007) ACM Trans Web. doi:10.1145/1232722.1232727
4. Castronovo C, Huang L (2012) J Mark Dev Compet. 6:117–134
5. Trusov M, Bucklin R, Pauwels K (2009) J Mark. doi:10.1509/jmkg.73.5.90
6. Clark M, Melancon J (2013) Int J Mark Stud. doi:10.5539/ijms.v5n4p132
7. Ferguson R (2008) J Consum Mark. doi:10.1108/07363760810870671

Chapter 28
A Network Structure of Emotional Interactions in an Electronic Bulletin Board

Haruka Adachi and Mikito Toda

Abstract As social network service (SNS) spreads all over the world, we can no longer live without its influence. The study of SNS would reveal how atmosphere of our society changes as we interact through SNS. Here, we focus our attention on emotions expressed in SNS and investigate how our emotions are affected by others in SNS. To reveal how positive/negative emotions are magnified and diffused through SNS, we analyze a network of emotional words in an electronic bulletin board based on the theory of complex networks.

28.1 Introduction

Recently, we use various social network service (SNS), and it is difficult to live without being influenced by SNS at all. An important part of information transfer in our society takes place through SNS, and the amount of data recording such human activities is increasing rapidly. Analysis of these data would reveal various aspects of our society, such as how atmosphere of our society changes as we interact through SNS. Moreover, universal features are observed in the statistical properties of SNS. For example, power law behavior is generally found in time evolution of the number of words which appear in blogs as various social events occur [1]. Such power law behavior shows close resemblance to those phenomena observed in nonequilibrium statistical physics. Therefore, we expect that statistical physics could unravel such universal characteristics in SNS, and that its methodology would enable us to understand not only Nature but also human society.

The development of SNS has positive and negative impacts on our society. As for positive ones, spontaneous collaboration could be formed through SNS and would be effective to solve social problems without participation of big organizations. As for negative ones, groundless rumors and unlawful activities can spread very easily through SNS, and such an incident could pose a serious threat for our

H. Adachi • M. Toda (✉)
Nara Women's University, Nara 630-8506, Japan
e-mail: adachi@ki-rin.phys.nara-wu.ac.jp; toda@ki-rin.phys.nara-wu.ac.jp

H. Takayasu et al. (eds.), *Proceedings of the International Conference on Social Modeling and Simulation, plus Econophysics Colloquium 2014*, Springer Proceedings in Complexity, DOI 10.1007/978-3-319-20591-5_28

society. Moreover, emotional interaction in SNS plays a crucial role for social security and fruitful usage of facilities provided by SNS. One of the typical cases in which negative reactions focus on a specific site is the so-called "blog under fire". Challenged by these features of SNS, a social experiment is done to see how negative/positive reactions affect other ones by artificially reducing the numbers of such reactions [2]. Their experiments manifest the fact that there exists emotional contagion in SNS. On the other hand, this study has sparked controversy concerning whether it is appropriate to do such a social experiment or not [3].

The purpose of our study is to analyze how positive/negative emotions spread over SNS. However, we do not attempt to conduct any social experiments. Instead, we focus on emotional interaction by analyzing correlation in usage of emotional expressions in SNS. We construct a network of emotional words using those messages with explicit reference to others in an electronic bulletin board, and study characteristics of the network based on the theory of complex networks. In Sect. 28.2, we explain our data and methods of analyzing emotional interaction in SNS. In Sect. 28.3, we analyze properties of the network of emotional words based on the theory of complex networks. In Sect. 28.4, we conclude our study and discuss future prospects of our study.

28.2 Data and Methods

We explain our data and methods of analyzing emotional interaction in SNS. In order to see how positive/negative emotions spread over SNS, we focus on how emotional words are used in SNS, especially in those messages which are referred to and in those ones which refer to others. In the following, we call those messages with specific reference to others as comments. Then, we construct a network of emotional words by drawing a link between a pair of emotional words used in these messages. We expect that the analysis of the network thus constructed will reveal how emotional interaction takes place among these messages in SNS. In Fig. 28.1, we show a schematic picture explaining the data and the methods of our analysis for emotional relation based on a network of emotional words.

We analyze the data of an electronic bulletin board in Japanese on-line encyclopedia "Nico Nico Pedia" [4]. This data is provided by Mirai Kensaku Brazil Co., Ltd. through National Institute of Informatics in Japan. For our analysis of Japanese sentences, we use the morphological analyzer "MeCab" [5]. It can divide Japanese sentences to words based on its own dictionary. For classification of emotional words, we use "Japanese Dictionary of Appraisal—attitude—" by Gengo Shigen Kyokai [6]. It enables us to classify emotional words according to whether they are positive or negative. It also provides us with their types based on the appraisal theory.

First, we choose those messages with specific reference to others and those which are referred to by others. In the electronic bulletin board "Nico Nico Pedia" [4], some messages contain a specific symbol indicating that these messages are

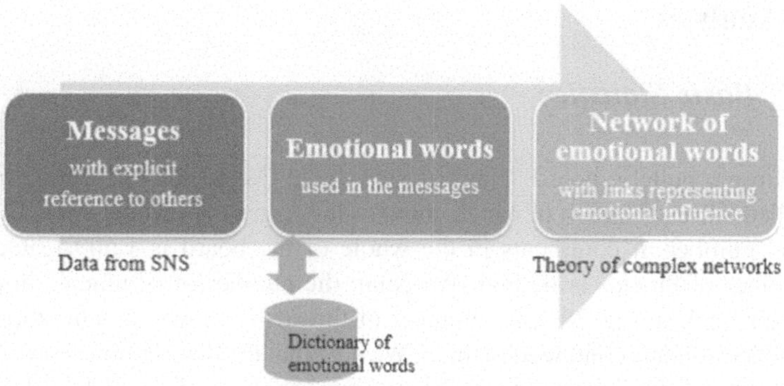

Fig. 28.1 A schematic explanation of the data and methods of our analysis for emotional relation based on a network of emotional words. First, we choose in SNS those messages with specific reference to others and those which are referred to by others. Second, we choose emotional words used in these messages based on a dictionary. Third, we construct a network of emotional words by drawing a link, from emotional words in the message referred to, to emotional words in the comments referring to the original message. We then analyze characteristics of the network thus constructed based on the methodology of the theory of complex networks [7]

comments to others identified by their index numbers. Thus, we can collect pairs of messages with explicit reference from one to the other.

After choosing emotional words used in these messages, we construct a network of emotional words by drawing a link, from emotional words in the message referred to, to emotional words in the comments referring to the original message. Note that the direction of the link from one emotional word to another indicates the direction of emotional influence from the original message to the comment. The network thus constructed is a weighted directional one where the weight of the link is defined by the number of times the pair of the words appears in the whole data we analyze.

We then analyze characteristics of the network thus constructed based on the methodology of the theory of complex networks [7]. In particular, we are interested in the relation of how positive/negative emotions affect emotions of others. The distribution of degrees of nodes enables us to single out important nodes, suggesting that those words are more influential or that they are more frequently used under the influence of other words. We also estimate the quantity called "modularity" to reveal groups of nodes with frequent mutual reference, showing that multiple types of emotional exchanges take place in the SNS. Thus, analysis of the network of emotional words will be fruitful to understand how people interact in SNS.

28.3 Analysis

28.3.1 Basic Statistics

The electronic bulletin board "Nico Nico Pedia" [4] consists of separate multiple parts called "threads", each of which contains messages concerning a specific topic. The total number of comments in the whole of the board is 1,509,179, and the total number of threads is 62,864. We count the number of comments in each of the threads, and, in Fig. 28.2, the number of threads is shown as a function of the number of comments contained in them. We note that the distribution shows a power law behavior, a feature often seen in the statistical analysis of social data. The power of the distribution is approximately $\gamma \approx -3/5$ with the distribution $P(x)$ represented as $P(x) \propto x^{\gamma}$ where x is the number of comments in each of the threads. It would be interesting to model a process of making a comment to messages as a growing tree showing the power law behavior. This will be a future topic of our study.

Among those threads with larger numbers of comments, we choose the one which the managing company of the board recognizes as "under fire". The number of comments in this thread is 7624, the seventh largest in the board. The reason of choosing it for our analysis is as follows. First, we think that the number of comments in it is large enough to apply statistical analysis. Second, we expect that, by comparing the results of this thread with those of others, we can obtain

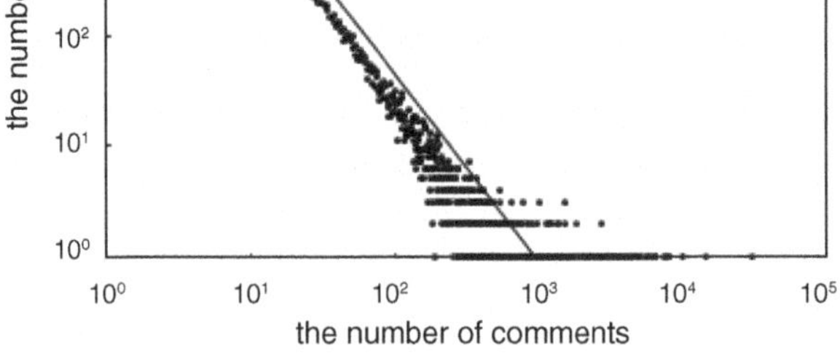

Fig. 28.2 The number of threads is shown as a function of the number of comments contained in them. The total number of comments in the whole of the board is 1,509,179, and the total number of threads is 62,864. The distribution exhibits a power law behavior. The power of the distribution is approximately $\gamma \approx -3/5$ with the distribution $P(x)$ represented as $P(x) \propto x^{\gamma}$ where x is the number of comments in each of the threads. The *solid line* is eye guide for power law with the exponent $-3/5$

characteristics of emotional exchange which are specific to those threads "under fire".

From comments of the thread, we choose those which include emotional words. The number of those comments is 787, about a tenth of the whole of the comments. Then, we construct a network of emotional words contained in these comments and the messages referred to by them. The network of emotional words in this thread is shown in Fig. 28.3. It is a weighted directional network. For visualization of the network, we use the tool "Gephi" [8]. The nodes represent emotional words used in the messages which are referred to or in the comments which refer to other messages. The link shows a pair of emotional words, one in the message referred to and the other in the comment referring to the message. The direction of the link indicates the direction of influence, from the word in the message to the one in the comment. The weight of the link shows the number of times the pair of emotional

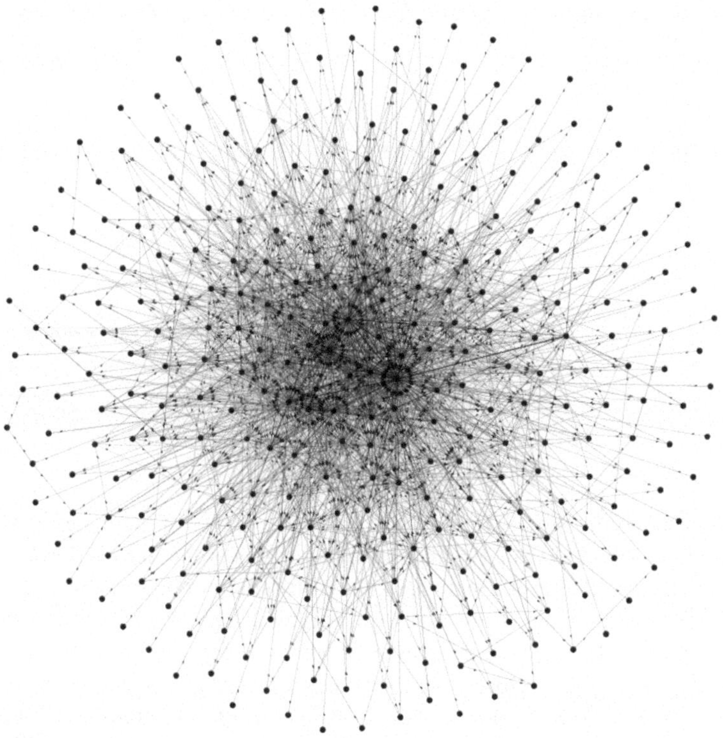

Fig. 28.3 The network of emotional words. The nodes represent emotional words and the link connects a pair of emotional words, from the one used in the message to the other used in the comment. The direction of the link means the direction of influence and its weight does the number of times the pair appears in the thread we analyze. The total number of comments which contain emotional words is 787. The total number of the nodes is 317. The total number of the links is 2479 and the total weights of the links is 6302

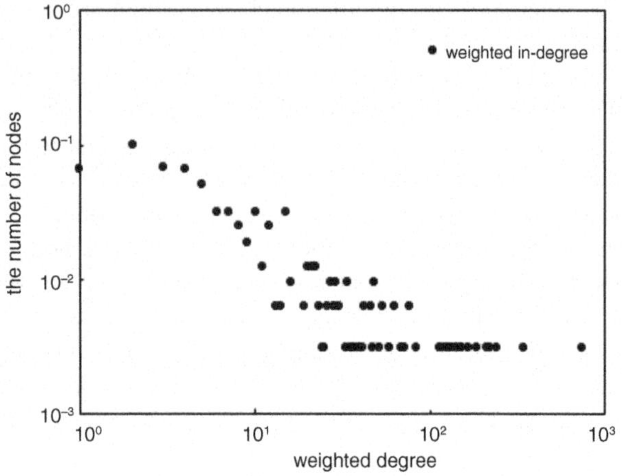

Fig. 28.4 The distribution of in-degrees. The distribution shows a close resemblance to a power law behavior though the plots are scattered

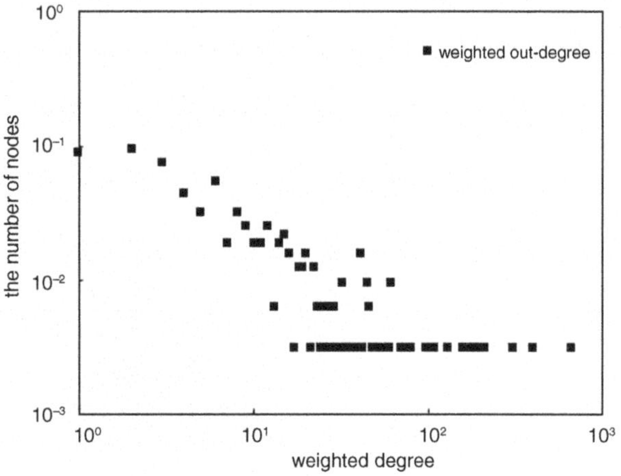

Fig. 28.5 The distribution of out-degrees. The distribution shows a close resemblance to a power law behavior though the plots are scattered

words appears in the thread. In Fig. 28.3, the total number of the nodes is 317. The total number of the links is 2479 and the total weights of the links is 6302.

In the analysis of complex networks, the distribution of degrees plays an important role. In the following, we analyze only weighted degrees. In Figs. 28.4 and 28.5, we show the distribution of in-degrees and that of out-degrees, respectively. Both of the distributions show a close resemblance to a power law behavior though their plots are scattered. We can interpret the distributions of in-degrees and out-degrees as follows. The larger the value of out-degree for a specific word is, the more influential this word is. The larger the value of in-degree for a specific word is, the more frequently this word is used under the influence of other words. Thus, the distributions of in-degrees and out-degrees show which emotional words play what kind role in emotional interaction. Moreover, clustering of nodes based on the weights of links joining them would reveal different types of emotional exchanges taking place in the thread.

28.3.2 Network Modularity

In order to classify nodes in the network, we use the quantity Q called modularity, which is originally introduced in [9] for non-weighted non-directed networks, and is extended to weighted directed networks in [10]. The point of introducing modularity is to reveal community structure of the network. Here, community structure means a partition of the nodes into different groups so that the links within each of these groups are more dense than those between different groups.

In the following, we consider modularity Q for a weighted directed network $\mathscr{N}$. First, we consider a partition $\mathscr{P}$ of the nodes of the network $\mathscr{N}$ into groups. We assume that each of the nodes belongs to a unique group and let C_i denote the group which the ith node belongs to. Next, we define a quantity $Q_{\mathscr{N}}$ as a function of the partition $\mathscr{P}$ as follows,

$$Q_{\mathscr{N}}(\mathscr{P}) = \frac{1}{2W} \sum_{i=1}^{N} \sum_{j=1}^{N} \left(W_{i,j} - \frac{W_i^{\text{out}} W_j^{\text{in}}}{2W} \right) \delta\left(C_i, C_j\right) , \tag{28.1}$$

where N is the total number of the nodes of the network, $W_{i,j}$ is the weight of the link from i to j, $W_i^{\text{out}} = \sum_{j=1}^{N} W_{i,j}$ is the total weight of the links from i, $W_j^{\text{in}} = \sum_{i=1}^{N} W_{i,j}$ is the total weight of the links to j, and $2W = \sum_{i=1}^{N} \sum_{j=1}^{N} W_{i,j}$ is the total weight of the links of the network. The quantity $\delta\left(C_i, C_j\right)$ takes the value 1 if the nodes i and j belong to the same group, and takes the value 0 otherwise.

The function $Q_{\mathscr{N}}(\mathscr{P})$ takes the value in the range $-1 \leq Q_{\mathscr{N}}(\mathscr{P}) \leq 1$, and characterizes to what extent the partition $\mathscr{P}$ reflects the community structure of the network $\mathscr{N}$. Note that the function $Q_{\mathscr{N}}(\mathscr{P})$ consists of the two terms, the first being the sum of the weights of the links within the same groups and the second representing that of a random partition for a random graph with the same total weights W_i^{out} and W_i^{in} for each of the ith node. If $Q_{\mathscr{N}}(\mathscr{P})$ is positive, the partition $\mathscr{P}$ has more dense links within the same groups than random partitions. On the other hand, if $Q_{\mathscr{N}}(\mathscr{P})$ is negative, it has less dense links within the same groups than random partitions. Thus, the larger the value of $Q_{\mathscr{N}}(\mathscr{P})$ is for the partition $\mathscr{P}$, the more closely it reflects the community structure of the network $\mathscr{N}$. As we vary partitions of the network $\mathscr{N}$, the partition $\mathscr{P}_{\text{M}}$ which attains the largest value of $Q_{\mathscr{N}}$ provides us with the best description of the community structure of the network $\mathscr{N}$. Then, we define the modularity Q of the network $\mathscr{N}$ by the following,

$$Q(\mathscr{N}) = Q_{\mathscr{N}}(\mathscr{P}_{\text{M}}) \tag{28.2}$$

For a given network $\mathscr{N}$, to estimate the largest value of $Q_{\mathscr{N}}$ is known to be NP-complete [11]. Therefore, we need to resort to an approximate method which provides us with a reasonably good estimation of the largest value. The tool "Gephi"

[8] also enables us to estimate the largest value of $Q_{\mathcal{N}}$ using one of the best known algorithms [12] for weighted directed networks.[1]

28.3.3 Community Structure

In Fig. 28.6, we show community structure of the nodes of the network. The color of the node indicates the community it belongs to. The color of the link indicates the community of the node it comes from. There, we show only those nodes which belong to the largest three communities. The total number of communities of the network is 32.

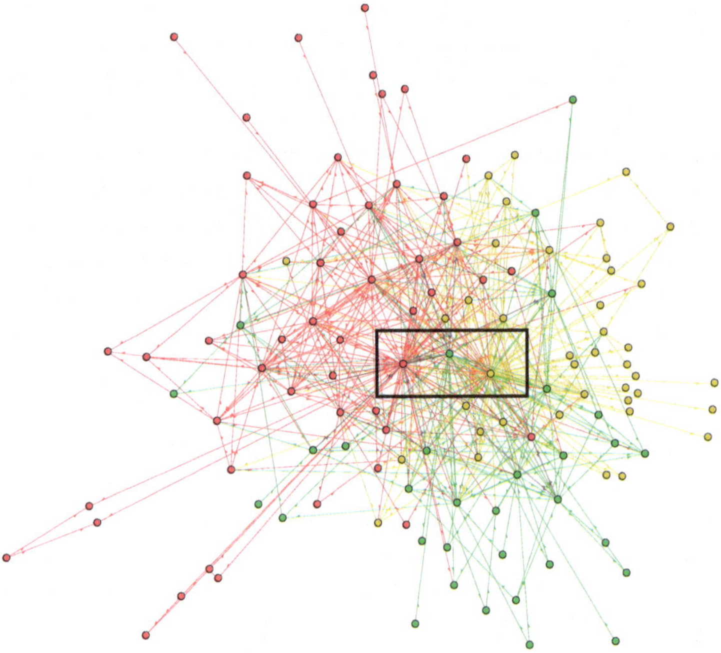

Fig. 28.6 Community structure of the nodes of the network shown in Fig. 28.3. Partition to communities is done by maximizing the function $Q_{\mathcal{N}}$ defined by Eq. (28.1). The color of the node indicates the community it belongs to. The color of the link indicates the community of the node it comes from. Here, we show only those nodes which belong to the largest three communities. Around the middle of the network, we can see three nodes with larger numbers of degrees than others. Their colors are *red*, *green*, and *yellow* from *left* to *right*, respectively. The *rectangle* indicates the locations of the hubs

[1]See the following web page, The Louvain method for community detection in large networks, http://perso.uclouvain.be/vincent.blondel/research/louvain.html.

Around the middle of Fig. 28.6, we can see three nodes with larger degrees than others. Their colors are red, green, and yellow from left to right, respectively. These nodes are the hubs of the communities in the sense that they have the largest values of both in-degree W_i^{in} and out-degree W_i^{out} in each of their communities. It is interesting that those nodes for the red and yellow communities have the largest absolute values of the difference $\delta_i = W_i^{\mathrm{out}} - W_i^{\mathrm{in}}$ in each of their communities. We also note that all of the words corresponding to these three nodes are classified as positive in the dictionary of emotional words.

In Table 28.1, we summarize these features of the community structure. These features mean the following. In each of the communities, there exists key emotional words. However, these words have different directions of influence. While the word "like" in the red community is most influential to arouse emotion of others, the one "idol" in the yellow community appears most under influence of other emotional words. In the green community, the word with the largest absolute value of the difference δ_i is different from the one with the largest values of in-degree and out-degree. Moreover, the word with the largest absolute value of the difference in the green community is negative while the one with the largest in/out-degrees there is positive.

Such differences in the three communities could imply how emotional exchanges differ among them. In order to see such differences in more detail, we show, in Table 28.2, a list of words with which the key emotional words have larger values of in-degrees/out-degrees. This table indicates how the key emotional words affect others or how they are used under influence of others within the same communities. In Table 28.2, for each of the three communities, three words are listed in the order of in-degree/out-degree which its key word has. For each of the word listed, we also show its order, its value of in-degree/out-degree, and its nature, i.e., positive or negative. The values of in-degree/out-degree of these words listed are not so large considering the total values of in-degree/out-degree of the key words. This

Table 28.1 We summarize characteristics of the three communities shown in Fig. 28.6

	Community1	Community2	Community3
Color	Red	Yellow	Green
The number of nodes	50	41	33
The word w_i with the largest in/out-degrees	Like (positive)	Idol (positive)	Expect(positive)
In-degree W_i^{in} of w_i	242	730	211
Out-degree W_i^{out} of w_i	304	653	194
The word w_j with the largest absolute value of difference $\delta_j = W_j^{\mathrm{out}} - W_j^{\mathrm{in}}$	Like (positive)	Idol (positive)	Refuse(negative)
The difference δ_j of w_j	+62	−77	+23

We show the words which have the largest values of both in-degree W_i^{in} and out-degree W_i^{out} in each of their communities. All of the words are classified as positive in the dictionary of emotional words. Those words for the red and yellow communities also have the largest absolute values of the difference $\delta_j = W_j^{\mathrm{out}} - W_j^{\mathrm{in}}$ in each of their communities

Table 28.2 A list of words with which the key emotional words in Table 28.1 have largest in-degrees/out-degrees

	Community1 Like (positive)		Community2 Idol (positive)		Community3 Expect (positive)	
Order	In-degree	Out-degree	In-degree	Out-degree	In-degree	Out-degree
1	Regret (negative) 5	Hatred (negative) 10	Joy (positive) 16	Permit (positive) 8	Insert (positive) 4	Relief (positive) 4
2	Sad (negative) 5	Love (positive) 6	Permit (positive) 9	Dislike (negative) 7	Refuse (negative) 3	Denial (negative) 4
3	Lovable (positive) 3	Will (positive) 6	Approval (positive) 8	Importance (positive) 4	Uneasiness (negative) 3	Laugh (positive) 3

For each of the three communities, three words are shown with which its key word has largest in-degree/out-degree. For each of the word listed, we also show its order, its value of in-degree/out-degree, and its nature, i.e., positive or negative

means that the key words are not linked with specific words. Rather, the key words influence many emotional words or they are used under influence of various ones. In this sense, they really play the role of hubs in the communities.

In Table 28.2, we note that the key word "like" has the largest out-degree with "hatred", a negative word. The word "like" also has the largest value of in-degree with "regret" and "sad"; both of them have a negative nature. This implies that, within the community in which the word "like" is the hub, negative or mixed emotions spread. Moreover, the word "like" has positive value of the difference $\delta_i = W_i^{\text{out}} - W_i^{\text{in}}$, implying that the usage of "like" stimulates spreading of such mixed emotions. On the other hand, within the community in which the word "idol" is the hub, five among six words listed are positive. We also note that "idol" has negative value of the difference $\delta_i = W_i^{\text{out}} - W_i^{\text{in}}$. Such features suggest that, within the community, positive emotions are exchanged where the word "idol" is used in response to others.

Thus, the community structure characterized by modularity Q reveals that there exists multiple emotional exchanges taking place in this thread. Our analysis indicates that the analysis using modularity Q is a useful method to understand emotional relation in SNS.

28.4 Conclusion

In this study, we have analyzed emotional relation in SNS based on the network of emotional words. The network is constructed using explicit reference from comments to messages. We have analyzed the network relying on the theory of complex networks, especially the quantity called modularity. Based on modularity Q, we obtain the community structure of the network of emotional words. The community structure thus obtained reveals that there exist multiple emotional exchanges taking place in the thread. The analysis also shows key words which are most influential. Thus, our analysis indicates that the analysis using modularity Q is a useful method to understand emotional relation in SNS.

As a next step of our analysis, we are planning to compare the results for threads "under fire" with those which are not "under fire". We expect that such comparison would enable us to foresee the processes leading to the situation called "under fire" so that we could take a measure to prevent it from happening. We also take a closer look at emotional events taking place within communities to stimulate fruitful processes of emotional exchange.

In our future study, we will extend the present analysis towards the following directions. First, we will extend our analysis by including not only emotional words but also other types of words and expressions. In our data of comments, a large number of them do not contain any emotional words. In order to analyze emotional exchange in these comments, we need more sophisticated methods to extract emotions expressed implicitly there. We will also extend our analysis to observe not only emotional exchange but also more general exchange of information. Second, we will investigate emotional exchange without explicit reference to other messages. We expect that analysis of emotional relation between messages and comments with explicit reference to them would give a clue to understand implicit emotional influence in SNS. Third, we are interested in time series of messages. For example, we are currently analyzing the distribution of time intervals between the messages referred to and the comments referring to them. We will study questions such as if there is any time correlation between comments referring to the same message. These results will be published in near future.

Acknowledgements We would like to acknowledge Dr. Y. Sano of Tsukuba University for her informative discussion. M. Toda is financially supported by Grants-in-Aid for challenging Exploratory Research and Grants-in-Aid for Scientific Research (C) from Japan Society for the Promotion of Science (JSPS), and the Cooperative Research Program of "Network Joint Research Center for Materials and Devices".

References

1. Sano Y, Yamada K, Watanabe H, Takayasu H, Takayasu M (2013) Phys Rev E 87:012805
2. Kramer ADI, Guillory JE, Hancock JT (2014) Proc Natl Acad Sci 111:8788
3. Kramer ADI (2014) https://www.facebook.com/akramer/posts/10152987150867796

4. Mirai Kensaku Brazil Co., Ltd (2014) "Nico Nico Pedia", provided by National Institute of Informatics. http://www.nii.ac.jp/cscenter/idr/nico/nico.html
5. Kudo T (2005) MeCab: yet another part-of-speech and morphological analyzer. http://mecab.sourceforge.net/
6. Sano M (2012) Japanese Dictionary of Appraisal – attitude – Appraisal Dictionary ver1.2, Gengo Shigen Kyokai
7. Newman MEJ (2010) Networks - an introduction. Oxford University Press
8. Gephi 0.8 alpha (2011) http://oss.infoscience.co.jp/gephi/gephi.org/
9. Newman MEJ, Girvan M (2004) Phys Rev E 69:026113
10. Arenas A, Duch J, Fernandez A, Gómez S (2007) New J Phys 9:176
11. Brandes U, Delling D, Gaertler M, Goerke R, Hoefer M, Nikoloski Z, Wagner D (2006) http://arxiv.org/abs/physics/0608255
12. Blondel V, Guillaume J-L, Lambiotte R, Lefebvre E (2008) J Stat Mech Theory Exp 10:P10008

Chapter 29
Scale-Free Network Topologies with Clustering Similar to Online Social Networks

Imre Varga

Abstract In this paper I propose a novel method to model real online social networks where the growing scale-free networks have tunable clustering coefficient independently of the average degree and the exponent of the degree distribution. Models based on purely preferential attachment are not able to describe high clustering coefficient of social networks. Beside the attractive popularity my model is based on the fact that if a person knows somebody, probably knows several individuals from his/her acquaintanceship as well. The topological properties of these complex systems were studied and it was found that in my networks the cliques are relevant independently of the system size as usual in social systems.

29.1 Introduction

While networks are present everywhere in our everyday life, these complex systems attract considerable scientific interest. Researches showed that social networks are different from other networks in some sense. The reason of this was studied by Newman and Park [1]. The biggest difference is in average clustering coefficient. In social networks there is a high probability that two friends of a given individual will also be friends of each other thus the clustering coefficient is high. Opposite to non-social networks, where these triangles are rare.

Many models of networks appeared in the last decades, but most of them are not able to describe social networks directly. Models based on "small-world" networks of Watts and Strogatz [2] do not reproduce the power law degree distribution. Most of growing scale-free network models result low clustering coefficients [3–5]. There are some trials to create scale-free networks with tunable clustering [6–9], but in these models the desired value of clustering coefficient determines other properties of the networks. Avoiding this problem I wanted to create a model for online social

I. Varga (✉)
Department of Informatics Systems and Networks, University of Debrecen, Debrecen, Hungary
e-mail: varga.imre@inf.unideb.hu

H. Takayasu et al. (eds.), *Proceedings of the International Conference on Social Modeling and Simulation, plus Econophysics Colloquium 2014*, Springer Proceedings in Complexity, DOI 10.1007/978-3-319-20591-5_29

networks in which I can set the average clustering coefficient without affecting other properties (e.g. degree distribution exponent, average degree) of the network.

29.2 Basic Model

In order to achieve my goal I generalized the well-known Barabási-Albert (BA) model [3] modifying the linking method. The growing networks start from a small fully connected network of N_0 nodes where each nodes have $N_0 - 1$ links to others. Then I start to grow the network by adding more and more new nodes to it step by step. When a new node joins it is attached by $m = N_0 - 1$ links to existing nodes. These vertices are chosen by two different ways.

(a) Some nodes are chosen based on preferential attachment. The probability of a node to be chosen is proportional to the number of existing connections of it. Thus nodes with more neighbors have larger probability to get a new one. The number of these chosen nodes is denoted by π.
(b) In the second phase the new node is linked to ν number of neighbors of each previously chosen vertices. The neighbors of popular nodes have the same probability to be linked to the new node, independently from their degree.

The exact linking algorithm has the following steps:

1. *Create a new node,* $i = 1$.
2. *If* $i > \pi$*, then the linking method of this node is over.*
3. *Link the node to a probably large degree, popular one by preferential attachment.* $i = i + 1$ *and* $j = 1$.
4. *If* $j > \nu$*, go to step 2.*
5. *Link the node to one of the neighbors of ith popular neighbor of this node with equal probability.* $j = j + 1$.
6. *Go to step 4.*

These steps are repeated until the number of nodes N reaches a desired value ($N \gg N_0$). The basic idea of this two-phase linking is that to have a popular friend is advantageous and then one gets to know some acquaintances of the popular friend. Finally the number of links of a new node can be written as $m = \pi(1 + \nu)$. This method is a kind of generalized version of BA model, if $\nu = 0$ the networks generated by these two methods are the same. Now the model has three independent parameters: N, π and ν.

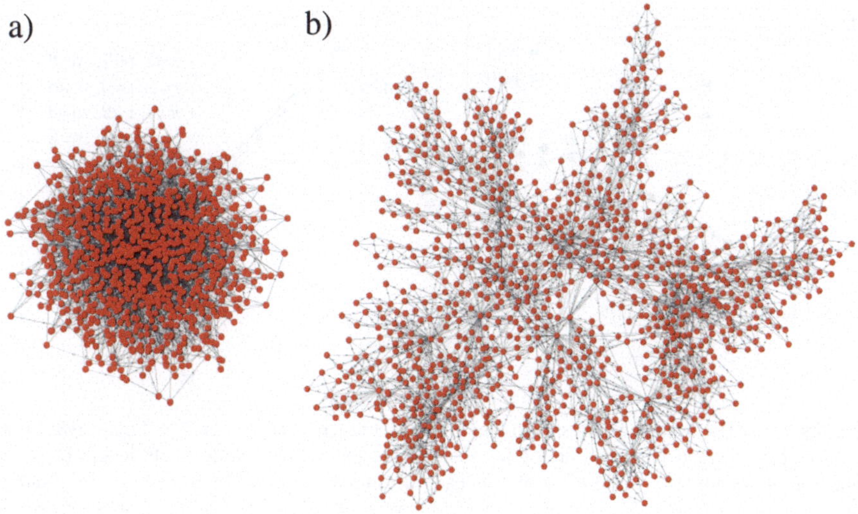

Fig. 29.1 The graphs of the model with the same number of nodes ($N = 1000$) and links ($m = 3$) using the same representation technique. On the *left side* the graph is a BA network ($\pi = 3, \nu = 0$). On the *right side* a completely different graph of my model ($\pi = 1, \nu = 2$) is presented

29.2.1 Properties of Generated Networks

This small change in the generation method leads to large differences in the network properties compared to BA-model. The differences can be seen right at the first sight even if the average degree and the density is the same (see Fig. 29.1).

In order to characterize the differences quantitatively I studied different properties first of all the average shortest path length $\langle L \rangle$ in the generated networks. It is small compared to the number of nodes and links. I found that $\langle L \rangle$ grows proportionally to the logarithm of N, so the networks have small-world property as expected. The coefficient of this proportionality depends on the parameters π and ν. BA-like networks have smaller average shortest path length than networks with high value of ν. The reason of this is the fact that in the latter case the graphs contain networks of small strongly connected groups of nodes due to the linking method. So increasing ν (at the same value of m) results networks where cliques are more important. Naturally larger number of links leads to smaller networks, where $\langle L \rangle$ obeys power law decay with parameter dependent exponent. Based on my simulation results curve fitting showed (see Fig. 29.2a) that the average shortest path length has the following functional form

$$\langle L \rangle \propto (\pi(\nu + 1))^{-F(\pi,\nu)} \ln N. \tag{29.1}$$

However initially nodes have the same amount of neighbors finally their degree varies in a wide range. Based on the growing algorithm one can analytically

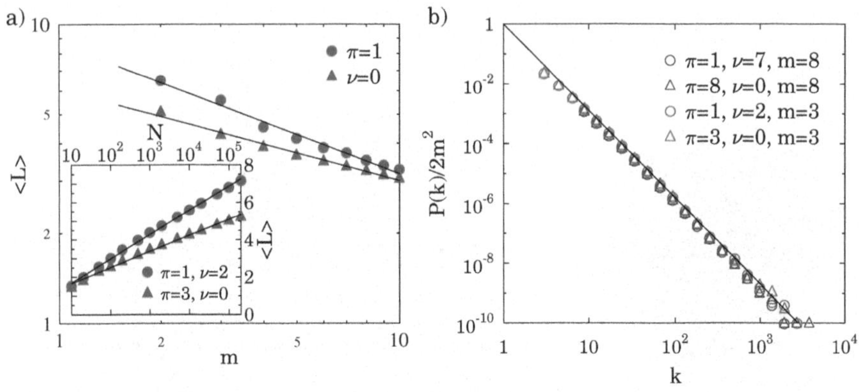

Fig. 29.2 (**a**) The average shortest path length $\langle L\rangle$ as a function of $m = \pi(\nu + 1)$. *Straight lines* indicate power law dependency on log-log scale. *Inset*: the average shortest path length $\langle L\rangle$ as a function of system size N on lin-log plot. In case of same density the BA-like graphs are smaller than generalized graphs. *Straight lines* indicate fits with Eq. (29.1). (**b**) All the graphs generated by this method have power law degree distribution. Rescaling the degree distribution data collapse occurs independently of π and ν. The exponent of the *solid line* is 2.9 as in BA model

determine the average degree of nodes

$$\langle k\rangle = 2m = 2\pi(1 + \nu). \tag{29.2}$$

The degree distribution can be well fitted by a straight line on log-log scale indicating scale-free networks with power law degree distribution with form $P(k) \propto k^{-\gamma}$. The curves with different values of π and ν can be rescaled by $2m^2$ to get data collapse as it is shown in Fig. 29.2b. This means that the exponent is independent from m in all cases not only for BA networks. The exponent γ of the degree distribution is independent of the number of nodes connected in the first step π and in each secondary step ν as well, its value is $\gamma = 2.895 \pm 0.038$ as expected. The value of the exponent is obtained by averaging the exponents of systems at different input parameter combinations. This independence needs some explanations. Let's see for example the $\pi = 1$ and $\nu = 9$ system. Only 10 % of the links based on purely preferential attachment and 90 % just randomly connected to the neighbors of popular nodes. How can this network be scale-free? As a matter of fact the 90 % also preferred, because sooner or later these neighbors also become popular as they popular neighbor gets more and more links.

To characterize the networks from the point of view of the cliques I calculated the clustering coefficient of nodes in my undirected graphs. Local clustering coefficient C of a node is the ratio of the number of existing links between neighbors of this node and the number of possible connection between them. In a general case C is proportional to the reciprocal of the degree of node, which indicates small degree nodes are mainly members of cliques while hubs of the networks connect them together.

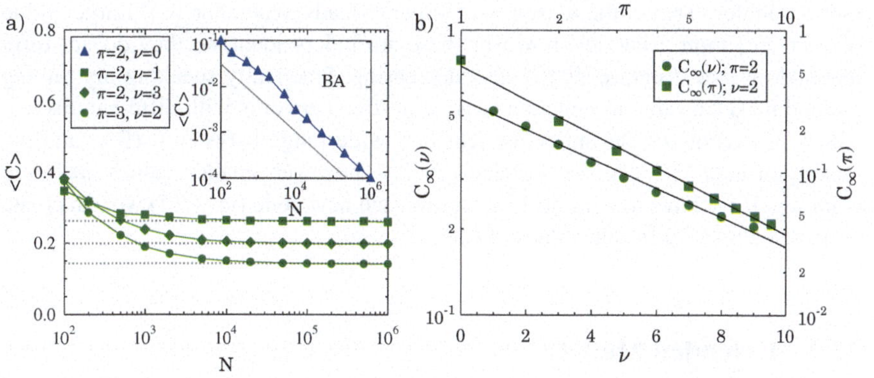

Fig. 29.3 (**a**) The average clustering coefficient $\langle C \rangle$ is decreasing with the increasing number of nodes N in the system, but it tends to zero only in BA networks (*inset*). When parameter $\nu > 0$ in a large network the value of $\langle C \rangle$ is constant. Values of C_∞ (obtained by curve fitting) which are determined by π and ν are indicated by *dashed lines*. (**b**) C_∞ as a function of ν on log-lin plot and C_∞ as a function of π on log-log plot fitted by Eq. (29.4)

The most interesting feature of my graphs can be seen if we analyze their average clustering coefficient $\langle C \rangle$. When a network is growing, $\langle C \rangle$ is decreasing. I found this can be written in the following functional form

$$\langle C \rangle \propto N^{-3/4} + C_\infty, \tag{29.3}$$

where N is the number of nodes and C_∞ is a constant at given parameter set. In case of BA network ($\nu = 0$) the value of $C_\infty = 0$, so we get back the well-known power law form. In this systems the formation of neighbor-triangles is random. Increasing the system size the degree of nodes is increasing as well so the chance of a node to belong mainly link-triangles is continually decreasing. This leads to small clustering coefficient. In generalized cases Eq. (29.3) means that $\langle C \rangle$ tends to finite values, not to zero. If $\nu > 0$, new nodes mainly compose triangles (independently from system size) due to the linking algorithm, so a given part of the system always have large clustering coefficient. One can see it on Fig. 29.3a. It indicates that when $\nu = 0$ in a large network cliques are negligible, while in the generalized networks they remains important at any system sizes. Large number of simulations were performed to discover how the constant value in $\langle C \rangle$ depends on the input parameters. I found that

$$C_\infty \propto \pi^{-A} e^{-B\nu}, \tag{29.4}$$

if $\pi > 1$ and $\nu > 0$, where A and B are constants. More links lead to smaller average clustering coefficient, where both types of linking methods (π and ν) have influence on C_∞ but they act in different ways. (See Fig. 29.3b.) Generally preferential links do not compose new triangles, so increasing π results just larger degree, but not more triangles. That is the reason why larger π leads to smaller $\langle C \rangle$. Larger value of ν creates more triangles, however these are independent, so they do not form tetrahedron-like structure. $\langle C \rangle$ is also decreasing. Practically speaking my linking method makes us able to generate large scale-free networks with different discrete values of average clustering coefficient in a wide range between 0 (BA) and the maximum at $\pi = 1, \nu = 1$ namely 0.739, however smaller values are more common. If we have maximum 15 edges to each new node ($m \leq 15$) we can create networks with 45 different values of C_∞.

29.3 Extended Model

At this point we are able to adjust the average clustering coefficient by the input parameters. However the values of π and ν determine the average degree of nodes as well. In order to model different real world networks we must tune $\langle C \rangle$ and $\langle k \rangle$ independently. That is the reason why my model has been extended. To change the number of links a reduction process is applied. After the growing period the system undergoes a destroying procedure where independently chosen nodes and their connections are removed. I used the so called *general attack* process [5] which means that all the nodes has the same probability to be removed. The strength η of this reduction process can be characterized by the ratio of number of removed nodes ΔN and the original number of nodes at the end of growing phase, so $\eta = \Delta N/N$. Thus finally the extended model has four parameters: N, π, ν and η. This reduction process has significant influence to the topological properties of the network.

29.3.1 Properties of the Reduced Networks

Remaining nodes loose connections by removing their neighbors. The final average degree in the system is determined by three things which can be expressed as

$$\langle k \rangle = \frac{\sum_i k_i - \sum_j k_j - \sum_l k_l}{N - \Delta N}, \tag{29.5}$$

where $i = 1, 2, \ldots, N$, j runs over removed nodes and l runs over the remained neighbors of removed vertices. The first term in the numerator is the sum of original degree of nodes before reduction. The second one is the loss of degree of the removed nodes. The third term describes the loss of degree due to the fact that remained nodes lose the links to removed neighbors. While removed nodes can have links to other removed nodes as well, the last two terms are not equal, their ratio is $(1 - \eta)$. In this way the Eq. (29.5) can be written as follow using mean field

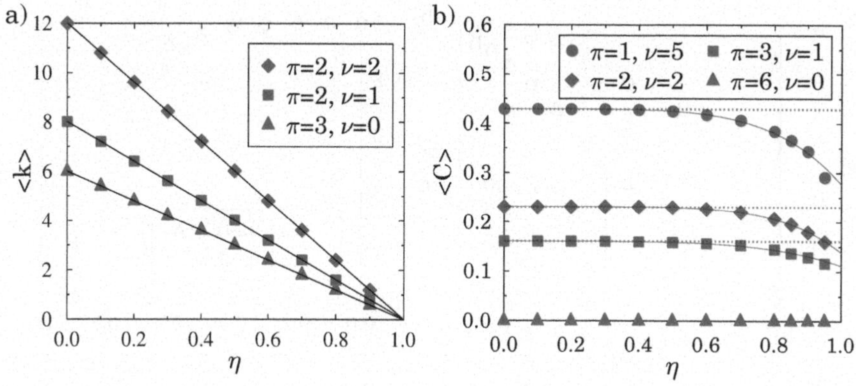

Fig. 29.4 (**a**) The average degree $\langle k \rangle$ is decreasing linearly with the reduction strength η. (Fitted by Eq. (29.7).) (**b**) The average clustering coefficient is decreasing very slowly during the reduction process. For small reduction it remains almost constant. In case of BA network ($\nu = 0$) $\langle C \rangle$ is always close to zero. *Dotted lines* denote C_∞ and *grey fitted curves* represent Eq. (29.9), where R^2 coefficient is above 0.96 for all $\nu > 0$ data sets

approximation

$$\langle k \rangle = \frac{2mN - 2m\Delta N - 2m\Delta N(1-\eta)}{N - \Delta N}. \tag{29.6}$$

Using Eq. (29.2) and the definition of η the Eq. (29.6) can be simplified to

$$\langle k \rangle = \frac{2m(1-\eta-\eta(1-\eta))}{1-\eta} = 2m(1-\eta) = 2\pi(1+\nu)(1-\eta). \tag{29.7}$$

In my simulations the average number of links of nodes decreases linearly with increasing reduction strength as predicted analytically. The effect of the reduction process on $\langle k \rangle$ is illustrated in Fig. 29.4a.

The reduction has only minor influence on average clustering coefficient, which is negligible even if half of nodes are removed. Stronger reduction leads to a bit smaller value of $\langle C \rangle$. I determined the functional form of this dependency which can describe as

$$C_\infty - \langle C \rangle \propto \eta^D \tag{29.8}$$

for large networks, where exponent D determines how fast the average clustering coefficient decreasing. (See Fig. 29.4b.) Using Eqs. (29.3), (29.4), and (29.8) finally we can write the average clustering coefficient as a function of input parameters of the model if $\pi > 1$ and $\nu > 0$

$$\langle C \rangle \propto KN^{-3/4} + K'\pi^{-A}e^{-B\nu} - K''\eta^D, \tag{29.9}$$

where K, K', K'', A, B and D are coefficients and exponents of the model.

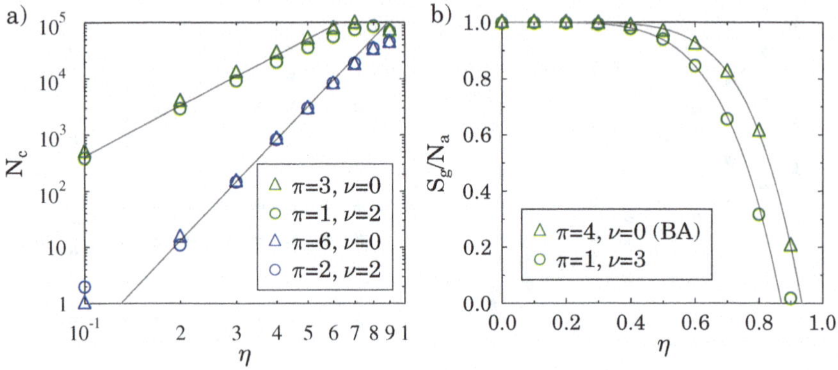

Fig. 29.5 (**a**) Number of clusters N_c as a function of reduction strength η on log-log scale. *Straight lines* indicate power law behavior, where the exponent depends only m, but independent from π and ν. (**b**) Strong reduction destroys giant component, it disappears faster in generalized networks. The decay can be described by Eq. (29.11) illustrated by *grey curves*

The values of $\langle k \rangle$ and $\langle C \rangle$ in my network are independently tunable with the reduction process, which has other side effects. The originally connected networks fall into pieces. Separate clusters appear, which are smaller networks without connections to other parts of the system. Increasing the reduction strength η the number of clusters N_c is increasing according to power law, where the exponent depends on the number of links only, independently from their role in the growing process (Fig. 29.5a). Large number of clusters can occur depending on η and the system size N. Based on the simulation results the value of N_c can be characterized by the following form

$$N_c \propto \frac{N}{\pi(\nu+1)} \eta^{\pi(\nu+1)}, \tag{29.10}$$

if the reduction is not negligible. When the reduction is very strong the number of clusters N_c saturates.

If the reduction strength is smaller than approximately 0.4 clusters are negligible except one which gives almost 100 % of the system. It is called giant component in the literature. It can be still dominant even if more than 75 % of the nodes are removed. After this the dominancy of giant component disappears fast in case of strong reduction. The speed of this process depends on the growing period. Not only the number of links of a new node m are important, but also the parameters π and ν separately. The size of giant component S_g can be written by the form

$$S_g \propto N_a(1-\eta^{E(\pi,\nu)}) = N(1-\eta)(1-\eta^{E(\pi,\nu)}), \tag{29.11}$$

where $N_a = N - \Delta N$ is the number of nodes in the reduced system. (See Fig. 29.5b.) The exponent E depends not only on the value of m, but also π and ν, however larger m results smaller exponent, so larger giant component. In BA networks ($\nu = 0$) the giant component is always larger than in generalized networks at a given link number. This shows that BA networks are strongly connected while if $\nu > 0$ the system is a weakly connected set of densely linked groups of nodes. Since the number of clusters is independent from π and ν at a given value of m, but the size of giant component is smaller for larger ν, clusters (excluding the giant component) are larger. The average cluster size is much smaller in BA networks then in the generalized case. These are also proofs of presence and importance of cliques. These clusters have a power law size distribution with a parameter dependent exponent. Number of clusters $n(S)$ of size S can be expressed as

$$n(S) \propto S^{-\tau(\pi,\nu)}. \tag{29.12}$$

29.4 Model of Real Online Social Network

Due to the discussed topological properties my networks are appropriate candidates for modeling real world online social networks. I managed to get a set of data of almost 60 million Facebook users [10]. This network has small world property, its degree distribution can be characterized by two power law regimes (see Table 29.1), so it is a kind of scale-free network. The quite high average clustering coefficient indicates the presence of cliques of users.

Based on my presented results I found a set of input parameters which leads to a very similar network. The values of input parameters in my Facebook model are: $\boldsymbol{\pi = 3}$, $\boldsymbol{\nu = 1}$ and $\boldsymbol{\eta = 0.72}$ ($N = 10{,}750{,}000$). This final sample contains more than three million nodes. In this size scale N has not got influence to the network properties, so not necessary to create larger system. The properties of the real social network and my model network are summarized in Table 29.1. As one can see the values of the main quantities $\langle C \rangle$ and $\langle k \rangle$ well describe the real case and other properties give quite good qualitative description (e.g. presence of separate clusters or power law degree distribution) as well.

Table 29.1 Comparison of my extended model network and the Facebook data set

		Facebook	Extended model
Average shortest path length	$\langle L \rangle$(N)	Logarithmic	Logarithmic
Degree distribution	$P(k)$	Power law	Power law
Degree distribution exponent	γ	1.32, 3.38	2.96
Average degree	$\langle k \rangle$	**3.13**	**3.24**
Dominance of giant component	S_g/N_a	0.99	0.90
Cluster size distribution	$N(S)$	Power law	Power law
Average clustering coefficient	$\langle C \rangle$	**0.16**	**0.15**

The features of the two networks are in good agreement

29.5 Conclusion

In summary, I proposed a simple method for generating scale-free networks where the average clustering coefficient is tunable in a broad range and determined by the input parameters π and ν. The method is a kind of generalized version of growing Barabási-Albert model where the links of a new node play different roles. Beside the preferential attachment some links obey the so called "friend of my friend is my friend" philosophy. After the growing process a reduction process was used in order to create large variety of networks changing $\langle k \rangle$ and $\langle C \rangle$ independently. This reduction process means random removal of nodes. The strength of reduction is characterized by parameter η. A detailed study of the model was presented proofing that in these scale-free networks the cliques have very important role which cannot be described by the original BA model. Comparing a real online social network and the graphs generated by the proposed algorithm I found very good agreement. For clarity my model does not describe the time evolution of real social networks just generate graphs topologically similar to a given state of real online social networks. In the near future the model networks are being subjected to agent-based simulation of information spreading using the model of Kocsis and Kun [11]. This model can be a good base of later study of effectiveness of advertising in online social networks.

Acknowledgements The publication was supported by the TÁMOP-4.2.2.C-11/1/KONV-2012-0001 project. The project has been supported by the European Union, co-financed by the European Social Fund.

The computations were performed on the HPC centre provided by the Hungarian National Information Infrastructure Development Institute at the University of Debrecen.

Special thanks to my colleague Gergely Kocsis for fruitful discussions.

References

1. Newman MEJ, Park J (2003) Why social networks are different from other types of networks. Phys Rev E 68:036122
2. Watts DJ, Strogatz SH (1998) Collective dynamics of "small-world" networks. Nature 393(6684):440–442
3. Barabási A, Albert R (1999) Emergence of scaling in random networks. Science 286:509–512
4. Lee HY, Chan HY, Hui PM (2004) Scale-free networks with tunable degree distribution exponents. Phys Rev E 69:067102
5. Varga I, Németh A, Kocsis G (2013) A novel method of generating tunable underlying network topologies for social simulation. In: Proceedings of the 4th IEEE international conference on cognitive infocommunicaitons, pp 71–74
6. Holme P, Kim BJ (2002) Growing scale-free networks with tunable clustering. Phys Rev E 65:026107
7. Newman MEJ (2009) Random graphs with clustering. Phys Rev Lett 103:058701

8. Heath LS, Parikh N (2011) Generating random graphs with tunable clustering coefficients. Phys A 390:4577–4587
9. Samalam VK (2013) A model for generating tunable clustering coefficients independent of the number of nodes in scale free and random networks [arXiv:1311.6401]
10. Gjoka M, Kurant M, Butts CT, Markopoulou A (2010) Walking in Facebook: a case study of unbiased sampling of OSNs. In: Proceedings of the IEEE INFOCOM '10, pp 1–9
11. Kocsis G, Kun F (2011) Competition of information channels in the spreading of innovations. Phys Rev E 84:026111

Chapter 30
Identifying Colors of Products and Associated Personalized Recommendation Engine in e-Fashion Business

Keiichi Zempo and Ushio Sumita

Abstract One of the important factors ignored in the literature in e-marketing is "the color" of a product. While one may be able to identify the dominating color of a product based on the overall impression, it is not easy to mechanize the process to determine the dominating color. Accordingly, in many applications, the color of a product is defined subjectively by those who enter the data. Consequently, the color of a product has been a missing link in e-marketing. The purpose of this research is to fill this gap by developing an algorithmic procedure for identifying the dominating color of a product by analyzing a digital image of the product. The algorithmic procedure enables one to reveal color preferences of consumers by analyzing the digital images of the products obtained from the purchasing records. A recommendation engine is also developed based on color class preference vectors of individual consumers.

30.1 Introduction

In recent years, as pointed in [1], there has been an increasing interest in Web usage mining as a means to capture Web user behavioral patterns and to derive e-business intelligence. In [2–4], for example, automatic personalization was proposed based on clustering of user transaction and page-views. A prevalent alternative approach for building personalized recommendation engines would be collaborative filtering. Given a record of activity of a target user, the collaborative filtering approach compares that record with the historical records of other users so as to find the top users who have similar taste or interest. However, it is known that the collaborative filtering approach has some deficiency, see e.g. [5–8], and some optimization strategies have been proposed in [9–11] to overcome such shortcomings. More

K. Zempo (✉) • U. Sumita
Faculty of Engineering, Systems and Information, University of Tsukuba, Tsukuba, Ibaraki 305-8573 Japan
e-mail: zempo@iit.tsukuba.ac.jp

H. Takayasu et al. (eds.), *Proceedings of the International Conference on Social Modeling and Simulation, plus Econophysics Colloquium 2014*, Springer Proceedings in Complexity, DOI 10.1007/978-3-319-20591-5_30

recently, a personal browsing assistant system is developed in [12], where the pre-fetched resources from the hyper-linked Web pages are compared so as to recommend which Web page should be requested next. As an application, there is recommendation engine specialized to fashions, e.g. [13, 14]. To the best knowledge of the authors, however, the color information has not been incorporated in the literature for developing better personalized recommendation engines.

The reason why colors of products has been ignored in e-marketing can be found in that a product typically involves many different colors. While one dominating color of a product may be identified in the eye of human based on the overall impression, it is difficult to mechanize the process for identifying the dominating color. Accordingly, in many applications, the color of a product is defined subjectively by those who enter the data. Furthermore, terms for describing a color are often quite vague and too many. Consequently, the color of a product has been a missing link in e-marketing. The purpose of this paper is to fill this gap by developing an algorithmic procedure for identifying the dominating color of a product by analyzing a digital image of the product. The algorithmic procedure enables one to reveal the color preference of a consumer by analyzing the digital images of the products purchased by the consumer. A recommendation engine is also developed based on color class preference vectors of individual consumers as shown in Fig 30.1.

Throughout the paper, vectors and matrices are indicated by underbar and doubleunderbar respectively, e.g. $\underline{\xi}$, $\underline{\underline{P}}(t)$, etc.

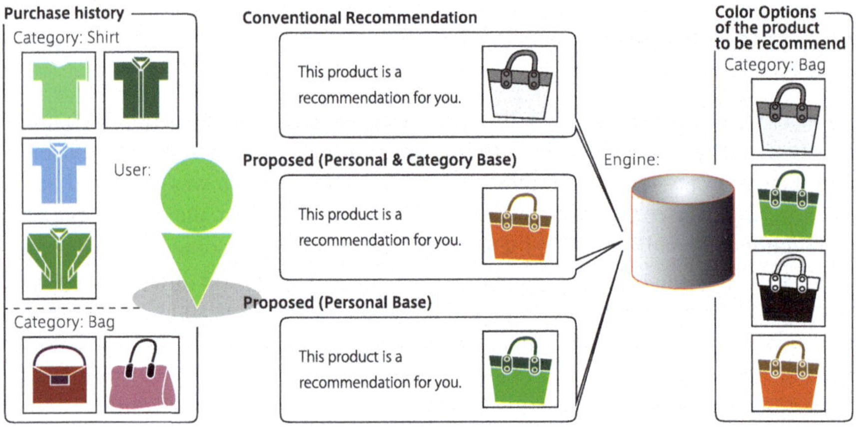

Fig. 30.1 Overview of the proposing personalized recommendation engine

30.2 Personalized Recommendation Engine Based on the Color of the Product

30.2.1 Development of Algorithm for Identifying the Dominating Color of a Product

A typical digital image of a product used in e-fashion business consists of a number of pixels, which would be too many to define the single dominating color of the product. In order to overcome this difficulty, we introduce $\underline{\Phi}(v_p) \in \mathscr{R}^6$, which we call a CCPV (Color-Class Profile Vector) of a digital image v containing a product p. In the eye of human, however, the Euclidean distance in RGB does not necessarily reflect the way humans differentiate different colors sensuously. Because of this reason, CIE (Commission International de l'Éclairage), the international commission on illumination, proposed the space denoted by CIE-L*a*b* in 1978. In CIE-L*a*b* space, RED, GREEN, YELLOW, BLUE, WHITE and BLACK are extremums of the axes as representative colors[15–17]. By defining the closeness of their representative six colors, we converted each pixel to facilitate by clustering. Based on this idea, we transform a set of pixels constituting a digital image of a product, denoted by v_p in CIE-L*a*b*, into the set of six dimensional vectors. By measuring the Euclidean distances between each of the transformed vectors and six fixed points in CIE-L*a*b* representing RED, GREEN, YELLOW, BLUE, WHITE and BLACK and then taking the average over the pixels in v_p, the sensuous color of the product in the eye of human is represented by a vector $\underline{\Phi}(v_p)$ in CIE-L*a*b*. $\underline{\Phi}(v_p)$ are calculated through the following steps.

Step 1: **Extraction of the pixels of the product image from the background**
Every digital image obtained from the data has the background constructed by the unique pixel for representing "NON-COLOR". This pixel is different from the pixel corresponding to "WHITE" and never appears in digital images of products. Accordingly, the set of pixels exactly constituting the digital image of the product p can be extracted. The resulting set of pixels is denoted by v_p, and the number of pixels in v_p is written as N_{v_p}.

Step 2: **Transformation of RGB vectors into CIE-L*a*b* vectors**
In Step 2, this transformation is conducted. Transformation $\mathscr{T}_{\mathrm{I}}$ of the pixel $\underline{\gamma} = {}^{\mathrm{t}}(\gamma_{\mathrm{R}}, \gamma_{\mathrm{G}}, \gamma_{\mathrm{B}}) \in \mathrm{RGB}$ into $\underline{\eta} = {}^{\mathrm{t}}(\eta_{\mathrm{L}}, \eta_{\mathrm{a}}, \eta_{\mathrm{b}}) \in$ CIE-L*a*b* is constructed in three stages. In the first stage, $\underline{\gamma}$ is mapped into an intermediate vector $\underline{X} = {}^{\mathrm{t}}(X_1, X_2, X_3)$ via the liner transformation defined by,

$$\begin{pmatrix} X_1 \\ X_2 \\ X_3 \end{pmatrix} = \begin{pmatrix} 0.4125 & 0.3576 & 0.1804 \\ 0.2127 & 0.7151 & 0.0722 \\ 0.0193 & 0.1192 & 0.9502 \end{pmatrix} \times \begin{pmatrix} \gamma_{\mathrm{R}} \\ \gamma_{\mathrm{G}} \\ \gamma_{\mathrm{B}} \end{pmatrix}. \quad (30.1)$$

The second stage constructs $\underline{f} = {}^t(f_1, f_2, f_3)$ from $\underline{X} = {}^t(X_1, X_2, X_3)$ through the following definition. For $i = 1, 2, 3$, let f_i be defined by,

$$f_i = \begin{cases} X_i^{\frac{1}{3}} & \text{if } X_i > 0.008856 \\ \dfrac{903.3X_i + 16}{116} & \text{else} \end{cases} . \tag{30.2}$$

Finally, $\underline{f}$ is mapped into $\underline{\eta}$ by ,

$$\begin{pmatrix} \eta_L \\ \eta_a \\ \eta_b \end{pmatrix} = \begin{pmatrix} 0 & 116 & 0 \\ 500 & -500 & 0 \\ 0 & 200 & -200 \end{pmatrix} \times \begin{pmatrix} f_1 \\ f_2 \\ f_3 \end{pmatrix} + \begin{pmatrix} -16 \\ 0 \\ 0 \end{pmatrix} . \tag{30.3}$$

Step 3: **Construction of a CCPV**

Given $\underline{\eta} \in$ CIE-L*a*b*, we consider another transformation $\mathscr{T}_{\rm II}$: CIE-L*a*b* $\rightarrow$ CC $\subset \mathscr{R}^6_+$, where $\mathscr{R}^6_+$ is the set of nonnegative vectors in $\mathscr{R}^6$. The space CC, standing for "Color Class", is introduced so as to develop several different color classes as we will see. For constructing CC, the transformation $\mathscr{T}_{\rm II}$ is defined by measuring the inverse of the squared Euclidean distances between $\underline{\eta}$ and six fixed points in CIE-L*a*b* representing RED, GREEN, YELLOW, BLUE, WHITE and BLACK. More formally, we consider the following six fixed points in CIE-L*a*b*.

$$\begin{array}{llll} \underline{\eta}_{\rm R} = {}^t(\ 50, 50,\ 0), & \underline{\eta}_{\rm G} = {}^t(\ 50, -50,\ \ 0), \\ \underline{\eta}_{\rm Y} = {}^t(\ 50,\ 0, 50), & \underline{\eta}_{\rm B} = {}^t(\ 50,\ \ 0, -50), \\ \underline{\eta}_{\rm W} = {}^t(\ 100,\ 0,\ 0), & \underline{\eta}_{\rm BK} = {}^t(\ 0,\ \ 0,\ \ 0), \end{array} \tag{30.4}$$

where each color in RGB are represented as,

$$\begin{array}{ll} \mathscr{T}_{\rm I}^{-1}(\underline{\eta}_{\rm R}) = {}^t(\ 0.59, 0.06, 0.18), & \mathscr{T}_{\rm I}^{-1}(\underline{\eta}_{\rm G}) = {}^t(\ -0.04, 0.25, 0.16), \\ \mathscr{T}_{\rm I}^{-1}(\underline{\eta}_{\rm Y}) = {}^t(\ 0.29, 0.17, 0.01), & \mathscr{T}_{\rm I}^{-1}(\underline{\eta}_{\rm B}) = {}^t(\ \ 0.04, 0.19, 0.55), \\ \mathscr{T}_{\rm I}^{-1}(\underline{\eta}_{\rm W}) = {}^t(\ 1.25, 0.95, 0.91), & \mathscr{T}_{\rm I}^{-1}(\underline{\eta}_{\rm BK}) = {}^t(\ \ \ 0,\ \ 0,\ \ 0). \end{array} \tag{30.5}$$

Given $\underline{\gamma} \in \upsilon_p$, let $\underline{\eta} = \mathscr{T}_{\rm I}(\underline{\gamma}) \in$ CIE-L*a*b* and define $\underline{\phi}(\underline{\gamma}) = \mathscr{T}_{\rm II} \circ \mathscr{T}_{\rm I}(\underline{\gamma}) = \mathscr{T}_{\rm II}(\underline{\eta})$ by

$$\underline{\phi}(\underline{\gamma}) \stackrel{\rm def}{=} c \begin{pmatrix} ||\underline{\eta}_{\rm R} - \underline{\eta}||^{-2} \\ ||\underline{\eta}_{\rm G} - \underline{\eta}||^{-2} \\ ||\underline{\eta}_{\rm Y} - \underline{\eta}||^{-2} \\ ||\underline{\eta}_{\rm B} - \underline{\eta}||^{-2} \\ ||\underline{\eta}_{\rm W} - \underline{\eta}||^{-2} \\ ||\underline{\eta}_{\rm BK} - \underline{\eta}||^{-2} \end{pmatrix} , \tag{30.6}$$

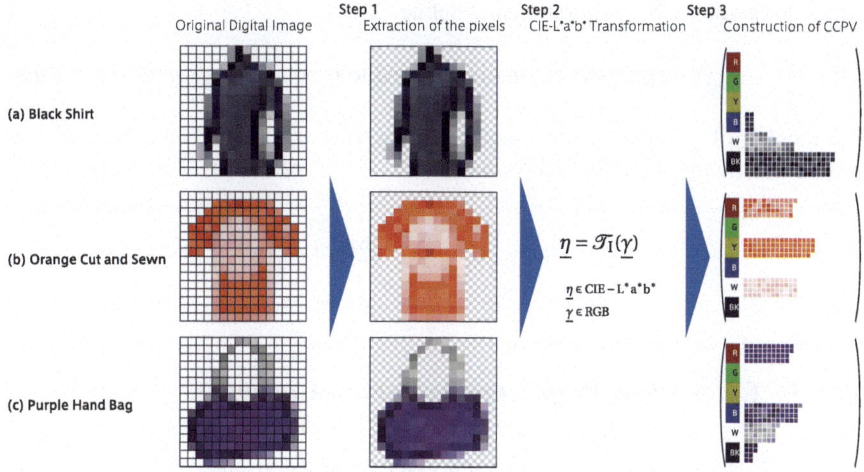

Fig. 30.2 Algorithm of CCPV construction for typical digital image

where $||\underline{x}||$ denotes the Euclidean norm of $\underline{x}$, and c is the normalization constant. It should be noted that $\underline{\phi}(\underline{\gamma})$ is a probability vector, where each component describes the how a typical person would sense the pixel represented by $\underline{\gamma}$ to the corresponding color in RED, GREEN, YELLOW, BLUE, WHITE and BLACK.

The schematic diagram of the above steps are shown in Fig. 30.2. The color-class profile vector of v_p can now be defined by,

$$\underline{\Phi}(v_p) \stackrel{\text{def}}{=} \frac{1}{N_{v_p}} \sum_{\underline{\gamma} \in v_p} \underline{\phi}(\underline{\gamma}) . \tag{30.7}$$

We may say that $\underline{\Phi}(v_p)$ describes how a typical person would sense the six different colors RED, GREEN, YELLOW, BLUE, WHITE and BLACK from the overall impression of the digital image v_p of product p.

30.2.2 Development of Color-Classes via Clustering of CCPVs

The algorithmic procedure described in Sect. 26.2.1 enables one to represent each digital image v_p of product p by the corresponding CCPV, $\underline{\Phi}(v_p)$. The data obtained from X Corporation contain 5665 such digital images, to each of which one of 425 colors was assigned by X Corporation. The purpose of this section is to develop a reasonable number of color classes by clustering these 425 colors, so that the effects of color in marketing can be analyzed efficiently. For this purpose, we represent each color defined by X Corporation by a CCPV in CIE-L*a*b*. More specifically, let x

be a color given by X Corporation and define,

$$V(x) = \{v_p : \text{the color } x \text{ is assigned to product } p\} . \tag{30.8}$$

The number of elements in $V(x)$ is denoted by $N(x) = |V(x)|$. The color x is then represented by $\underline{\Phi}_x \in$ CIE-L*a*b* where,

$$\underline{\Phi}_x = \frac{1}{N(x)} \sum_{v_p \in V(x)} \underline{\Phi}(v_p) . \tag{30.9}$$

30.2.3 Color Class Preference Vectors of Customer

In order to define a color class preference vector of a consumer, we introduce the following sets.

$CUST = \{i : 1 \le i \le N_c\}$: the set of customers
$S = \{j : 1 \le j \le N_s\}$: the set of product categories
$S(j)$: the number of products in the product category $j \in S$
$q_r(j)$: the set of products which are identical having the same product ID but belong to different color classes in the rth product in the product category $j \in S$
$Q(j) = \{q_1(j), \cdots, q_{S(j)}(j)\}$: the set of product groups in $S(j)$, where each group consists of identical products having different color classes
N_{CC} : the number of color classed to be combined
$CC = \{1, \cdots, N_{CC}\}$: the set of color classes
$n(i, j, x)$: the number of products, purchased by consumer $i \in CUST$, which belong to $Q(j)$ having color class $x \in CC$

For $l \in CC, l = 1, \cdots, m$, let the color class distribution vector, $\underline{\theta}(i, j)$, be defined by

$$\underline{\theta}(i,j) = [\theta(i,j,1), \cdots, \theta(i,j,N_{CC})]; \quad \theta(i,j,l) = \frac{n(i,j,l)}{\sum_{k=1}^{N_{CC}} n(i,j,k)} . \tag{30.10}$$

The corresponding mean and variance vectors, $\underline{\mu}(j)$, $\underline{\sigma}(j)$, can be obtained as

$$\underline{\mu}(j) = \frac{1}{N_c} \sum_{i \in CUST} \underline{\theta}(i,j) , \tag{30.11}$$

$$\underline{\sigma}(j) = [\sigma(j,1), \cdots, \sigma(j,N_{CC})]; \ \sigma(j,l) = \sqrt{\frac{1}{N_c - 1} \sum_{i \in CUST} \{\theta(i,j,l) - \mu(j,l)\}^2} . \tag{30.12}$$

Then the color class preference vector of consumer $i \in CUST$ for the product category $j \in S$ can be defined in the following manner.

$$\underline{z}(i,j) = [z(i,j,1), \cdots, z(i,j,N_{CC})]; \;\; z(i,j,l) = \frac{\theta(i,j,l) - \mu(j,l)}{\sigma(j,l)} \,. \tag{30.13}$$

Let $CCQ\left(j,\check{j}\right)$ be the set of color classes which products in $q_{\check{j}}(j) \in Q(j)$ possess. If consumer i is to purchase a product $p \in q_{\check{j}}(j) \in Q(j)$, then the color $\tilde{x}(i,j)$ to be recommended is determined by

$$\tilde{x}(i,j) = \arg \max_{x \in CCQ(j,\check{j})} \{z(i,j,x)\} \,. \tag{30.14}$$

In the approach discussed above, the color class preference vector of consumer $i \in CUST$ is defined for each product category $j \in S$. As an alternative approach, the single color class preference vector of consumer $i \in CUST$ may be employed for all the products in $Q = \bigcup_{j=1}^{N_S} Q(j)$. In this case, in place of Eq. (30.10), we define

$$\underline{\theta}(i) = [\theta(i,1), \cdots, \theta(i,N_{CC})] \; ; \;\; \theta(i,l) = \frac{\sum_{j=1}^{N_S} n(i,j,l)}{\sum_{j=1}^{N_S} \sum_{k=1}^{N_{CC}} n(i,j,k)} \,. \tag{30.15}$$

Then the color class preference vector of $i \in CUST$ for all the product in Q can be defined by

$$\underline{z}(i) = [z(i,1), \cdots, z(i,N_{CC})] \; ; \; z(i,l) = \frac{\theta(i,l) - \mu(l)}{\sigma(l)} \,, \tag{30.16}$$

where the mean and the variance vectors are also changed accordingly as $\underline{\mu}$ and $\underline{\sigma}$. Let $CCQ(j,\check{j})$ be defined as before. If consumer $i \in CUST$ is to purchase a product $p \in q_{\check{j}}(j) \in Q(j)$, then the color $\tilde{x}(i)$ to be recommended is determined by,

$$\tilde{x}(i) = \arg \max_{x \in CCQ(j,\check{j})} \{z(i,x)\} \,. \tag{30.17}$$

The latter approach may work better because of the larger data volume involved in constructing $\underline{z}(i)$. If the color class was not defined because of the lack of the purchase history for the customer or the lack of the color options, the engine would recommend the default choice of the color.

30.3 Numerical Experiments Based on Real Data

30.3.1 Data Description

Sumita Research Laboratory at the University of Tsukuba has been working with a TV shopping company, hereafter called X Corporation, for developing a CRM (Customer Relationship Management) support engine based on real data. X Corporation has been in retail business worldwide, offering a variety of products ranging from Apparel products, Jewelries, and Home electronics appliances to foods. A typical digital image used in the e-business consists of $400 \times 400 = 160{,}000$ pixels, which would be too much to define the dominating color-class of the image.

The data obtained from X corporation consist of demographic information of those consumers who purchased at least one product during the period between September 1st, 2004 and August 31st, 2007, as well as their purchasing records and channels, product records and TV programs during the period. The amount of consumers, N_c, was 455,415, the amount of product categories, N_s, was 34 and the data consisted of about 2.3 million records. The average number of purchase occasions per customer and purchased quantity per customer were 3.70 and 5.33, respectively. The digital images collected from the data obtained from X Corporation amount to 6762, involving 1782 types of products spread over 34 small categories. The structure and the key components of these records are in Fig. 30.3. The database of X Corporation defines 430 colors appear for the products corresponding to the 6762 digital images. However, five of them are clearly useless (e.g. NON-COLOR, CLEAR) and eliminated. Consequently, the data to be used

Genre Category

1 Fashion Wear
- 10 Inner-Lingerie, 11 Jacket-Coat, 12 Shirt-Blouse, 13 Skirt, 14 Knit, 15 Pants, 16 Men's, 17 Ensemble, 18 Cut and Sewn, 19 Suit-One piece

2 Bag
- 20 Hand Bag, 21 Shoulder Bag, 22 Tote Bag, 23 Boston Bag, 24 Racksack/Organizer Bag, 25 Pouch/Pochette, 26 Bag etc.

3 Fashion Accessory
- 30 Ring, 31 Necklace/Pendant/Brooch, 32 Earring/Ear Clip, 33 Bracelet/Anklet, 34 Set etc.

4 Brand Accessory
- 30 Ring, 31 Necklace/Pendant/Brooch, 32 Earring/Ear Clip, 33 Bracelet/Anklet, 34 Set etc.

5 Fashion Gadget
- 50 Presbyopia glasses, 51 Watch, 52 Stole/Scarf/Belt, 53 Wallet/Key Purse, 54 Shoes/Umbrella/Hat, 55 Wig/Hair Accessory, 56 Others

Fig. 30.3 Category structure

for our analysis contain 425 colors (corresponds to 5665 digital images) defined by X Corporation. In what follows, these 425 colors are categorized into several number of newly defined color-classes by analyzing the 6762 digital images. The algorithmic procedure used to establish the color-classes can be applied to a digital image of any product with one of the 425 colors, identifying the dominating color-class of the product automatically. In turn, the algorithmic procedure enables one to canalize the consumers from the perspective of color preferences, thereby filling the missing link in e-marketing.

In order to cluster 425 colors, each represented by $\underline{\Phi}_x$, we employ the group average method in hierarchical clustering [18, 19]. In this approach, a set of vectors would be grouped together one by one based on the nearest Euclidean distance until the predetermined number of clusters would exhaust the original set. In each grouping, the resulting cluster is represented by one vector which can be generated as the weight center of the two clusters to be merged. We terminated the grouping just before the six representative color (RED, GREEN, YELLOW, BLUE, WHITE, BLACK) combined to the other six representative color.

For each cluster generated by the above algorithm, the histogram is constructed by 425 colors over the digital images involved in the cluster. Namely if a cluster consists of $\underline{\Phi}_{x(1)}, \cdots, \underline{\Phi}_{x(T)}$, then the histogram is constructed over the products in $\bigcup_{l=1}^{T} V(x(l))$. The grouping resulted into generate 14 color-classes, (i.e. $N_{CC} = 14$), named as BLACK, BEIGE, WHITE, PINK, BROWN, GRAY, BLUE, NAVY, GREEN, PURPLE, RED, ORANGE, SAXE-BLUE and YELLOW.

30.3.2 Accuracy Test for Color Class Recommendation Engine

In this subsection, we examine the accuracy of the color class recommendation engine developed in Sect. 30.2.3. The data set obtained from X Corporation is decomposed into ten subsets of equal size randomly. Based on the cross validation approach, nine subsets are used to construct $\underline{z}(i,j)$ in Eq. (30.13) and $\underline{z}(i)$ in Eq. (30.16), while the remaining subset is used for testing accuracy. In order to provide a basis for comparison, the following random estimation accuracy is considered.

Random Estimation:

> If consumer i is to buy a product $p \in q_{\check{j}}(j)$ and a color class is chosen randomly, the probability of its correctness is given by $|CCQ(j,\check{j})|^{-1}$ where $CCQ(j,\check{j})$ is the set of color classes which products in $q_{\check{j}}(j)$ possess.

In Table 30.1, the results for testing accuracy based on $\underline{z}(i,j)$ in Eq. (30.13) and the results for testing accuracy based on $\underline{z}(i)$ in Eq. (30.16) are exhibited respectively. One can observe that the color class recommendation engine outperforms the random estimation consistently with only one exception for "51 Watch" in row of table $\underline{z}(i,j)$. However, even for this product, the color class recommendation

Table 30.1 Accuracy test of the recommendation engine based on $\underline{z}(i,j)$ and $\underline{z}(i)$ for customer i and category j ("ratio" notes acc./rand.)

	Based on $\underline{z}(i,j)$				Based on $\underline{z}(i)$			
	Records	Acc.	Rand.	Ratio	Records	Acc.	Rand.	Ratio
Fashion wear								
10 Inner-lingerie	565,088	0.43	0.32	1.36	551,873	0.48	0.32	1.53
11 Jacket-coat	34,482	0.34	0.30	1.11	56,984	0.45	0.30	1.49
12 Shirt-blouse	50,832	0.36	0.32	1.13	66,679	0.44	0.32	1.36
13 Skirt	25,641	0.40	0.33	1.21	38,999	0.52	0.33	1.54
14 Knit	44,945	0.27	0.25	1.06	63,412	0.33	0.26	1.30
15 Pants	64,242	0.43	0.35	1.22	77,971	0.54	0.36	1.51
16 Men's	168	0.47	0.31	1.51	351	0.50	0.31	1.60
17 Ensemble	29,206	0.34	0.30	1.14	48,895	0.42	0.30	1.42
18 Cut and sewn	136,540	0.58	0.27	2.18	159,947	0.34	0.27	1.25
19 Suit-one piece	24,460	0.39	0.32	1.22	39,179	0.53	0.33	1.61
(Sub total)	975,604	0.43	0.31	1.39	1,104,290	0.45	0.31	1.45
Bag								
20 Hand bag	13,918	0.25	0.20	1.25	29,218	0.48	0.21	2.26
21 Shoulder bag	17,250	0.20	0.17	1.21	35,931	0.39	0.19	2.04
22 Tote bag	27,320	0.25	0.23	1.05	51,454	0.35	0.24	1.47
23 Boston bag	17,318	0.17	0.16	1.06	36,155	0.31	0.17	1.82
24 Rucksack/organizer bag	4153	0.24	0.21	1.18	11,580	0.40	0.22	1.85
25 Pouch/pochette	20,771	0.21	0.19	1.11	34,361	0.33	0.19	1.70
26 Bag all	2865	0.30	0.22	1.37	8183	0.48	0.20	2.42
(Sub total)	103,595	0.22	0.20	1.13	206,882	0.37	0.20	1.83
Fashion accessory								
30 Ring	3288	0.82	0.44	1.86	8152	0.50	0.45	1.12
31 Necklace/pendant/brooch	7058	0.84	0.36	2.31	14,913	0.58	0.38	1.52
32 Earring/ear clip	2066	0.59	0.43	1.36	6246	0.53	0.36	1.47
33 Bracelet/anklet	1241	0.87	0.31	2.86	3187	0.43	0.27	1.59
34 Set etc.	5247	0.58	0.42	1.38	12,740	0.36	0.30	1.18
(Sub total)	18,900	0.74	0.39	1.87	45,238	0.49	0.36	1.35
Brand accessory								
40 Ring	1110	0.72	0.40	1.80	3900	0.61	0.39	1.56
41 Necklace/pendant/brooch	4366	0.79	0.36	2.17	10,699	0.50	0.36	1.38
42 Earring/ear clip	1895	0.56	0.43	1.30	5640	0.59	0.38	1.56
43 Bracelet/anklet	358	0.72	0.24	2.97	1066	0.41	0.25	1.64
44 Set etc.	131	0.60	0.47	1.27	842	0.49	0.36	1.38
(Sub total)	7860	0.72	0.38	1.90	22,147	0.54	0.37	1.47

(continued)

Table 30.1 (continued)

	Based on $\underline{z}(i,j)$				Based on $\underline{z}(i)$			
	Records	Acc.	Rand.	Ratio	Records	Acc.	Rand.	Ratio
Fashion gadget								
50 Presbyopia glasses	2098	0.17	0.17	1.03	7749	0.40	0.20	1.99
51 Watch	2881	0.25	0.36	0.69	6528	0.73	0.36	2.02
52 Stole/scarf/belt	25,785	0.31	0.27	1.18	42,137	0.41	0.26	1.55
53 Wallet/key purse	26,031	0.28	0.18	1.52	45,128	0.38	0.18	2.12
54 Shoes/umbrella/hat	20,126	0.40	0.24	1.68	35,385	0.43	0.23	1.85
55 Wig/hair accessory	876	0.50	0.49	1.00	7,554	0.95	0.49	1.95
56 Others	7046	0.37	0.24	1.55	16,023	0.36	0.25	1.40
(Sub total)	84,843	0.32	0.24	1.37	160,504	0.44	0.24	1.81
Total	1,190,802	0.41	0.29	1.40	1,539,061	0.44	0.29	1.53

engine based on $\underline{z}(i)$ supersedes the random estimation by a factor of two. It can be seen that, when the volume of test data is high, the color class recommendation engine based on $\underline{z}(i)$ outperforms the color class recommendation engine based on $\underline{z}(i,j)$. This implies that color preferences of consumers are reflected beyond product categories for products which are purchased rather often at modest prices, as represented by Fashion Wear (10 through 19) , Bag (20 through 26) and Fashion Gadget (50 through 56). For more expensive products which are likely to be purchased with less frequency, however, color preferences of consumers within the product category prevail over those derived from all products, as can be seen in Fashion Accessory (30 through 34) and Brand Accessory (40 through 44). This result is in agreement that one who have the color to prefer may buy the other color product as an accent color. In any case, one may select whichever the recommendation engine based on $\underline{z}(i,j)$ or $\underline{z}(i)$, by considering which is suitable for the genres of product.

30.4 Conclusion

One of the important factors ignored in the past analyses in e-marketing is "colors" of products. This is so because it is difficult to define a color of a product, which typically consists of many different colors. The purpose of this research is to fill this gap by developing an algorithmic procedure for identifying the dominating color of a product by analyzing a digital image of the product. Since humans tend to clearly distinguish RED from GREEN as well as YELLOW from BLUE, the Euclidean distance in CIE-L*a*b* is more consistent with the sensuous feeling of human for colors than the Euclidean distance in RGB. Accordingly, for analyzing color preferences of consumers in e-marketing, CIE-L*a*b* is more appropriate than RGB. Based on this idea, we proposed the CCPV (Color-Class Profile Vector) which represents the overall impression of a digital image containing a product. Since each product has its color in the data base, these vectors can be utilized to categorize many different colors, resulting in 14 color classes. This enables one to study color preferences of consumers by segments. Furthermore, it provides a basis for constructing a recommendation engine based on the color classes for enhancing e-commerce. We had also confirmed the effectiveness of personalized recommendation engine with CCPV from the numerical experiments based on real data. This study is still in its infancy. It would be necessary to combine the color analysis proposed in this thesis with other approaches, such as automatic personalization and collaborative filtering, so as to empower the existing recommendation engines. This line of research is underway and will be reported elsewhere in due course.

References

1. Srivastava J, Cooley R, Deshpande M, Tan P (2000) ACM SIGKDD Explorations Newsletter. doi:10.1145/846183.846188
2. Mobasher B, Dai H, Luo T, Nakagawa M (2002) Discovery and evaluation of aggregate usage profiles for web personalization. Data Min Knowl Disc 6(1):61–82. doi:10.1023/A:1013232803866
3. Pinyapong S, Kato T (2005) IEICE Trans Inf Syst. doi:10.1093/ietisy/e88-d.5.938
4. Liu K, Fang B, Zhang W (2011) IEICE Trans Inf Syst. doi:10.1587/transinf.E94.D.542
5. Sarwar BM, Karypis G, Konstan J, Riedl J (2001) Proceedings of the 10th international conference on World Wide Web. doi:10.1145/371920.372071
6. Jung K, Lee J (2002) E-commerce and web technologies. Springer, Heidelberg. doi:10.1007/3-540-45705-4_26
7. Inoue K, Urahama K (2004) IEICE Trans Inf Syst E87-D(7):1987–1988
8. Kang H, Yoo SJ (2007) IEICE Trans Inf Syst. doi:10.1093/ietisy/e90-d.12.2100
9. Aggarwal CC, Wof JL, Yu PS (1999) Proceedings of the 1999 ACM SIGMOD international conference on management of data. doi:10.1145/304182.304188
10. Sarwar B, Karypis G, Konstan J, Riedl J (2000) Proceedings of the 2nd ACM conference on Electronic commerce. doi:10.1145/352871.352887
11. Vartak M, Madden S (2013) Proceedings of the 2013 ACM SIGMOD international conference on management of data. doi:10.1145/2463676.2465270
12. Jung JJ, Lee K, Park S, Jo G (2005) IEICE Trans Inf Syst E88-D(5):843–850
13. Lin Y, Kawakita Y, Suzuki E, Ichikawa H (2012) International symposium on applications and the Internet. doi:10.1109/SAINT.2012.75
14. Huang C, Wei C, Wang Y (2013) IEEE international conference on multimedia and Expo workshops. doi:10.1109/ICMEW.2013.6618318
15. C.I.E. (1971) Recommendations on uniform color spaces, color-difference equations, psychometric color terms. Supplement No. 2 to CIE publication No. 15 (E.-1.3.1)
16. Fairchild MD, Berns RS (1993) Color Res Appl. doi:10.1002/col.5080180308
17. Connolly C (1997) IEEE Trans Image Process. doi:10.1109/83.597279
18. Kaufman L, Rousseeuw PJ (2008) Finding groups in data: an introduction to cluster analysis. Wiley, New York. doi:10.1002/9780470316801
19. Wiggerts TA (1997) Proc Conf Rev Eng. doi:10.1109/WCRE.1997.624574

The manufacturer's authorised representative in the EU is Springer
Nature Customer Service Centre GmbH, Europaplatz 3, 69115 Heidelberg,
Germany. If you have any concerns regarding our products, please
contact ProductSafety@springernature.com

Printed and bound by CPI Group (UK) Ltd, Croydon, CR0 4YY
30/09/2026
02225567-0017